The Mathematics Teacher in the Digital Era

MATHEMATICS EDUCATION IN THE DIGITAL ERA
Volume 2

Series Editors:
Dragana Martinovic, University of Windsor, ON, Canada
Viktor Freiman, Université de Moncton, NB, Canada

Editorial Board:
Marcelo Borba, State University of São Paulo, São Paulo, Brazil
Rosa Maria Bottino, CNR – Istituto Tecnologie Didattiche, Genoa, Italy
Paul Drijvers, Utrecht University, The Netherlands
Celia Hoyles, University of London, London, UK
Zekeriya Karadag, Bayburt University, Turkey
Stephen Lerman, London South Bank University, London, UK
Richard Lesh, Indiana University, Bloomington, USA
Allen Leung, Hong Kong Baptist University, Hong Kong
John Mason, Open University, UK
John Olive, The University of Georgia, Athens, USA
Sergey Pozdnyakov, Saint-Petersburg State Electro Technical University, Saint-Petersburg, Russia
Ornella Robutti, Università di Torino, Turin, Italy
Anna Sfard, Michigan State University, USA & University of Haifa, Haifa, Israel
Bharath Sriraman, The University of Montana, Missoula, USA
Anne Watson, University of Oxford, Oxford, UK

For further volumes:
http://www.springer.com/series/10170

Alison Clark-Wilson • Ornella Robutti
Nathalie Sinclair
Editors

The Mathematics Teacher in the Digital Era

An International Perspective on Technology Focused Professional Development

Editors
Alison Clark-Wilson
London Knowledge Lab
Institute of Education
University of London
London, UK

Ornella Robutti
Dipartimento di Matematica
Università di Torino
Turin, Italy

Nathalie Sinclair
Faculty of Education
Simon Fraser University
Burnaby, BC, Canada

Additional material can be downloaded from http://extras.springer.com

ISSN 2211-8136
ISBN 978-94-007-4637-4
DOI 10.1007/978-94-007-4638-1
Springer Dordrecht Heidelberg New York London

ISSN 2211-8144 (electronic)
ISBN 978-94-007-4638-1 (eBook)

Library of Congress Control Number: 2013955236

© Springer Science+Business Media Dordrecht 2014
This work is subject to copyright. All rights are reserved by the Publisher, whether the whole or part of the material is concerned, specifically the rights of translation, reprinting, reuse of illustrations, recitation, broadcasting, reproduction on microfilms or in any other physical way, and transmission or information storage and retrieval, electronic adaptation, computer software, or by similar or dissimilar methodology now known or hereafter developed. Exempted from this legal reservation are brief excerpts in connection with reviews or scholarly analysis or material supplied specifically for the purpose of being entered and executed on a computer system, for exclusive use by the purchaser of the work. Duplication of this publication or parts thereof is permitted only under the provisions of the Copyright Law of the Publisher's location, in its current version, and permission for use must always be obtained from Springer. Permissions for use may be obtained through RightsLink at the Copyright Clearance Center. Violations are liable to prosecution under the respective Copyright Law.

The use of general descriptive names, registered names, trademarks, service marks, etc. in this publication does not imply, even in the absence of a specific statement, that such names are exempt from the relevant protective laws and regulations and therefore free for general use.

While the advice and information in this book are believed to be true and accurate at the date of publication, neither the authors nor the editors nor the publisher can accept any legal responsibility for any errors or omissions that may be made. The publisher makes no warranty, express or implied, with respect to the material contained herein.

Printed on acid-free paper

Springer is part of Springer Science+Business Media (www.springer.com)

Foreword

Integrated Circuits

David Pimm

It is 1982. I am 28 years old and two-and-a-half years into paid work as a neophyte researcher in mathematics education in the UK. I sit in a secondary mathematics classroom, armed with a coding sheet, watching a lesson based on a piece of software on a Research Machines 380Z, the whole class attending to a small monitor. My aim is to see what difference the computer's presence makes to the lesson, particularly to the teacher's practice.

I learn many things over the course of a term, including but not limited to:

- How restrictive coding sheets can be, how they serve to determine what can actually be documented as occurring;
- How the need to write a code every 10 s (an early form of digital sampling) fragments a lesson and, more subtly, how coding is not a real-time activity – for instance, in order to ascertain whether a teacher question has been asked is not simply a matter of the grammatical form of an utterance, it involves waiting to see how it is taken by the students;
- How looking for altered patterns of interaction requires a strong grounding in prior norms; and,
- How hard it can be to see the really new.

Reading across the chapters in this volume brought back these earlier memories, not least because of the intertwined discussions of classroom structures and teacher 'orchestrations', as well as both the regularities and distinctions among them that occur in different settings. A staggering amount has changed in the past 30 years, with regard to the richness, range and sophistication of the mathematical devices and their encodings that are available for school use in the teaching and learning of mathematics. How far we have come from 'Computer-Aided Learning (CAL)' as it was then known. What is much more complicated to address, and problematic in a

D. Pimm
Simon Fraser University

variety of senses, is the scope and depth of actual use of this plethora of electronic possibilities in actual school classrooms and other learning contexts.

In a chapter of a book for teachers on ICT (information and communication technology) and mathematics, Sinclair and Jackiw (2005) attempt to look to the future: in a prognosticating vein, they identify three waves in technology-for-mathematics-education. The first focuses on the individual student (the site of the 'L' in CAL), the second develops the context of classroom learning (including the teacher and the curriculum), and the third looks outward to the world (wide web). Issues they explore include the relation between curricular specificity and mathematical expressivity with any particular instance of technology-and-software. I mention their piece here in part as an introduction to my thinking across the various chapters of this collection, given the latter's more or less common focus on and concern with the *teacher* in a mathematics classroom and the use of pertinent digital technology.

In Sinclair and Jackiw's terms, then, this book is primarily second wave. But also in terms of research on technologised mathematics classrooms, this feels to me a second-wave collection. Many early studies in this area focused on what teachers did with the technology (for more on the broader study mentioned above, see, for example, Ridgway et al. 1984) and were largely descriptive and untheorised (certainly the one I was involved in was: in an important sense the coding scheme both summarised and operationalised the extent of theorisation of the act of mathematics teaching – and it was not technology specific). "What is it like teaching mathematics with technology present?" such studies asked. We and many others learnt how hard it is (but not much about why). Others also learnt, among other things, about the potential for the 'worksheetification' of expressive technology, as well as the tension between programming and small program use [1].

But even given such commonality, this book is still quite plural. There is no one single focus, neither in terms of object of attention, phenomenon of interest, nor method of examination. Actual classrooms are looked at, often; teachers (present in various incarnations, novice and expert, experienced with technology or not, etc.) are considerably more in focus than students (computer-aided teaching?) and are represented by their views via questionnaires and interviews and by their reported classroom actions. We are told about what is and also about what might be. Mathematics is the presumed common concern of all authors, but this book is only cursorily about technology's effect on or interaction with mathematics. Technological devices are widespread across the chapters, but even they are not in continuous view.

In terms of foreground, I find a marked concern with exploring the complexity of being the teacher in a technologised classroom, as well as beginning to address the intricate 'teacher education' question about what manner and nature of support to offer teachers in their work (whether beginning or ongoing) in such settings. Some chapters report on more narrowly delineated research studies (some attempt to document the state of play while others opt for case studies, then, both in terms of the closeness of focus to a very few teachers and necessarily I feel in terms of examining specific forms of technology). Others still attempt to propose or harness a far more theorised frame and then look at some teaching settings through it. And some do both and some look across a range of studies. In terms of method and manner of exploration, there is again appreciable diversity. At times, for me, I felt a sense of struggle to find a suitably helpful grip size that is sufficiently specific to offer traction on the particular while rising sufficiently above it to improve the view before disappearing into the clouds.

Returning briefly to my opening personal vignette, we are far, far along the road from that naïve encounter. Mathematics education has produced much more informed and sophisticated classroom watchers. But despite this enhanced attention, there are still some fundamental and significant questions about operating as a teacher in a technologised classroom that do not seem to have gone away. And that, I find, is very, very interesting.

For instance, back in the older days, there used to be far greater concern with what used to be called 'screen effects', most particularly problematic features visible on screens that might mislead learners about the mathematics, due in considerable measure to issues of pixel size and the way certain mathematical phenomena were generated or implemented. While those schooled in mathematics via different means and media had little difficulty discerning which was 'something mathematical' from a 'screen effect', concern was raised about how students, some of whose first experiences – for instance, of graphs of functions – might be screen mediated, could develop into more discerning viewers of such screen displays. For me, this can helpfully be thought about in terms of device transparency (see below).

Once again, the phenomenon is not entirely new (for example, my generation of successful mathematics students, like those before us, had to become past masters at working with 'faulty' geometric drawings), but the 'new' devices amplify the effects and their significance enormously. Yet, Bernard Parszyz' (1988) interesting account distinguishing figure from diagram only arose in large part from early work with dynamic geometry software. So I wish to think more about digital devices as (among many other things) massive if perhaps temporary amplifiers of didactic phenomena. I venture 'temporary' because one thing I have noticed is that there is far less discussion/concern in the literature currently about screen effects 'misleading' students, only in part because of better pixelation and cleaner algorithms both simulating and dissimulating the mathematics. Why else, I wonder, might this 'problem' have dissipated or shifted?

Once again, this seems to me a first-wave difficulty, one for which the computer might have been blamed for leading to mistaken understanding. Now, it and its fellows have shifted to being seen as second-wave phenomena, ones where we seem more willing to lay blame at the teacher's door. The teacher 'should' now know what to do with the 'opportunity' of the mistaken pixel or the rounding error or whatever else arises from attempting to look with novice eyes through the e-screen (to use Mason's interesting term) to the mathematics.

In relation to this book, I wonder about the pedagogic equivalent of screen effects and their link, among other things, to Clark-Wilson's work on classroom 'hiccups'. These can, in part, be seen in terms of 'glitches' in classroom functioning that parallel screen effects – only this time the concern (and responsibility) is centrally placed with the teacher rather than the student. And a parallel worry may also hold with regard to novice teachers, whose initial experience increasingly comes in device-richer settings, namely that they may not be able to distinguish clearly, at least initially, between 'pedagogic screen effects' and somehow more permanent or fundamental didactic challenges arising from teaching mathematics.

I do not wish to fall into the yawning trap of essentialising the latter and consequently 'peripheralising' the former, to say nothing of examining the interesting ways in which the two interact (cue mention of the older version of TPCK, perhaps). Perhaps all didactic phenomena are 'temporary', in this sense of being conditioned by and to a considerable extent produced by the varied devices that have always graced the mathematics classroom (see, for instance, Kidwell, Ackerberg-Hastings and Roberts's (2008) account of 200 years of mathematical materials). But the size and frequency of effects are much greater now.

These observations bring me to consider once more one of the background framings of this book, more evident in some chapters than others, in regard to professional development, namely how to bring new teachers into contact with and increase awareness of these amplification and perturbation effects (some strong, some weak; some resolving, some mutating), as well as the potential for generating fundamentally new instances, of digital didactic phenomena. And all of these bring the hope of greater understanding of this very general second-wave focus and concern, namely what it means to teach mathematics in a classroom setting, however temporary (cue a third wave) such settings themselves may be. But this book really underscores for me the fact that one profound interest for researchers in mathematics education of all backgrounds, over and above a specific concern with digitised classrooms themselves, comes from what highlights and sharp reliefs arise from their study in relation to the general topic of teaching mathematics in schools.

The professional development issue also brought to mind a tension in teacher education work around taking such observed regularities from research and incorporating them into a curriculum for teacher development (for example, Paul Drijvers et al.'s seven orchestrations from Chapter 'Technology Integration in Secondary School Mathematics: The Development of Teachers' Professional Identities', bearing in mind their proposed expansion when looking at very young children, provided in Chapter 'Teaching Roles in a Technology Intensive Core Undergraduate Mathematics Course'). This process can contribute both to their normalisation and indeed, in some settings, to their institutionalisation. One need only recall the infamous, imposed 'three-part lesson' of the UK national numeracy strategy, which became a focus for external teacher evaluation. A key question when observing such norms is to ask what forces have combined to normalise them, as well as perhaps looking for ways to subvert them, believing as I do that the particular is always richer than the general.

Frames of Conceptualisation

There are two theoretical (analytic/conceptual) frames which crop up to a greater or lesser extent in many chapters of this book, sometimes in the same piece: TPACK and instrumental orchestration. In the penultimate chapter, Ken Ruthven does sterling work in attempting a comparison across them, before offering his own third, with its interesting shift from 'knowledge' to 'expertise'. This is important work, especially when carried out so cleanly and carefully as here. I take advantage of my geographic location in this book to add a couple of further observations.

First, I believe the 'metaphor' of Venn diagram provides a problematic image: it 'forces' TPACK to be a subset of PCK. It misses the fundamentally new (as well as framing everything, as Ruthven points out, as knowledge rather than know-how, a point I return to at the end). It is a flat image – as scrupulously even-handed as set theory itself – and has no scope for change or growth, no dynamic, no becoming. The categories themselves seem to be the preoccupying feature, giving rise to demarcation disputes, and not the reality they are supposed to capture, help analyse or explain.

Second, the addition of TPACK's 'and' [2], as well as changing all of its component terms into nouns, seemingly changes TPCK into a straightforward list (it also risks losing the alphabetic/historical link to PCK). It unbundles the notion and risks shedding the implied interaction among the components. My parallel is with the mathematical notion of a topological vector space: this is not simply a vector space that is also a topological space; the two descriptors interact, in that the vector space operations must also be continuous with respect to the topology. Likewise, it is not simply technology, pedagogy and content knowledge in shopping-list isolation, but their mutual interactions and shapings which are of interest and significance in this book.

With instrumental orchestration [3], the focus is on patterns of classroom pedagogic organisation in the presence of technological devices. Its goal is to help examine one of the key elements of this didactic difficulty, namely how to manage gaining facility with the device while maintaining a focus on the mathematics. Such devices in a mathematics classroom are nevertheless still means to an end, not ends in themselves (whether or not it is a device which has wider currency in the outside world or is solely for pedagogic use in a classroom setting).

These chapters brought to mind the work of Jill Adler (2001) on multilingual classrooms in the South Africa context and her productive use of Lave and Wenger's notion of transparency to discuss the effect of language as a resource that needs at different times to be both visible and invisible. I wonder whether these notions have a helpful role to play in discussions of technology in the mathematics classroom, not least when one approach seems to be to frame digital devices (conservatively) in terms of classroom or curricular 'resources'. And there is the general question of how such transparency transitions are effected [4].

When I was a UK school student in the mid-1960s (pre-calculators), there were still occasions when such split (double) focus was involved, whether in being taught how to use a pair of compasses, a slide rule or logarithm tables (and why we were told to say 'bar one' not 'minus one'), so this is not a completely new phenomenon. But it certainly seems to be one where such a difference in degree actually becomes a difference of kind.

And one of the ways in which studies of classrooms-with-technology can enhance the general study of teaching is by bringing to light altered forms of pedagogic phenomena or even more importantly new phenomena that have not been encountered before. Thus, in terms of the future, I expect the pertinent phenomena will be clearer and theorisations more settled, as the absorption of digital devices continues its complicated and intricate trajectory.

Some Closing Words About 'Technology'

Words matter. This is a tenet that has shaped my career in mathematics education. With regard to the last bullet in my second opening paragraph, namely our frequent inability (individually and culturally) to perceive the radically novel, Marshall McLuhan (1964) had some interesting observations about the effect of metaphors, their conservative, backward-looking quality in search of resemblance, of similarity. In consequence, we can miss the new. In one of his examples, 'the train is an iron horse', he draws attention to how, unlike with horses, trains do not 'get tired' (even if they do require 'feeding'). The automobile as a 'horseless carriage' is another instance McLuhan discusses. And Brian Rotman (2008), in his engaging and challenging book *Becoming Beside Ourselves*, reminds us that it is not even perhaps the mathematics that will change most from digital pressures so much as the mathematician. It is in this frame of mind, one of looking at the implicit effects of the choice of certain words, that I end this brief foreword in an etymological vein.

Technology is a peculiar word, not least that its suffix ('-ology' [5]) seems to promise the study of something, namely *techne*, 'know-how' or craft, in contrast to *episteme*, which might be rendered 'knowledge'. (Recall my earlier comments on TPACK and Ruthven's attempt to substitute 'expertise' for 'knowledge'.) And in the Arzarello et al. chapter, the word 'praxeology' is used in this same sense of 'technology'.

But technology actually refers to devices, to stuff – stuff into which mathematical and other know-how has been stuffed. Now we have 'digital' as the widespread preferred adjective, including 'digital technology', superseding 'new technology' or 'information technology' (to me one of the least informative) or 'information and communication technology'. What is being signalled by this plurality of and rapid drift in naming – for change of terminology is seldom either neutral or innocent? In particular, is it drawing attention to the most educationally salient aspect of these devices?

And in the same way that in the nineteenth-century 'geometry' and then 'algebra' morphed into 'a geometry' (which also permits 'geometries') and 'an algebra' (likewise seemingly authorising 'algebras'), so 'technology' has, arguably faced with proliferation and fragmentation, shattered into 'a technology' (and 'technologies'). One thing this pluralisation both reflects and perhaps achieves is a destruction of a sense of naïve uniformity, that there is a monolithic thing called 'technology' that has constant and predictable effects and that internal differences are minor or insignificant. Perhaps this has never been the case and 'technology' itself might have been used early on to provide an umbrella term for 'calculators and computers'. These are words and expressions whose history could be explored to document an aleatory and forgetful field (see Tahta and Love 1991; I believe 'technology' could be a significant word to add to their closing list).

I end by drawing attention to the term that is unvaryingly used to frame the central problem of this book, and that is 'integration': the unacknowledged but seemingly universal verb of pedagogic technological desire, an unanalysed good. Yet 'integration', to which obeisance is always made, is never actually interrogated – or possibly even noticed. I recall a very telling conversation I had a decade ago with French didactician Aline Robert, when I was engaged in an exploration of new mathematics teacher induction in France. I had absorbed that the transitive verb commonly used for teacher education was *former* (literally, to shape or mould someone) and was using it fluently, if unthoughtfully, when she asked me: *Qu'est-ce que ça veut dire de former quelqu'un*? This direct and powerfully simple question set me back on my heels. I draw on it here: What does it actually mean to *integrate* technology into a classroom?

In their three-wave chapter that I mentioned at the outset, Sinclair and Jackiw (2005, p. 236) ask:

> Will it [school mathematics] ever reach the point where the phrase 'innovation in ICT' reflects not just the presence of ICT in a classroom but something positive or exceptional about its *use* in the school, its involvement in the teaching and learning of mathematics?

One final thing strikes me about the word 'integration': it usually reflects a directionality, an attempt to integrate *this* into *that*. And the *that* (in this case, the classroom) usually has a prior integral and bounded existence. Is the goal a classroom in which the presence of digital devices passes unnoticed, unremarked? If so, will there then be a new boundaried whole into which the next 'innovation' will need to be 'integrated'?

Just as the seemingly innocuous term 'collaboration' as a universal classroom good has significant negative undertones for anyone with a memory of mid-twentieth-century European occupation (one only 'collaborates' with an enemy), so too anyone who has experienced social, cultural, political or especially racial 'integration' (think civil rights and forced bussing, think immigration) cannot be purely sanguine about its challenges, prejudices and resistances by all concerned. And from psychotherapy (Clarke 1988), there is an uncomfortable awareness that someone can only resist if someone else is pushing.

Notes

[1] Interestingly, 30 years on, programming as a mathematical activity worthy of the attention of school students seems to have all but disappeared (though see Chapter 6 by Buteau and Muller). For a further take on this, see: http://www.salon.com/2006/09/14/basic_2/.
[2] Possibly simply to change it from an initialism into an acronym – and perhaps as a play on rapper 2pac? My fingers will keep typing 'teachnology', my own hybrid second-wave form for TPCK.
[3] In passing, I am slightly uneasy about the use of 'Sherpa' in one of these forms, given that it is the name of a people, not a job description. And 'orchestrations' are what composers do, not conductors.
[4] The notion of meta-commenting (see Pimm 1994) may provide a useful means of detecting teacher attempts at shifting the degree of transparency.

[5] There is an old advert for British Telecom which has a doting grandmother (played by Maureen Lipman) cooing that her grandson had managed to pass an '-ology' in the UK's public exams at 16: 'you get an 'ology and you're a scientist'. And then there is the broad misuse of the term 'methodology' (the study of method), almost always when the plain term 'method' will suffice.

References

Adler, J. (2001). *Teaching mathematics in multilingual classrooms*. Dordrecht: Kluwer.
Clarke, R. (1988). Gestalt therapy, educational processes and personal development. In D. Pimm & E. Love (Eds.), *Teaching and learning school mathematics* (pp. 170–182). London: Hodder and Stoughton.
Kidwell, P., Ackerberg-Hastings, A., & Roberts, D. (2008). *Tools of American mathematics teaching, 1800–2000*. Baltimore: Johns Hopkins University Press.
McLuhan, M. (1964). *Understanding media: The extensions of man*. Toronto: New American Library of Canada.
Parszyz, B. (1988). "Knowing" vs "seeing": Problems of the plane representation of space geometry figures. *Educational Studies in Mathematics, 19*(1), 79–92.
Pimm, D. (1994). Spoken mathematical classroom culture: Artifice and artificiality. In S. Lerman (Ed.), *Cultural perspectives on the mathematics classroom* (pp. 133–147). Dordrecht: Kluwer.
Ridgway, J., et al. (1984). Investigating CAL?. *Computers & Education, 8*(1), 85–92.
Rotman, B. (2008). *Becoming beside ourselves: The alphabet, ghosts and distributed human being*. Chapel Hill: Duke University Press.
Sinclair, N., & Jackiw, N. (2005). Understanding and projecting ICT trends in mathematics education. In S. Johnston-Wilder & D. Pimm (Eds.), *Teaching secondary mathematics with ICT* (pp. 235–251). Buckingham: Open University Press.
Tahta, D., & Love, E. (1991). Reflections on some words used in mathematics education. In D. Pimm & E. Love (Eds.), *Teaching and learning school mathematics* (pp. 252–272). London: Hodder and Stoughton.

Contents

Introduction 1
Alison Clark-Wilson, Ornella Robutti, and Nathalie Sinclair

Interactions Between Teacher, Student, Software and Mathematics:
Getting a Purchase on Learning with Technology 11
John Mason

Part I Current Practices and Opportunities for Professional Development

Exploring the Quantitative and Qualitative Gap Between
Expectation and Implementation: A Survey of English
Mathematics Teachers' Uses of ICT 43
Nicola Bretscher

Teaching with Digital Technology: Obstacles and Opportunities 71
Michael O.J. Thomas and Joann M. Palmer

A Developmental Model for Adaptive and Differentiated
Instruction Using Classroom Networking Technology 91
Allan Bellman, Wellesley R. Foshay, and Danny Gremillion

Integrating Technology in the Primary School Mathematics
Classroom: The Role of the Teacher 111
María Trigueros, María-Dolores Lozano, and Ivonne Sandoval

Technology Integration in Secondary School Mathematics:
The Development of Teachers' Professional Identities 139
Merrilyn Goos

Teaching Roles in a Technology Intensive Core Undergraduate
Mathematics Course 163
Chantal Buteau and Eric Muller

Part II Instrumentation of Digital Resources in the Classroom

Digital Technology and Mid-Adopting Teachers' Professional Development: A Case Study ... 189
Paul Drijvers, Sietske Tacoma, Amy Besamusca, Cora van den Heuvel, Michiel Doorman, and Peter Boon

Teaching Mathematics with Technology at the Kindergarten Level: Resources and Orchestrations ... 213
Ghislaine Gueudet, Laetitia Bueno-Ravel, and Caroline Poisard

Teachers' Instrumental Geneses When Integrating Spreadsheet Software ... 241
Mariam Haspekian

A Methodological Approach to Researching the Development of Teachers' Knowledge in a Multi-Representational Technological Setting ... 277
Alison Clark-Wilson

Teachers and Technologies: Shared Constraints, Common Responses ... 297
Maha Abboud-Blanchard

Didactic Incidents: A Way to Improve the Professional Development of Mathematics Teachers ... 319
Gilles Aldon

Part III Theories on Theories

Meta-Didactical Transposition: A Theoretical Model for Teacher Education Programmes ... 347
Ferdinando Arzarello, Ornella Robutti, Cristina Sabena, Annalisa Cusi, Rossella Garuti, Nicolina Malara, and Francesca Martignone

Frameworks for Analysing the Expertise That Underpins Successful Integration of Digital Technologies into Everyday Teaching Practice ... 373
Kenneth Ruthven

Summary and Suggested Uses for the Book ... 395
Alison Clark-Wilson, Ornella Robutti, and Nathalie Sinclair

Glossary ... 403

Index ... 407

Introduction

Alison Clark-Wilson, Ornella Robutti, and Nathalie Sinclair

Teacher education is an important issue for society and it is framed within cultural, social, political and historical contexts. In recent years international research in mathematics education has offered a range of theoretical perspectives that provide different and interrelated frames and viewpoints (Ball and Bass 2003; Clark and Hollingworth 2002; Davis and Simmt 2006; Jaworski 1998; Wood 2008). The role of digital technologies within this discourse has an increasing relevance as the society and government place demands on teachers to integrate technology into their classroom practices so that students can experience its potential as a powerful learning tool (Drijvers et al. 2010; Lagrange et al. 2003; Trouche 2004).

Many of the chapters in this book open by stating how, despite over 20 years of research and curriculum development concerning the use of technology in mathematics classrooms, there has been relatively little impact on students' experiences of learning mathematics in the transformative way that was initially anticipated. The direct response to this has been an increase in research that focuses on the role of the teacher within technology-mediated lessons, in addition to the need for governments and schools to justify their expenditure on educational digital technologies. Many researchers in the past have discussed the role that teachers might or could play in the technology-rich classroom – exploring, for

---

A. Clark-Wilson (✉)
London Knowledge Lab, Institute of Education, University of London,
23-29 Emerald Street, London WC1N 3QS, UK
e-mail: a.clark-wilson@ioe.ac.uk

O. Robutti
Dipartimento di Matematica, Università di Torino,
Via Carlo Alberto 10, Turin 10123, Italy
e-mail: ornella.robutti@unito.it

N. Sinclair
Faculty of Education, Simon Fraser University, 8888 University Drive,
Burnaby V5A 1S6, BC, Canada
e-mail: nathsinc@sfu.ca

example, how the computer might blur the distinction between teacher and students (Papert 1980) or how the computer might even *become* the teacher (Pimm 1983) – but it is only recently that more systematic study of the unique demands and opportunities of the teacher teaching with digital technology had been undertaken.

Research on the use of technology in the mathematics classroom has traditionally focused very strongly on the affordances of particular software environments (in addition to hardware configurations) as well as on the ways such software affects or even changes the nature of mathematical objects and relations (see Sinclair and Jackiw 2005). When the research lens is trained on the classroom teacher, however, the emphasis can shift away from the technology, the associated tasks for students, and even the mathematical concepts at play, as it is hard to maintain a deep focus on multiple aspects. This will be evident in the chapters that follow. However, many of the authors have provided more information about the tasks that featured in their research as additional materials that can be accessed on the Springer webpages associated with the book.

In compiling an edited volume such as this, which features research from Australia, Canada, England, France, Italy, Netherlands, Mexico, New Zealand and the United States, it is inevitable that there is a broad range of terminology adopted. For example, the word technology is used to mean software, programs, applets, applications, courseware, display technology and hardware. Similarly, this research domain has its own set of vocabulary and constructs. To assist the wider understanding of this domain within an international context, we offer a Glossary chapter, which can be found on the Springer website (http://extras.springer.com[*]), compiled by the editors and authors through a collaborative process of exchange of ideas.

Many of the chapters within the book make reference to examples of particular teachers using particular tasks that have incorporated technology in some way. In each case it will be important to consider the particular context for the associated research as, in many cases, the researchers are focusing on teachers' existing or developing practice, rather than exemplifying forms of 'best practice'. As such, each of the classroom tasks described within each chapter must be interpreted within its specific context and it is for the reader to actively question the appropriateness of the task or the technology under scrutiny. Alongside this, the relevance of the cultural, social, political and historical context relating to each particular classroom cannot be overlooked.

A Journey Through the Text

The book has been divided into three main parts, which are preceded by an opening chapter that plays the dual role of inviting the reader into the context of doing and teaching mathematics with digital technologies and of alluding to some of the opportunities and tensions that such work entails. In his chapter 'Interactions between teacher, student, software and mathematics: Getting a purchase on learning with technology', John Mason shifts your attention away from the classroom-based

---

[*]Log in with ISBN 978-94-007-4638-1.

contexts contained within the remaining chapters of the book and invites you to engage with three different *e-screens* that are accessible from the Springer website (http://extras.springer.com*). Mason's chapter will challenge you to consider how to make sense of your mathematical experiences from the perspectives of several contrasting structural frameworks. He offers these in the spirit of his own work with pre-service and in-service teachers, which is centrally about helping them learn to attend to the mathematical ideas at play in the technological setting.

Following this opening chapter, the structure of the book unfolds as follows: Part I consists of six chapters that draw on a range of research perspectives (including grounded theory, enactivism and Valsiner's zone theory) and methodologies (including questionnaires, interviews, video analyses and longitudinal observations) that provide an overview of current practices in teachers' use of digital technologies in the classroom and explore possibilities for developing more effective practices. Part II gathers six chapters that share many common constructs (such as *instrumental orchestration, instrumental distance* and *double instrumental genesis*) and research settings that have emerged from the French research community, but have also been taken up by other colleagues. The two papers in Part III provide more meta-level considerations of research in the domain by contrasting different approaches and proposing connecting or uniting elements.

Part I

The first two chapters in this part both provide a snapshot of the ways in which large numbers of teachers are currently using digital technologies. In her chapter 'Exploring the quantitative and qualitative gap between expectation and implementation: A survey of English mathematics teachers' uses of ICT', Nicola Bretscher surveys 188 secondary teachers' technology use in England with the aim of exploring the gap between the reality and the potential of ICT use in the mathematics classroom. Rather than taking ICT broadly as the unit of analysis, Bretscher's survey made important distinctions between software and hardware use, as well as between teacher-centred use of technology (whole classroom settings with data projectors and interactive whiteboards) and student-centred use (on laptops or in computers labs). Bretscher follows Remillard's (2005) socio-cultural approach in her study of teachers' technology use; this approach focuses more on how technology gets used as a resource in teaching, amongst other resources, which have institutional, contextual and historical dimensions, and not just cognitive ones. The results of her survey show a predominant use of IWBs in teacher-centred classroom environments, with relatively little use of mathematical software (such as graphing software and dynamic geometry software). Bretscher discusses the three factors that lead to the statistically significant differences between using IWBs in a whole-class context and giving students direct access to ICT in a computer suite: teachers' confidence in using ICT; teachers' perception of the difficulty of classroom management; and the amount of curriculum material covered in ICT lessons. These factors should be

---
*Log in with ISBN 978-94-007-4638-1.

very important for professional development, and are worth bearing in mind in the following chapters in which more qualitative approaches are used to study the conditions of digital technology integration.

In the chapter entitled 'Teaching with digital technology: Obstacles and opportunities', Michael Thomas and Joann Palmer consider the types of obstacles – which they define as "anything that prevents an affordance-producing entity being in a classroom situation" (p. 72) – to research secondary teachers' implementation of digital technologies. In order to study the indicators of teacher progress in implementation of technology use, the authors introduce the construct of Pedagogical Technology Knowledge (PTK), which they define as including not only proficiency in using technology but also understanding of the techniques required to build didactical situations incorporating it. Unlike TPACK and instrumental genesis, PTK attends also to the personal orientations of teachers and, thus, to the role that beliefs and attitudes play within technology integration. With a focus on this aspect of teacher orientation, Thomas and Palmer report on two studies that sought to examine the importance of teacher confidence in the growth of PTK. One study involving 22 secondary teachers shows that, for these teachers, there is a correlation between strong confidence in one's ability to teach with graphing calculators and a more positive attitude toward technology and its use in the learning of mathematics. A follow-up study involving 42 female secondary teachers confirmed the strong correlation between confidence in using technology in the mathematics classroom and PTK. These findings have important implications for professional development, which the authors outline in their final section.

The rate at which technology evolves is such that there is an abundance of practice-based theories, which are offered to the community as a set of ideas that might prompt further research and discourse. The contribution by Allan Bellman, Wellesley Foshay and Danny Gremillon in the chapter 'A developmental model for adaptive and differentiated instruction using classroom networking technology', is offered in this context; they suggest a progression in teachers' development concerning their uses of a particular technology that can make students' learning outcomes more visible in the classroom. Drawing on grounded theory, and using extensive longitudinal observations, this chapter offers a model that could be used to consider how teachers might become masters of a given technology that has been designed to treat assessment differently.

In the chapter 'Integrating technology in the primary school mathematics classroom: The role of the teacher', María Trigueros, María-Dolores Lozan and Ivonne Sandoval examine the role of the teacher when integrating technology in the primary mathematics classroom. Guided by an enactivist perspective, this chapter examines the three different uses of technology that can occur in the classrooms – *replacement*, *amplification* and *transformation* – and relate them to five aspects of the role of the teacher in terms of communication of mathematics, interaction with students, validation of mathematical knowledge, the source of mathematical problems, and the actions and autonomy of students. The authors provide three case studies, each of which tends toward one of the three uses of technology (though all move between at least two different uses). They also emphasise the way in which the characteristics of the technology used has a strong influence on the role of the teacher. Thus, their chapter shows also how any approach to technology integration for teachers must be attentive to the kinds of digital technologies being used.

In the chapter 'Technology integration in secondary school mathematics: The development of teachers' professional identities', Merrilyn Goos uses the construct of *teachers' pedagogical identities* to prompt our thinking about the process through which two Australian secondary mathematics teachers (who are new to the profession) develop the afore mentioned identities as they begin to use technology in their classrooms. She defines technology in a broad sense to include display technologies in the classroom and various mathematical programs and applications. Goos' theory emanates from a socio-cultural view of both teachers' and students' learning and it adapts Valsiner's zone theory, as well as her earlier work, to provide a framework that takes account of the way that the technology alters the teachers' role and the factors that influence how the teachers adopt the technology. Goos' expansion of Valsiner's zones lead her to define a teacher's *zone of free movement* and *zone of free action* as a complex system that overlaps with the teacher's *zone of proximal development*, which can be used to explain how beginning teachers are able to develop innovative practices that involve technology.

We move to a university mathematics department setting in the chapter 'Teaching roles in a technology intensive core undergraduate mathematics course', as Chantal Buteau and Eric Muller describe the on-going development of a computer programming course for undergraduate mathematicians in the Canadian setting, with an emphasis on the roles of the course tutors and the department in which they operate. The mathematics department's goal was to achieve "an education of mathematics majors and prospective teachers of mathematics that would empower them to develop, implement, and use their own interactive mathematical objects". The authors use the context of the *Mathematics Integrated with Computers and Applications* courses, which have been in development since 2001, to raise issues around the associated course design and implementation.

Part II

Most of the chapters in this part are inspired in one way or another by Trouche's (2004) notion of *instrumental orchestration*, which is a construct that is based on the ergonomic framework of the instrumental approach. It articulates teachers' work before and during their activity with students, which is described in terms of schemes of action. In the chapter 'Digital technology and mid-adopting teachers' professional development: a case study', Paul Drijvers, Sietske Tacoma, Amy Besamusca, Cora van den Heuvel, Michiel Doorman and Peter Boon combine instrumental orchestration with elements of TPACK (a framework that is described in more detail also in the chapter 'Frameworks for analysing the expertise that underpins successful integration of digital technologies into everyday teaching practice') to describe the practices that teachers may develop when they use technology in their classrooms, and how these practices change over time when working in a community of teachers. These practices are described in terms of the use of seven different orchestrations. The authors stress the importance of supporting mid-adopting teachers in their professional development concerning technology, rather than focusing on the introduction of new technologies to non-expert teachers. Many of the orchestrations they describe are used in subsequent chapters of this part, and further developed to better suit different contexts.

In the chapter 'Teaching mathematics with technology at the kindergarten level: Resources and orchestrations', Ghislaine Gueudet, Laetitia Bueno-Ravel and Caroline Poisard seek to adapt Drijvers' et al. orchestrations to the context of the kindergarten mathematics classroom. Indeed, one of their aims is to study the kinds of orchestrations that might be specific to this level of schooling, in contrast to those described in the context of secondary school mathematics teaching. The authors provide case studies of three kindergarten teachers using two different applications, one a digital abacus and the other a game focusing on using number as position. They find new orchestrations in these kindergarten classrooms that are related to the differences in the resource system in the kindergarten classroom as well as in the importance of verbalisation for these children who do not yet read or write. Their chapter shows the need for theory development that is attentive to context – in this case the grade level of the students.

The chapter entitled 'How do teachers integrate technology in their practices? A focus on their instrumental geneses' moves to the context of the secondary mathematics classroom, where Mariam Haspekian describes a new study that expands upon her doctoral work in which she introduced two constructs, *instrumental distance* and *double instrumental genesis*, focusing on the use of spreadsheets within mathematics education. These constructs, originating from French research based on the activity theoretic approaches first proposed by Vygotsky, offer insight into the processes through which teachers appropriate the spreadsheet for use as a mathematical and pedagogical resource. Her case study of an experienced secondary mathematics teacher illuminates aspects of the teachers' practices.

There is resonance between the work of Haspekian and that of Alison Clark-Wilson, whose research also concerns the professional learning of experienced secondary mathematics teachers. Clark-Wilson's longitudinal study of two teachers focuses on aspects of what and how the teachers learnt as they began to use a complex multi-representational technology within an English school setting. In her chapter 'A methodological approach to researching the development of teachers' knowledge in a multi-representational setting', Clark-Wilson illuminates the methodology through which her construct of the *hiccup* became evident, highlighting the challenges that studies into teachers' knowledge development present to researchers. Her approaches, which are framed within the instrumental perspective, offer insight into a researchers' thinking in designing systematic and objective research protocols.

By contrast, in the chapter 'Teachers and technologies: Shared constraints, common responses', Maha Abboud-Blanchard adopts a meta-level approach in her cross-analysis of the outcomes of three different French studies that each researched aspects of the process of teachers' integration of technology within mathematics classrooms.

Two of these studies involved empirical research in secondary classrooms concerning the use of a computer algebra system, dynamic geometry program and web-based resource, whilst the third concerned a meta-analysis of over 600 individual publications on this theme. Abboud-Blanchard, who uses the construct of the *double approach* (Robert and Rogalski 2002), shows how it is possible to identify the characteristics of ordinary teachers' uses of technology and contrast the resulting outcomes in terms of the different components of their practice.

The chapter 'Didactic incidents: A way to improve the professional development of mathematics teachers', by Gilles Aldon, connects to the other chapters in this part through some shared theoretical perspectives. It is based on the instrumental approach, emanating from the viewpoints of *documentational genesis*, and the theory of didactic situations (*milieu*). Aldon introduces the ideas of *didactic incidents* and *perturbations*, theoretical constructs that help to explain the dynamic relation between teaching and learning from the perspective of documentary genesis that is contextualised within a milieu. These ideas support the analysis of teaching and learning as it underpins the construction of knowledge in a dynamic way and as processes that evolve over time, with the possibility of enhancing teachers' professionalism. The data and examples are taken from the European Union Comenius funded project EdUmatics, and provide evidence of this dynamism. The transformation of a resource into a document, or of an artefact into an instrument, when schemes of utilisation are activated, are not stigmatised once and for all, but they are seen as an interrelated set of on-going processes.

Part III

In the chapter 'Meta-didactical transposition: A theoretical model for teacher education programme', Ferdinando Arzarello, Ornella Robutti, Cristina Sabena, Annalisa Cusi, Rossella Garuti, Nicolina Malara and Francesca Martingone highlight the need to take the complexity of teacher education into account with respect to the institutions in which teaching operate, alongside the relationships that teachers must have with these institutions. This chapter also considers the evolution of the professional role of a teacher both as an individual within classes, and as member of a community of teachers. To address this need, the authors use Chevallard's (1985, 1992, 1999) *Anthropological Theory of Didactics* (ATD), which is mainly centred on the transposition of mathematics created by the teacher with the students. It is applied to teacher education, that is, to teachers as learners in a community, in which they improve their professionalism using technologies and resources, discussing among themselves and other communities, particularly that of researchers. The result of this study is the presentation of a model entitled the *Meta-Didactical Transposition,* which describes the evolution of teacher education over time, by analysing the different variables involved: *components* that change from external to internal; *brokers* who support teachers interacting with them; and *dialectic interactions* between the community of teachers and researchers. This chapter is particularly useful to help researchers interpret the process of teacher education, not only in a particular national context, but in various situations, because this process is usually characterised by praxeologies that are adopted and evolve over time, as well as increases in teachers' awareness when they come into contact with the community of education researchers. Moreover, the model usefully describes the role that institutions may have in this process, especially as, in every country, teacher education programme often start from specific directives given by a central institution that is linked to schools or to the academy.

The chapter entitled 'Frameworks for analysing the expertise that underpins successful integration of digital technologies into everyday teaching practice', by Kenneth Ruthven, compares and contrasts some of the main frameworks that are currently being used in research on teacher expertise in the context of digital technology integration (and within this book). These frameworks are all relatively new and each attends to a somewhat different aspect of the teaching activity. In this chapter, Ruthven analyses each of these frameworks and shows, through examples, how they function differently – both in terms of their epistemological assumptions and their intended unit of analysis – as research tools. All three approaches that Ruthven discusses emerge out of research focused at the secondary level but involving a range of different digital technologies.

Ruthven begins with the Technology, Pedagogical Content Knowledge approach (TPACK), which complements, with the addition of the 'technology component', the widely-used PCK framing of the knowledge that mathematics teachers need to teach. Using examples from research, Ruthven shows how this framework is used to "signal the need to consider technological, pedagogical and epistemological aspects of the knowledge underpinning subject teaching and their interaction in general terms" (p. 380) as part of the other components of knowledge, but that it provides only a "rather coarse-grained tool for conceptualising and analysing teacher knowledge" (p. 380). He then considers the Instrumental Orchestration approach, which also draws on an existing framework (instrumental genesis) focusing on student use of digital technologies (rather than on teacher expertise, as in PCK). This approach has been successful in providing a fine-grained analysis of the organisation of classroom activity around the use of a tool. Researchers using this approach have developed a typology of orchestrations that have enabled the identification of more general classroom patterns as well as comparison across different teachers and/or classrooms. As Ruthven writes, the typology "makes visible an important dimension of the professional knowledge that teachers participating in trialling had employed or developed in order to incorporate use of these digital technologies into their practice" (p. 384). While the Instrumental Orchestration approach used to study the integration of new technologies depends on the development of certain knowledge-for-teaching, Ruthven's third framework, the Structuring Feature of Classroom Practice approach, has a different purpose, which is to support the identification and analysis of teaching-with-technology expertise. This approach has evolved from prior research attending not only to teacher mathematics knowledge but also to classroom organisation and integration. It offers five structuring features of classroom practice that shape the ways in which teachers integrate new technologies. This approach provides a more differentiated characterisation of some of the key features of the instrumental orchestration approach.

As you read through the chapters and encounter these different approaches (or, indeed, other ones), you might consider some of the particular orientations toward the phenomenon of technology integration in mathematics teaching that Ruthven has outlined, asking yourself perhaps why a particular approach was chosen and what it offers as well as obscures in terms of understanding the phenomena at hand. For beginning researchers, this chapter offers an (often missing) critical comparison between competing frameworks that should make it easier to select the one more appropriate for your research questions and goals.

In the 'Conclusion', the editors make some concluding comments about the similarities and differences between the theoretical constructs, contexts and implications of the book's chapters, for example the ways in which the various constructs might be used to help to shape future research (and its associated methodologies) concerning the appropriation of mathematical digital technological tools in a range of educational settings. In addition, the editors look holistically at the implications of the various constructs on the design, content and implementation of professional development for teachers. Finally, the editors suggest how the individual chapters or combinations of chapters might be used within teaching sessions aimed at the intended readership of this book, that is to say researchers, Masters' or Doctoral students and pre-service and in-service teachers.

References

Ball, D. L., & Bass, H. (2003). Toward a practice-based theory of mathematical knowledge for teaching. In B. Davis & E. Simmt (Eds.), *Proceedings of the 2002 annual meeting of the Canadian Mathematics Education Study Group Edmonton* (pp. 3–14). Alberta: CMESG/GDEDM.

Chevallard, Y. (1985). *Transposition Didactique du Savoir Savant au Savoir Enseigné*. Grenoble: La Pensée Sauvage Éditions.

Chevallard, Y. (1992). Concepts fondamentaux de la didactique: perspectives apportées par une approche anthropologique. *Recherches en Didactique des Mathématiques, 12*(1), 73–112.

Chevallard, Y. (1999). L'analyse des pratiques enseignantes en théorie anthropologique du didactique. *Recherches en Didactique des Mathématiques, 19*(2), 221–266.

Clark, D., & Hollingworth, H. (2002). Elaborating a model of teacher professional growth. *Teaching and Teacher Education, 18*, 947–967.

Davis, B., & Simmt, E. (2006). Mathematics-for-teaching: An ongoing investigation of the mathematics that teachers (need to) know. *Educational Studies in Mathematics, 61*(3), 293–319.

Drijvers, P., Kieran, C., & Mariotti, M. A. (2010). Integrating technology into mathematics education: Theoretical perspectives. Mathematics education and technology-rethinking the terrain. *New ICMI Study Series, 3*(Part 2), 89–132.

Jaworski, B. (1998). Mathematics teacher research: Process, practice and the development of teaching. *Journal of Mathematics Teacher Education, 1*, 3–31.

Lagrange, J.-B., Artigue, M., Laborde, C., & Trouche, L. (2003). Technology and mathematics education: A multidimensional study of the evolution of research and innovation. In A. J. Bishop, M. A. Clements, C. Keitel, J. Kilpatrick, & F. K. S. Leung (Eds.), *Second international handbook of mathematics education* (pp. 239–271). Dordrecht: Kluwer Academic.

Papert, S. (1980). *Mindstorms: Children, computers and powerful ideas*. London: Harvester Press.
Pimm, D. (1983). Self images for a computer. *Mathematics Teaching, 105*, 40–43.
Remillard, J. T. (2005). Examining key concepts in research on teachers' use of mathematics curricula. *Review of Educational Research, 75*(2), 211–246.
Robert, A., & Rogalski, J. (2002). Le système complexe et coherent des pratiques des enseignants de mathématiques: une double approche. *Revue canadienne de l'enseignement des sciences, des mathématiques et des technologies, 2*(4), 505–528.
Sinclair, N., & Jackiw, N. (2005). Understanding and projecting ICT trends in mathematics education. In S. Johnston-Wilder & D. Pimm (Eds.), *Teaching secondary mathematics effectively with technology* (pp. 235–252). Maidenhead: Open University Press.
Trouche, L. (2004). Managing complexity of human/machine interactions in computerized learning environments: Guiding students' command process through instrumental orchestrations. *International Journal of Computers for Mathematical Learning, 9*, 281–307.
Wood, T. (Ed.). (2008). *The international handbook of mathematics teacher education* (Vol. 1–4). West Lafayette, USA: Purdue University/Sense.

Interactions Between Teacher, Student, Software and Mathematics: Getting a Purchase on Learning with Technology

John Mason

Abstract In this chapter three examples of teacher-guided use of ICT stimuli for learning mathematics (screencast, animation and applet) are critically examined using a range of distinctions derived from a complex framework. Six modes of interaction between teacher, student and mathematics are used to distinguish different affordances and constraints; five different structured forms of attention are used to refine the grain size of analysis; four aspects of activity are used to highlight the importance of balance between resources and motivation; and the triadic structure of the human psyche (cognition, affect and enaction, or intellect, emotion and behaviour) is used to shed light on how affordances may or may not be manifested, and on how constraints may or may not be effective, depending on the attunements of teachers and students. The conclusion is that what matters is the way of working within an established milieu. The same stimulus can be used in multiple modes according to the teacher's awareness and aims, the classroom ethos and according to the students' commitment to learning/thinking. The analytic frameworks used can provide teachers with structured ways of informing their choices of pedagogic strategies.

Keywords Interaction • Teacher-guided • Ways of working • e-screens • Screencast • Animation • Activity

---

J. Mason (✉)
Open University (Prof Emeritus), Milton Keynes, UK

University of Oxford (Senior Research Fellow), Oxford, UK
e-mail: j.h.mason@open.ac.uk

Introduction

What roles can teachers play in using e-screens[1] to support interactions with students and mathematics? How might teachers' pedagogic choices be informed? The questions being explored in this chapter concern the affordances of teacher-guided use of ICT for stimulating interactions between teacher, student and mathematics. Attention is restricted to interactions which begin with the teacher taking initiative, either because the applet itself is a screencast of a tutor, or because the use of the applet is directed by the teacher.

Ordinarily one expects to find a description of theoretical constructs before being told the method undertaken for the collection of data and the theoretical frame(s) for its subsequent analysis. However my approach is fundamentally experiential, which means that the data being offered are what arises in the reader through what they notice (what comes-to-mind) while reading and undertaking task-exercises. The analysis consists of a narrative to account for observations that I have made which may resonate with what others have observed, or as in the case here, to give an account of affordances based on experience in multiple settings. The empirical aspect of these studies lies not in my presenting my data here, but in generating recent experience in the reader. The analysis is informed by experience. Note the parallel with teaching and learning mathematics: experience can inform action-in-the-moment without being used to try to convince others using extra-spective data collected in some other situation.

The following assumptions provide an overview of the theoretical constructs being used, but these are only elaborated after you have had some exposure to the specific stimuli being considered.

A0: The human psyche involves cognition, affect, behaviour and attention-will.
A1: Teaching takes place in time and learning takes place over time.
A2: Action requires three roles to be filled: initiating, responding and mediating, and each of these roles can be played by the teacher, the student and the content.
A3: Effective activity requires a balance between motivation and resources.
A4: One thing that we do not seem to learn from experience is that we do not often learn from experience alone. Tasks are provided for students to initiate activity, which provides experience and, in order to learn effectively from experience, it helps to adopt a reflexive stance.
A5: Aligning teacher and student attention improves communication.

The affordances of the three forms of e-screen stimuli arise from the form of relations amongst discerned details in what is experienced. These relations are suggestive of general properties, which apply to many situations, being instantiated in the particular. Validity of these general properties can be tested by considering whether the proposed narrative fits or resonates with recent personal experience; whether the distinctions made help make sense of personal past experience; and most importantly, whether this articulated experience informs future practice through being sensitised to notice opportunities to act freshly and more effectively (Mason 2004).

---

[1] I use 'e-screens' to refer to electronic screens, as distinct from the mental 'screen' which is the domain of mental imagery.

Three Studies

The studies offered here all involve the use of an e-screen to initiate activity, in the form of a screencast, an animation, and an applet. Little will be achieved simply by reading the accounts however, since it is necessary to experience the stimuli for yourself, perhaps sensitised by assumptions A0 through A5.

ScreenCasts

With Jing and related software it is easy to record short videos showing work on mathematical problems or conceptual animations. There is a set of them at www.maths-screencasts.org.uk (set up July 2011; accessed Feb 2012) or Khan Academy (www.khanacademy.org accessed Mar 2012).

> Pick one of the screencasts, say the one on Lagrange multipliers:
> http://www.maths-screencasts.org.uk/scast/LagrangeMult.html
> What are you attending to as the screencast proceeds?
> What learning is afforded by watching the screencast?
> What would a student have to do to learn something from the screencast?

On the surface, the task for students is to make mathematical sense of what is presented, and to increase their confidence that they can tackle a similar problem effectively in the future. The question of what constitutes a 'similar problem' might need to be discussed explicitly. During the screencast your attention may have shifted between what was on the screen and what was being said, and drawn to the symbols being written and spoken at the same time. Would a student watching this know how the presenter knew to perform the actions she does?

Rolling Polygons

At the heart of this task is an animation, however the presentation begins with setting the scene by inviting the use of mental imagery.

> Imagine a point P moving in a circle centred at point C. Imagine a finite number of lines (at least 3) being drawn through C. From P drop perpendiculars onto your lines and mark their feet as $F1, F2, F3, \ldots$. Now join $F1, F2, \ldots$ in sequence to form a polygon. What happens to the polygon as P moves around the circle?

Changing *P* mentally is pretty difficult, and so the task proper begins with an animation involving a triangle (downloadable from ref http://extras.springer.com[*]; double click on right hand figure to see animation).

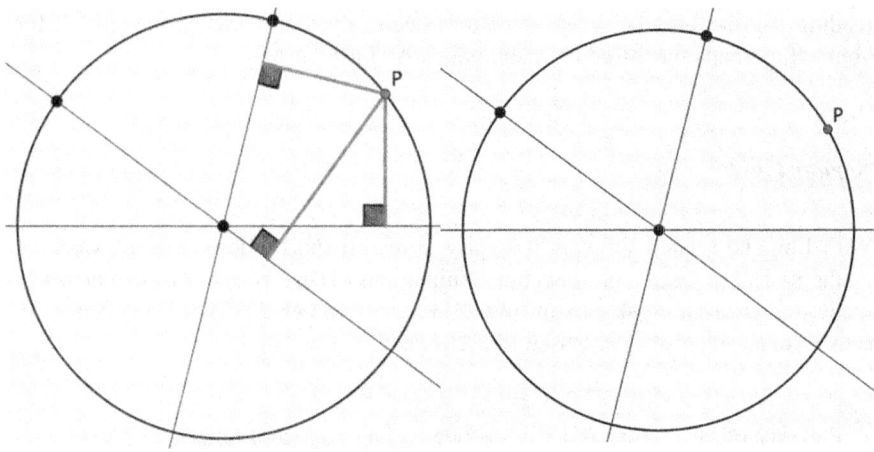

What role is played by the initial mental imagery?
What are you attending to as the animation proceeds?
What actions are stimulated by watching the animation?
What would a student have to do to learn something from the animation and the applet?

On the surface, the initial task for students is to become aware of and be at least somewhat surprised by a phenomenon, and then to begin to seek an explanation for that phenomenon. This in turn is likely to call upon students' powers to make deductions about angles using previously encountered fact such as the effects on angles of rotating lines through 90°, or angles in a quadrilateral with two right angles and angles subtended at a circle on the same side of a chord, etc..

The initial mental imagery is intended to contribute to the 'reality' of the task through exercising a fundamental human power and evoking curiosity as to what might happen. It sets the scene. This is a pedagogic strategy that can be used in many situations, because imagining ourselves doing something in the future is the basis for planning.

Secret Places

This task and its applet support is intended for teacher-led exploration, though it can be used by individuals or small groups working without the teacher.

---

[*]Log in with ISBN 978-94-007-4638-1.

Initially there are five places around a table, and one of them has been selected as a 'secret place'. You can probe any place, but all you will be told is whether it is 'hot' (meaning either it or one of the adjacent places on either side of it is the secret place), or 'cold'. How can you most efficiently (least number of probes) discover the secret place?

In the applet (http://extras.springer.com*), if you click on a place, it will show either 'red' or 'blue'. Red signals that the secret place is either the one chosen, adjacent to it, whereas blue signals that this is not the case.

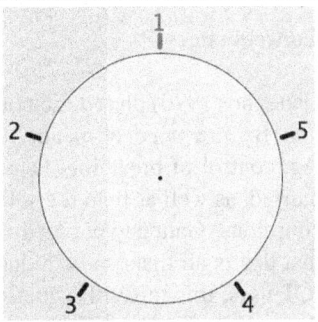

What are you attending to as you explore the effect of clicking on places? What actions are stimulated by predicting, justifying and then clicking? What would a student have to do to learn something from a teacher-led search for a strategy?

Applet available for download at
http://mcs.open.ac.uk/jhm3/Applets%20&%20Animations/Reasoning/Secret%20Places/Secret%20Places%201D.html (Set up Feb 2011; accessed Oct 2012).

On the surface, the task for students is to locate the 'secret place' as efficiently as possible, thus drawing on their natural powers to imagine (what could happen if they click somewhere) and to reason (possible consequences of clicking and whether that would be helpful).

For all three stimuli, what matters is what students do next, having encountered the stimulus; what ways of working have been established; and what sort of atmosphere students are used to.

Elaboration of Assumptions

The assumptions that follow make no direct reference to the use of technology. However, later in the chapter the descriptions of interactions with the e-screens include this necessary elaboration.

*Log in with ISBN 978-94-007-4638-1.

A0: The Human Psyche Involves Cognition, Affect, Behaviour and Attention-Will

This first assumption is implicit in Western psychology, but has its roots in ancient Eastern philosophy-psychology (Ravindra 2009; Mason 1994b). Despite this, it is all too easy to forget to engage the whole of students' psyche.

Consequences

Gattegno (1970) placed the notion of *awareness* at the core of his *science of education*. By *awareness* he meant 'that which enables action', which includes the somatic (eg. control of breathing, heart-rate, perspiration etc.) and the automated or internalised, as well as both the subconscious (eg. Freudian and other impulses) and the conscious. Gattegno claimed that it is awareness that can be educated, and indeed that that is all that can be 'educated': *only awareness is educable*. With the sample ICT uses, this raises the question of what awarenesses are available for educating due to the affordances of the ways of working and the medium used.

By contrast, *only behaviour is trainable*. This conforms with an image found in several of the Upanishads (Rhadakrishnan 1953, p. 623; Mason 1994), in which the human psyche is seen as a chariot. The chariot itself is seen as a metaphor for the body and hence for behaviour.

from website: http://members.ozemail.com.au/~ancientpersia/page8a.html

The horses drawing the psyche-chariot represent the emotions (affect). These are the source of energies which are made available to the psyche. Thus *only emotion is harnessable*. Emotion is the way that we access energy which acts through the disposition of various selves that take charge in the individual. All of these contribute to the setting in which attention acts, which many philosophers equate with the will, since as William James (1890, p. 424) observed, "each of us literally *chooses*, by his way of attending to things, what sort of a universe he shall appear to himself to inhabit". It is not simply what you are attending to, but how you are attending to it that matters (Mason 1998, see A5). A mixture of surprise and curiosity draws on psychic energy accessed through the emotional-affective charge channelled according to the habits of the active 'self'. The issue is developing a milieu in which mastering the mathematical aspect of a situation matters to students.

Trained behaviour is essential, being the manifestation of automated functioning and habits, but on its own it can be limiting and inflexible, whereas coupled with awareness the two together can stimulate and exploit the energy that is often called creativity. None of the studies offered here are directly intended to train behaviour concerning the carrying out of a mathematical procedure, though they could be used to train behaviour concerning collective and individual mathematical thinking.

Assumption *A0* can be used to probe implications of the adage 'practice makes perfect' which is the foundation stone of behaviourist theories of how learning takes place. Certainly it is necessary to integrate behaviours into psycho-somatic functioning. However, repetition alone is no more likely to lead to internalisation than is constant exposure to the same idea. Even stimulus–response (Skinner 1954) is only effective in certain circumstances. As Piaget (1970) pointed out under the label *genetic epistemology*, the individual is an active agent, constructing her own narrative. From a Vygotskian perspective, narrative construction is based in and takes place within socio-cultural milieu. The role of a teacher is to direct attention towards appropriate narratives which constitute conceptual understanding (Bruner 1990; Norretranders 1998), and to provoke students to integrate appropriate action or functioning through subordinating attention (Gattegno 1970; Hewitt 1994, 1996). 'Integration through subordination' is achieved by withdrawing the attention initially required to carry out an action so that the action can be carried out in future while absorbing a minimum of attention.

Henri Poincaré (1956) expressed surprise that people find mathematics difficult to learn, because from his perspective mathematics is entirely rational, and humans are rational beings. Jonathon Swift (1726) had already challenged this notion, proposing that human beings are at best 'animals capable of reason'. If rational reasoning is not activated, mathematical thinking is likely to be experienced as mysterious. The success of behaviourist strategies based on stimulus–response combinations shows that Swift was correct: people can be trained and enculturated into certain types of behaviours and this can be partly conscious and partly unwitting on their part. They can be successful in routine situations such as tests, but their success is short-term unless routines are frequently rehearsed. Such training only takes you so far. Once will is activated, attention wanders, different selves with different energy flows and dispositions come into play, and learning becomes much more complex. Hence the need to educate awareness as well as to train behaviour.

To be responsible for your own learning is a commonplace sentiment that fits with Western democratic values. The word *responsible* has roots in parallel with the Italian *spondere* which means 'to be able to justify actions' (to respond). Jürgen Habermas (1998) began from a similar position to Poincaré's assumption of rationality, but he focused on responsibility, which he cast in terms of justification:

"The rationality of a person is proportionate to his expressing himself rationally and to his ability to give account for his expressions in a reflexive stance. A person expresses himself rationally insofar as he is oriented performatively toward validity claims: we say that he not only behaves rationally but is himself rational if he can give account for his orientation toward validity claims. We also call this kind of rationality *accountability* (Zurechnungsfähigkeit)." (Habermas 1998, p. 310 emphasis in the original, quoted in Ascari 2011, p. 83)

He delineated three different domains or types of rational justification:

Epistemic: factual; assertive (epistemic rationality of knowledge)
Teleological: intentions behind actions (teleological rationality of action)
Communicative: attempts to convince, requiring listener acquiescence (communicative rationality of convincing), with both a *weak* and a *strong* form.

The point is that justification is an essential core component of mathematical thinking, as well as involvement in society. A successful practitioner without a narrative by means of which to justify choices on the basis of explicit criteria is at the mercy of habits in the face of changing conditions (Mason 1998).

The psyche operates within a socio-cultural-historical milieu with its undoubtedly important influences, most especially the atmosphere or ethos developed in the classroom or other setting and the social pressures from peers and from institutional norms (Brousseau 1997). One of the difficult things about online activity is that it is much harder to influence from a distance the atmosphere in which students are working than it is in face-to-face interactions.

A1: Teaching Takes Place in Time; Learning Takes Place Over Time

Despite the desire by government to have inspectors witnessing learning, learning is a maturation process. It requires time (Piet Hein 1966). Gattegno (1987) went so far as to suggest that learning actually takes place during sleep, when our brains choose what sense-impressions from the day to let go of. Thus memory is not about storing but about making and breaking links. Learning is reinforced not simply through re-encountering similar actions, activities and experience in fresh contexts, in what Bruner (1966) referred to as a spiral approach to the curriculum, but through developing an increasingly complex narrative to accompany the developing richness of connections.

Development

At the Open University (1981) we used the trio of *see–experience–master* (SEM) to emphasise that 'learning' does not take place on first encounter, nor even after some further experience. The triple can act as a reminder that encountering new ideas is a bit like being in a train station. An initial encounter is like seeing an express train go by when details are hard to make out and it all seems off-putting or complicated. With continued encounters there is growing familiarity, a bit like seeing a freight train rumble by. Eventually there is a degree of confidence and mastery, as when a passenger train stops and you get on and go-with the idea.

We translated Bruner's three modes of (re)presentation (enactive-iconic-symbolic) into a spiral of *manipulating-the-familiar, getting-a-sense-of,* and *articulating* that sense (*MGA*) as a reminder that it is natural to use what is familiar in order to get a sense of underlying relationships which, when articulated more and more succinctly eventually become confidence-inspiring and familiar for use in further manipulation.

MGA fits well with the principle of variation (Marton and Booth 1997; see also Watson and Mason 2005, 2006): learning a concept is becoming aware of what aspects of an example can be varied, and over what range, while remaining an instance of the concept. What is available to be learned is what is varied in a succession of experiences in contiguous space and time. Spiral learning and exposure to variation in key aspects is sometimes replaced by frequent repetition of nearly identical tasks in an attempt to train behaviour. However, if attention is not drawn (explicitly or implicitly) to carefully engineered variation of key aspects, the result may be successful performance on routine exercises, without educating awareness. It may also all too easily have a negative influence on disposition to engage, with students only willing to undertake what they know they can already succeed at.

Consequences

Learning, seen as educating awareness, training behaviour and harnessing emotion within a particular milieu, can be cast in terms of developing dispositions to attend in appropriate ways. A teacher cannot 'do the learning' for students. Indeed, the more they try to indicate to students the behaviour being sought as evidence of learning, the easier it is for students to display that behaviour without actually generating it for themselves, without educating their awareness (this is the *didactic tension* first articulated by Brousseau: see Brousseau 1997). What a teacher *can* do is participate in the various possible modes of interaction with students, without looking for evidence of 'learning' in too short a term (Piet Hein 1966). It often takes time to integrate a way of acting into your own functioning, even when this is stimulated by efficient and effective pedagogy (integration through subordination of attention).

Implications for Teaching

When choosing or designing task-sequences SEM and MGA can act as reminders when choosing or designing task-sequences to arrange for multiple encounters, and within each encounter, multiple instances with relevant variation. Learning is seen as a maturation process, like baking bread or brewing beer. It takes time. When rushed, the tendency is to revert to superficial success through routine exercises carried out using templates based on 'worked examples'.

SEM and MGA can also act as reminders that 'responsible learning', that is, having access to justifications for actions initiated is a gradual process, as complex narratives take time and multiple encounters with phenomena in order to come to articulation. As the articulation becomes more succinct and familiar, it becomes available as a component in yet further development.

A2: Action Requires Three Roles to be Filled: Initiating, Responding and Mediating

Following Bennett (1966, 1993) who developed a framework called *Systematics*, based on the quality of numbers, action has the quality of three-foldedness. Action requires an initiating impulse, a responding impulse and a mediating impulse. Without the mediator, there is nothing to bring or hold the initiating and responding together. Put another way, any action takes place within a context or milieu (Brousseau 1997) that enables the action to take place.

Consequences

From this perspective, interaction between a teacher-tutor, a student, and mathematics can take place in one of the six combinatorially distinct ways of arranging these three components in the three roles (Mason 1979). For convenience these six modes are known as the six ex's: Expounding, Explaining, Exploring, Examining, Expressing, Exercising, all within a milieu consisting of institutional affordances and constraints (including classroom and institutional social norms and demands). The milieu also includes the focal world(s) or spaces of the participants. Usually this consists of the mental worlds in which people dwell and from which they express their insights, but the presence of virtual screen-worlds provides a more explicitly taken-as-shared world of experience, namely the world of phenomena acted out on, and interacted with, a screen (Mason 2007).

The key feature for consideration here is the mediating or reconciling contribution of one of these roles so as to bring the other two into relation, and so as to sustain that relation for long enough for the action to reach fruition, leading to a result that can partake in further actions. The use of electronic screens associated with the tasks suggested above centres on the teacher as initiating impulse, and so draws particularly on the interactions summarised as *expounding* and *explaining*, although there are plenty of opportunities to shift into other modes from time to time.

Expounding is characterised by the presence (actual or virtual) of students bringing the teacher into contact with the mathematics in a special way. The term *pedagogic content knowledge* (Shulman 1986) has been used to describe what is needed in order to carry through this action effectively, while others try to capture it by describing the *knowledge needed for teaching* (Davis and Simmt 2006). Here the focus is more on the experience of the action as the teacher crafts tasks for students through contacting the didactic peculiarities of the topic, calling upon relevant pedagogic strategies. The effect of the action is to draw the students into the world and mind-set of the teacher. When the micro world or software plays the role of teacher, the quality of the action depends on the quality of the preparation of the software, which requires sensitivity to student experience and deep knowledge about didactic tactics and pedagogic strategies in relation to classic misunderstandings and misapprehensions within the particular topic.

Explaining is used in this way of thinking with a non-standard meaning. It is characterised by the teacher making contact with the thinking of the student, entering the student's world, centred on, made possible by, and hence mediated by the particular mathematical content. As soon as the teacher experiences "Ah that is where the difficulty lies", there is likely to be a shift into expounding. Staying with the world of the student involves 'teaching by asking' and 'teaching by listening' (Davis 1996) rather than teaching by telling. The more usual sense of *explain* as 'to make plain' is highly idiosyncratic, because what is 'plain' to the speaker may not be 'plain' to the audience. Thus the usual meaning of *explaining* is usually an instance of the action of *expounding*.

In relation to the previous axiom concerning teaching taking place in time, calling upon modes of interaction in which students play the initiating role, and those in which the content plays this role, can at least balance the student experience of modes of interaction, and can provide opportunity for *exploring* the ideas (teacher mediates between content and student); *expressing* (students feeling the need to construct their own narrative, so the student mediates between the content and the teacher); *exercising* through practising what needs to be practised (the teacher mediates between the student and the content by providing exercises); all in preparation for *examining*, in which students' own developing criteria for whether they are understanding and appreciating appropriately are tested against the expert's criteria (the content mediates between student and teacher).

Implications for Teaching

Perceiving actions in which one participates as involving three impulses within a milieu can transform teaching by altering what a teacher attends to, and how, and also how they see their contribution. Arranging the energies of the classroom so that as teacher you can dwell in mediating or in responding can be exhilarating as well as liberating for students. Provoking students into experiencing the desire to express promotes the maturation of their understanding and their appreciation of what they are integrating into their functioning, that is, the education of their awareness.

A3: Effective Activity Requires a Balance Between Motivation and Resources

Following Bennett (*op. cit.*), activity involves two axes: motivation and operation within a world of attention. Motivation in an activity has to do with the perceived

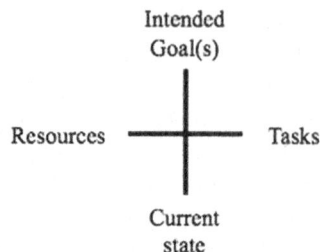

Fig. 1 Structure of activity in Systematics

gap between goals or aims and current state. It is 'what matters' to the student (Fig. 1). This conforms with a Vygotskian perception of activity but with the addition of a tension or gap between current state and goal, which plays the role of disturbance identified by Heidegger (1962) and many others (e.g. Festinger 1957; Piaget 1971) as what activates learning (leading to assimilation and accommodation).

The second axis concerns the resources available (both those brought by the students and those provided by the environment) and the tasks provided. If the resources available are inadequate for the gap between current state and goal, or if the tasks do not actually provide sufficient stimulus to reach the goal, then the activity will be ineffective.

Resources include student propensities and dispositions, and learner access to their natural powers such as stressing and ignoring, imagining and expressing etc. Where student powers are usurped by textbooks or modes of interaction with the teacher, students soon learn to park their powers at the door as not being required, and so become dependent on the teacher to initiate mathematical actions.

Tasks are inherently multiple by nature: as conceived by the author; as intended by the teacher; as construed by the student(s); as enacted by the students; and as recalled in retrospect by the student(s). Tahta (1981) pointed out that there are different aspects of a task: the outer task is what the task states (and is interpreted as by students), whereas the inner task is implicit, and has to do with mathematical concepts and themes that may be encountered, powers that may be used, and propensities that may come to the surface, all contributing to educating awareness.

In the language of affordances, constraints and attunements (Gibson 1979), affordances arise from the relationship between resources and tasks. The constraints are usually imposed from the tasks, for as is well known, creativity only takes place when there are constraints. Both student attunement and teacher attunement contribute to the motivational and the operational axes.

Ainley and Pratt (2002) distinguish between *purpose* of a task as the local context which gives learners a purpose in undertaking it, and the *utility* of a task or a technique in terms of the range of situations in which it can be used in the future. Both contribute to the development of positive or negative dispositions and propensities.

The two-axis structure of activity provides a richer structure than that provided by the adage 'start where the learners are'. Indeed, calling upon the whole psyche, and mindful of Vygotsky's distinction between *natural* and *scientific* knowledge, the effective teacher 'starts' where the learners could be rather than where they are, by invoking their energies through surprise or a sense of a gap so that they strive to move along the motivational axis, supported by access to appropriate resources and well judged tasks.

A4: One Thing We Don't Seem to Learn from Experience, Is that We Don't Often Learn from Experience Alone

Evidence for this is widely available, as you try to remember what you have read in the newspaper, what you saw on television, even what you set out to accomplish when you went into another room. What students get from engaging in an activity is highly variable, as Jaworski (1994) found when she asked students what a lesson had been about in which the task as set had been to draw and cut out copies of quadrilaterals and see if they would tessellate. Many students reported that the lesson was about 'cutting out quadrilaterals', 'using scissors', etc., and only a few mentioned tessellation. This reinforces the observation that different students attend to different things, stressing some things and ignoring others, and that even when they are attending to what the teacher intends, they may be attending in different ways (Mason 2003).

The student's stance towards learning, delineated by Marton and Saljö (1976) as a mixture of *surface*, *deep* and *strategic* approaches, colours all of learners' actions, and the closer they are to the strategic–surface, the more likely it is that task-completion characterises their epistemological stance. Even participation in suitable activity may not lead to the intended learning. Many students act as if their role is to attempt the tasks they are set, and that somehow those attempts will be sufficient to produce the expected learning. This epistemological stance is the basis of the *didactic contract* (Brousseau 1997). However tasks are supposed to generate activity, through which learners gain experience. Yet "one thing we don't seem to learn from experience, is that we don't often learn from experience alone" (Mason 1994a). A reflective stance, a withdrawing from the action in order to become aware *of* the action can make learning much more efficient than without it. To paraphrase William James (1890) "a succession of experiences does not add up to an experience of that succession". More is required. This is particularly hard to arrange when students are studying at a distance.

Evidence of learning is informed action in the future, which is what some call an *enactivist* stance (Varela et al. 1991) in which *knowing* is the same as *(en)acting*. This requires having an appropriate action come-to-mind (be-enacted) when needed, which brings us back to the education of awareness.

Implications for Teaching

To promote learning, including learning how to learn, it is useful to get learners to withdraw from activity and to reflect not only on which actions were successful and which were not, but on why. A further step is to prompt them to identify actions they would like to have come-to-mind in the future. This is how people most often learn from experience. Construction tasks (Watson and Mason 2005) are very useful for this purpose because they enrich personal example spaces while at the same time exercising techniques.

Stimulating effective reflection involves creativity and sensitivity, because the same prompts used over and over can lead to learners becoming dependent on the teacher rather than developing independence (Baird and Northfield 1992). In order not to train students to depend on the teacher to indicate appropriate behaviour, it is necessary to use both scaffolding and fading (Brown et al. 1989). Another way to express this is to say that the teacher needs to be alert to moving from directing behaviour (instruction) to increasingly indirect prompting as required, until students are spontaneously initiating that action themselves. This is what van der Veer and Valsiner (1991) suggest was intended by Vygotsky's notion of *zone of proximal development* (Mason et al. 2007).

A5: Aligning Teacher and Student Attention Improves Communication and Hence Affordances

Bringing what the teacher and what the students are attending to into alignment is only the beginning of effective teaching; alignment in how the teacher and the students are attending also matters. Different forms of attention include:

Holding Wholes: gazing in an unfocused manner, absorbing the overall, placing oneself in a state of receptivity towards a situation;
Discerning Details: distinguishing entities (which can then be held as 'wholes');
Recognising Relationships between discerned details in the situation;
Perceiving Properties as being instantiated as recognised relationships between discerned details; and,
Reasoning on the basis of agreed properties.

These five 'states' or structures of attention correspond closely with the 'levels' distinguished by Dina van Hiele-Geldof and Pierre van Hiele (van Hiele 1986) with the notable difference that rather than being seen as levels in a progression of development, attention is experienced as shifting rapidly between these states in no specific order.

Implications for Teaching

In order to be helpful to students it is necessary for teachers to be aware not only of what they are attending to in the moment, but how they are attending to it. This enables them to make use of an appropriate mode of interaction and to direct learner attention (however subtly or explicitly) so that either it comes into alignment with their own (cf. *exposition*) or it brings theirs into alignment with that of learners (cf. *explaining*).

Drawing Threads Together

Tasks are offered to students so that they engage in activity. The activity itself is not sufficient to generate learning. Rather, students need to participate in transformative action in which they experience shifts in the focus and structure of their attention. It is not simply a matter of agentiveness, of converting assenting into asserting (Mason 2009) but of relationship (Wan Kang and Kilpatrick 1992; Handa 2011), of playing various roles in different modes of action. Experience alone is not sufficient, and for most students, especially in order to stimulate the education of awareness as accompaniment to training of behaviour, an explicitly reflexive stance is required, as students become explicitly aware of actions that have proved fruitful and of actions that have not. Imagining themselves in the future initiating those actions can improve the chances that a relevant action will come-to-mind when needed, and this is how development takes place. This is what Vygotsky was getting at with the *zone of proximal development*: the actions that can be used when cued become actions that can be initiated by the student without explicit cues (van der Veer and Valsiner 1991; Mason et al. 2007).

Analytic Narrative Concerning the Three Studies

ScreenCasts

Background

The design of Open University Mathematics Summer Schools in the 1970s was based on a framework known as *Systematics* (Bennett *op cit.*). The format of one type of session introduced was called *Technique Bashing*: a tutor would publicly tackle an examination question, revealing as much as possible of their inner monologue and procedural incantations as they went. The idea was to draw the student into the world experienced by the tutor (a form of expounding). This was a real-time version of a mode of interaction based on tape-frames used in our distance taught courses, in which students listened to a tutor talking through a concept or a technique while directing their attention to a series of printed frames containing key phrases and whatever else needed to be written, together with space for students' own work. There were lots of stop instructions for students to switch modes and take initiative, either *exercising* or *expressing* but also *exploring*, in ways that are not possible in a face-to-face tutorial.

The idea of tape-frames was to have the tutor's voice in the student's head through the use of earphones, and we made use of BBC expertise to develop rules of thumb for linking the audio with the text so that students always knew what to be attending to, so we did not simply read the text out loud. Emphasis was placed on how the tutor knew what to do next, not just on what they did next, and this conforms with a plethora of subsequent research on worked examples indicating that what students want most is to know how the expert knows what to do next (see for example Renkl 1997, 2002).

With readily-available online video (Redmond 2012), students can now access screencasts of tutors displaying worked examples of concepts and techniques in this technique-bashing mode. What is not so clear is how, when recording a tutor's performance, student attention can be provoked to shift from dwelling in the particular (recognising relationships in the particular) so as to see the general through the particular (perceiving properties as being instantiated). Emphasis is on factual (A0) rather than teleological rationality; the person gaining most from bringing to articulation is the performing tutor (communicative rationality).

Questions

Students almost always ask for more examples, as if somehow exposure to sufficient examples will mean that they internalise or learn what is intended. This is a manifestation of the epistemological stance mentioned earlier. Having someone taking me sensitively through the steps, where I can stop and rewind whenever I want, looks like a powerful resource. Thus screencasts of a tutor 'working' typical problems are likely to be popular with students, as any teacher will surmise on the basis of what students ask them for. But what do students actually do with them, and what do students need to do so as to use them effectively and efficiently? How can initiative be shifted back to the student? These are important questions at any time when planning a lesson, but particularly when preparing a self-study resource such as a screencast.

Affordances

One question to be asked is what the student is attending to, and whether the resources required (student background, disposition and concern, and powers) are available. For example, what does the student think is 'typical' or generic about the particular example whose working is displayed in the screencast (A4)? Unless either the students have become used to asking this for themselves, or the tutor is explicit about it in the screencast, many students are likely to recognise at best a limited range of permissible change in the salient aspects that can be varied, and may even overlook some of those 'dimensions' (Marton and Tsui 2004).

Clearly some of the affordances are that the student can pause and back-up at will, as with a tape-frame but unlike a live lecture or tutorial. Constraints are that the examples worked are determined by the screencast. Even if students could choose the example, and a CAS could display the workings step by step, it would be difficult to insert the tutor commentary, especially the inner-incantations, which is what students appreciate (Jordan et al. 2011, p. 13). Of course students would also like to be able to stop and ask questions, but that involves a two-way interaction in real time, at least with current technology.

Here the action involves the tutor-screencast, the mathematical content and the student. The student begins the interaction when experiencing a sense of disturbance at not fully grasping or understanding a concept or the use of a technique (A5). They then choose to run the screencast seeking specific assistance. They may subsequently initiate a change of action by pausing, stopping or rerunning. However once running, the initiative immediately switches to the tutor-screencast in terms of the tutor's words and actions. The student attempts to follow. They need extra energy or initiative to shift from assenting to what they see and hear to asserting (trying their own version).

Commentary

If the student stance is 'watching and listening', then the interaction is typical of expounding (A2): the tutor has, by virtue of imagining the students watching and listening, been brought into contact with the content in a particular manner, presumably with awareness of typical stumbling blocks and sticking points experienced by students. Both mathematical and pedagogical content knowledge are required in order to be effective. Sensitivity to the nature and scope of one's own attention is necessary in order to be effective in aligning student attention with the tutor's attention (A5). One reason for not showing the tutor's face is to reduce distraction, to approximate the sense of the 'tutor in your head' being shown what to do. Even so students may be distracted by unfamiliar accent, turns of phrase, and a possible gap between them wanting 'the answer' and the tutor 'expounding'.

If the student is trying to make contact with the mathematical content, then there may be periods of time when the student is initiating and, if the tutor has focused on what the student seeks to find out, the tutor-screencast can act as the intermediary or mediating force to bring the student into contact with the mathematics concerning relevant issues. Typical of the interaction mode of *explaining* is the teacher trying to enter the world of the student; here the student enters the world of the tutor who is trying to act like a student, a form of pseudo-explaining. The use of short tightly focused screencasts is likely to contribute to their usefulness because students can pick and choose which ones might meet their needs most effectively. This leads to the need for an appropriate organisation of screencasts so that users can find what they are looking for and know what each contains without excessive effort, otherwise they will not be used.

There is a difficult issue of milieu-at-a-distance. It is hard enough to persuade live students that making and later modifying conjectures is preferable to keeping silent until you are certain that you are correct. On a screencast a tutor can display this behaviour, but always at the risk of students losing confidence in the tutor who, for example, might keep correcting themselves (explicitly and intentionally modifying previous conjectures). The tutor in a screencast is a role model for the doing of mathematics. If correct and clear mathematics flows out of a pen on screen then students will imagine that unless this happens for them, they are failing or deficient in some way.

Any explicit support the tutor offers in the screencast may not be detected by the student *as* support, because student attention is likely to be on making sense of the mathematical content. Consequently explicit attention to fading of any scaffolding support is going to be required, either through prompting reflection (A4) or by being increasingly indirect about the prompts and commentary in the screencast.

Ways of Working

There is a curious phenomenon with all screen-based activity, namely, "what does the student do when the show is over" (Mason 1985). The cessation of movement and sound creates a hiatus, not unlike the moment when you finish reading an engrossing novel. In that moment, attention shifts to the concerns of the material world, to what is to be done next; insights, relations and properties experienced during the session can evaporate all too readily. In order to learn from the experience of using a screencast (*A*4), students may need to be trained to pause at or near the end and to ask themselves what they have now understood that they did not before, and what they would now like to do in the future that they might otherwise not have done before. It is tempting to suggest that each screencast needs a linked set of exercises on which the student might be advised to work. However, it is not the doing of multiple exercises that leads to effective and efficient learning (*A*1/*MGA*), but rather the bringing to articulation for oneself of what makes a task belong to the space of exercises (Sangwin 2005) coped with by the technique, and the space of examples (Watson and Mason 2005) associated with a concept. The most powerful study strategy a student can use is to construct their own exercises and their own examples of concepts. Effective learning involves training students to 'learn how to learn' (Shah 1978; Claxton 1984).

It is less than clear how watching a screencast, however often, contributes to learning in the sense of the student having an appropriate action come-to-mind in the future as a consequence of interacting with the screencast. It seems that what matters is what activity the student engages in using the screencast as stimulus. Screencasts begin as the tutor *expounding* the use of a technique to solve a particular exercise (*A*2). The tutor is of course aware of the specific exercise as an example of a class of similar exercises. They see the particular as an instance of the general (*perceiving properties A5*). The students, however, see the particular. They may need extra stimulus to see the general through the particular, perhaps in the form of explicit meta-comments by the tutor who draws their own attention, and that of the students, out of the immediate activity so as to become aware of the actions being employed. There is a vast literature on the effectiveness of worked examples (see Atkinson et al. 2000) which could inform the way in which worked examples are presented on screencasts so as to maximise their usefulness and effectiveness for students.

For students who know what they do not know, screencasts could be very effective in clarifying the components of a technique, enriching a concept, or alerting students to mathematical powers, themes and heuristics. Their effectiveness will depend as much on the disposition and agency of the student as on the quality of the awareness exhibited in the screencast.

In some commercial collections such as the Khan Academy (*op cit.*) there is evident lack of sensitivity to classic student misapprehensions, such as for example, confusing the name of a person with their age when working on age related word problems (Word Problems 3). What the student encounters from the screencast is some measure of excitement-concern but manifested as behaviour without access to the thinking that brought that behaviour to mind as being appropriate to the situation.

Extensions

It would be useful to develop screen casts that display other aspects of learning and doing mathematics such as:

The mathematical use of various human powers (imagining and expressing, specialising and generalising, conjecturing and convincing: see Polya 1962 or Mason et al. 1982/2010) in a multitude of contexts;

The recognition of mathematical themes (such as doing and undoing, invariance in the midst of change, freedom and constraint see Gardner 1992, 1993a, b, c); and,

Example construction, including counter-example construction (see Watson and Mason 2005; Mason and Klymchuk 2009).

Rolling Polygons

Background

Mathematical animations have been used for over 50 years to introduce topics, to stimulate exploration and to provide a context for applying ideas to new contexts (Salomon 1979; Tahta 1981). A particularly effective way of working with animations, posters and mental imagery was developed by a group called Leapfrogs (1982) and involves watching (on an actual or a mental screen), then reconstructing what was seen, leading to mathematical interpretation and seeking justification for conjectures about relationships that were articulated.

The Rolling Polygon animation was made in order to offer experience of a range of 'ways of working' including a 'silent start' to a lesson or task, reconstruction, discussion, conjecturing, reasoning and justifying, and reflection (A1, A2, A3). These can all be used in many different contexts beyond animations.

Here the factual rationality is of little import, although one affordance is to bring to attention the way in which mathematical thinking depends on recognising factual relationships encountered in the past as being present. Put another way, relationships recognised in the current situation may be perceived as instances of more general properties.

Narrative

Attention is at first, naturally enough, directed towards the point P moving around the circle. Watching an animation and then reconstructing it proves to be an effective way of aligning student attention. The film invites the conjecture that the triangle remains the same shape independently of the position of P on the circle. This may or may not be experienced as the more technical description 'congruent'; it may not emerge until reconstruction of what was seen. The film also invites the conjecture that a point on the triangle traces an ellipse as P moves around the circle. There are implicit generalities which, if expressed as conjectures, give substance to conjectured relationships as properties of a whole class of phenomena. Thus the size of the circle, the angles between the lines and the position of the point on the triangle could all be varied.

In terms of variation theory, what is likely to stand out for most people is the invariance of the shape of the triangle. Astute observation may reveal that the angles of the triangle are the angles between the lines. Such an observation, treated as a conjecture, might lead to a shift in what is attended to, and how. The presence of the right-angles, for instance, could trigger the possibility of cyclic quadrilaterals or of diameters of a single circle. Choosing between alternative relationships to pursue is an important feature of mathematical problem solving.

Commentary

This task is typical of *phenomenal mathematics* (Mason 2004, 2008) in which a mathematical theorem or technique is introduced by displaying a phenomenon. When the phenomenon is surprising, many students are moved to want to explain it, to make sense of it and to explore possible variations which leave the phenomenon invariant (A0). At first the fact that the triangle appears to remain invariant in shape but not location is a surprise. The fact that a point on the triangle follows an ellipse is equally surprising, and leads to questions such as predicting the positions of the foci from the shape of the triangle, or determining under what conditions the locus will be a circle. If the triangle shape remains invariant, then it must be a rotation of the original, so one possibility is to seek the centre of that rotation, which could then lead to a justification of the first conjecture.

As with any challenging geometrical relationships, there are opportunities to catch shifts in both what is being attended to and what is being stressed (A5). Familiarity with stressing and consequent ignoring (Gattegno 1970) could open up questions about what is being ignored (and that might fruitfully be stressed!).

Affordances

The 'silent presentation' of the task, coupled with its surprise offers, students the opportunity to pose themselves problems as a way of making sense. It provides a task in which everyone can participate because it draws upon known resources (A3). Beginning with an invitation to imagine, before seeing the diagram, affords an opportunity to work on strengthening the power to form mental images (which may be pictorial, verbal, kinaesthetic or some combination of all three). The imagery instructions are in expounding mode, but as soon as surprise is experienced, there can be shifts to other modes such as exploring and expressing (A2). Describing how the film unfolded without recourse to a diagram or the film itself provides an opportunity to express what is being imagined or re-imagined (A5). Various possible approaches may begin to come to mind, so there is an opportunity to park ideas as they emerge so that an efficient and insight-generating approach can be selected.

Considering what can be changed while preserving the phenomenon is further opportunity to imagine and to express, and to conjecture various generalisations. Seeking a justification for the initial phenomenon may lead to recognition of relationships that are expressed as properties in some standard geometrical theorems. Reflecting on that reasoning can lead to increasing the scope of generality of the phenomenon itself.

Trapping the intentions (teleological rationality A0) behind approaches taken is really only possible by intentional withdrawal from action and reflection upon that action.

Ways of Working

Animations lend themselves to a way of working in which individuals collectively experience a phenomenon, then mentally re-play it for themselves, before joining others to try to reconstruct the 'plot', the sequence of images. This in turn alerts attention to critical details that can be examined on a second viewing. Thus shifts of mode of interaction can be rapid and multiple, providing a range of roles for students, teacher and mathematics (A2). Once a reasonable account of what was seen begins to develop, people naturally want to account-for the phenomenon, but it is particularly valuable to try to separate accounts-of and accounting-for, if only because that is vital when interpreting classroom video (Mason 2004) or when cooperating in a collaborative peer group.

Extensions

After thinking about the problem, students might feel moved to use a dynamic geometry package to explore for themselves. Alternatively, an applet (available on the website) can be provided which enables you to vary different constraints, such as the number of lines and the angles between the lines. You can also release the moving point from being confined to a circle to being confined to an ellipse, or even allow it to be completely free in the plane. There is also the question of what role the perpendiculars play: it they were replaced with lines of given slope, perhaps parallel to some given lines, would the triangle remain invariant, and would the locus remain an ellipse?

Implications for Teaching

Seeing this task as an instance of a class of tasks under the general heading *phenomenal mathematics* could transform tasks used with students. A reasonable conjecture is that every topic and every technique in school mathematics and in at least the first few years of undergraduate mathematics could be introduced through generating a phenomenon that surprises many people and invites or invokes attempts to explain what lies behind the phenomenon.

Secret Places

Background

Tom O'Brien (2006) demonstrated that children as young as 9 and 10 are capable of reasoning mathematically when number calculations are not required. The applet was produced to enable primary teachers and teacher educators to experience their own use of mathematical reasoning, in order to sensitise them to possibilities for children. The applet is designed to be used in a tutor-led mode rather than individuals by themselves.

Most people with whom this has been used rise immediately to the challenge. There is an initial sense that it should not be too difficult, however people often discover that they need to re-think what the blue and red information is telling them. Despite several decades of human computer interaction there is still some emotional arousal due to the machine responding to probes (as distinct, say, from a person playing the role of the computer).

Some people display a propensity to want to start clicking without thinking, so the role of the tutor is to act as a brake, getting people to park their first impulse and think more deeply. Participants find themselves imagining what will happen one or more steps ahead, with some resorting to notation in order to keep track of the possibilities. This could provide an instance of 'reasoning by cases' and of being systematic. Attention tends to be on resolving the particular at first, rather than developing a general strategy, so again the role of the teacher is to promote movement to the general.

Narrative

People seem to respond to the challenge very quickly, despite an absence of 'purpose' or evident 'utility' (A3). It seems that the challenge appears tractable, and the dissonance of not-knowing but finding out stimulates emotions which are then harnessed (A0). One or more initial forays with the applet involving rapid clicking develops discernment of pertinent screen details and a sense of the task. Attention then shifts to what information is revealed by different choices, which invokes relationships. There are of course differences in subsequent actions depending on the result of the first click (A5).

People quickly work out that it does not matter which place you try first, so it becomes a practice to click on place 1 to start with. However there is often a split of opinion about what the colour actually means.

Some people want to know how they will know if/when they get the correct place. However the software never confirms the location of the secret place(s). Under most conditions (sufficient places given the number of secret places chosen) there is no need for the applet to validate secret locations, since that 'knowing', with certainty, is the result of reasoning. Even after several 'games', confusion comes to the surface regarding what a blue place tells you about the adjacent places. This is amplified where people work in groups of two or more in an ethos which values conjecturing and justification (communicative rationality A0).

After a few random trials to *get-a-sense-of* what is going on (A1), people usually want to shift into individual or small group work. The initiating impulse has changed, either into an *exploratory* mode in which the teacher and presence of the software introduce and maintain the students in contact with the mathematical reasoning, or into an *exercising* mode in which the desire to try examples initiates student activity (A2). Integral to Pólya's advice (*op cit.*) but unfortunately sometimes overlooked, is the role of specialising (manipulating, exercising) not simply to collect data, but in order to get a sense of underlying structure, leading to a conjectured generality. Put succinctly, *doing* $\neq$ *construing*; something more is required (A4).

Working individually or in small groups, people usually recognise the need for case by case analysis. Sometimes it takes a while to realise that the number of clicks you have to make before you can be certain (for one secret place) depends on what colours show up when you make choices. For many this is an unexpected situation. Bringing to articulation a method for locating the secret places most efficiently can take some time, even when it can be done in practice: *doing* is not the same as *saying* and that again is not the same as *recording* succinctly (A4). Communicating with yourself, then a friend, then a sceptic (Mason et al. 1982/2010) is useful for prompting clarification and experience of locating and distilling the underlying essential relationships forming the structure of the situation.

While the initial or outer task is to 'find the secret place' the implicit inner cognitive task is to develop an efficient method or algorithm for succeeding given what happens with a specified number of places and what is revealed in successive clicks, and to justify this as the best possible strategy in dealing with all possible situations for that number of places. A great deal depends on how teachers prompt reasoning by requiring justifications for choices of places to click, and all that depends on past experience the class has had of mathematical thinking, conjecturing, justifying, etc..

In order to be able to support desired shifts, for example between resolving the particular and seeking a general strategy, it is useful for the teacher to be aware of differences in goals (A3) and, over time, to direct student attention into alignment with the larger educational goal. Emotional commitment (harnessed emotion) may be so strong that students are locked into a simplistic version of the didactic contract (doing what is required will produce expected learning), whereas the teacher is aware that although the outer task is to find the secret location, the inner task (A3) is for the students to educate their awareness about the use of reasoning, becoming aware of possible actions (clicking, deducing, anticipating, conjecturing, …), analysis of cases, ruling out ineffective actions, and developing a general strategy. This is the teacher's teleological rationality, but needs to be picked up by students if they are to gain substantially from the activity. Clearly the factual rationality is of little import in itself.

Adding a second secret place among five places produces an ambiguity because in some configurations there is not enough information to locate them. This can lead to seeking the minimum number of places for which a given number of secret places can be located, or what is equivalent, the maximum number of secret places among a given number of places for which the secret places can always be located.

Affordances

The initial task offers opportunity to encounter and use the notion of symmetry, to realise the importance of considering different possible cases, to break the situation down into all possible distinct cases and to embark on a systematic examination of them all in turn. It also offers opportunity to imagine an action and its consequences, to make conjectures and to modify them in the face of contrary evidence, and to reason about what information is provided by discovering a 'hot' or a 'cold' place. Finding a 'method' which works with a minimal number of clicks is one form of generality (over all choices of location of the secret place).

Maintaining a plenary mode interspersed with individual and small group reconstruction and exploration allows for multiple modes of interaction, and exposure to aspects of mathematical thinking that can be called upon in the future when working on core curriculum topics, informed by the teacher's awareness of the affordances, inner tasks and goals of the activity (A3).

Effectiveness depends greatly on the working ethos and atmosphere of the social setting. It can work well in generating mathematical reasoning in a conjecturing atmosphere in which everything asserted is treated as a conjecture and expected to be modified unless and until it is satisfactorily justified, and in which those who are confident question and support those who are not so confident. It does not work well in an ethos of striving to get the right answer.

The extended task promotes a sense of generality through relating the number of places with the number of clicks required (with one secret place) and then to extend this further to deal with several secret places. It also offers repeated exposure to similar forms of reasoning in multiple situations which can contribute to students integrating these actions into their repertoire of available actions (exercising).

Ways of Working

The applet was designed to be used in plenary so that the teacher is in charge of when buttons get pressed. Ever since electronic screens came into use in classrooms, it has been appreciated that requiring agreement as to what buttons to press next is a powerful stimulus to communication and reasoning, and some projects are based almost solely on this idea (Dawes et al. 2004). In fact the applet has a 'locking feature' to restrict what users can access if it is to be used in small group mode by students. The resistance to acting upon the first idea that comes to mind is one of the contributions that a teacher-led plenary mode can contribute to the education of student awareness (A1), by blocking the first impulse and calling upon more considered thinking. Learning to 'park' an idea and look for a different or better one is an important contribution and part of the potential 'inner task'(A3).

The point of the applet is not actually to find the secret place but to convince yourself and others (friends and sceptics) that your method will always find the secret place(s) in no more than the number of clicks that you claim. Satisfaction and other effective rewards arise from personal use of reasoning powers, and agreement from peers and an expert (teacher). Note however that there is no 'purpose' offered apart from the arising of curiosity, the activating of desire to find the location, and an initial sense that it cannot be too difficult. No one has ever dismissed the task as "well just click all the places ... who cares?".

In order to bring justification through reasoning (reasoning on the basis of agreed properties) to the fore, the teacher needs to manage the discussion, creating and maintaining a conjecturing atmosphere, providing thinking time as well as time and space for expressing ideas and insights, and for rehearsing and challenging the conjectures of others. Opportunities abound for constructing configurations for which a conjectured 'method' does not always find the secret place in the minimum number of clicks.

The applet itself at best provides an introduction to or on-going experience of reasoning by considering and eliminating cases. Unless it is used as part of a programme of experience of activities involving similar types of reasoning, with appropriate drawing of attention to effective and ineffective actions, use of the applet would be mere entertainment.

Extensions

The applet permits changes to the number of places at the table (numbers from 4 to about 25 are distinguishable), the number of secret places, and the spread of the 'hot' information (default value is 1 place each side of the secret place).

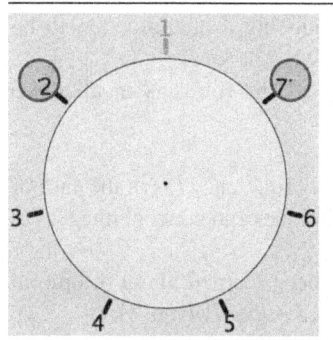

Here position 1 has been clicked and found to be 'cold'. Deductions have been made that positions 2 or 7 could not be the secret place, and have been marked 'cold' by the users to assist their reasoning.

But there is potential ambiguity in this additional notation: interesting things happen when it emerges that some people interpret the cold-marker to mean that clicking there would necessarily give a 'cold' response!

There is a second version that takes the same idea (locating a secret place) into two dimensions, where the space of activity is ostensibly a finite grid of squares. 'Hot' means adjacent horizontally or vertically, but the grid can be turned into a cylinder, torus, Mobius band or Klein bottle, and some of these can have displacements. The intention is to introduce 'as a matter of course' rather than as an object of explicit attention, different surfaces that can be constructed by identifying edges of a rectangle, as multiple contexts in which to exercise similar reasoning. The 2D version offers opportunity to encounter and explore topological notions of 'nearness' on familiar and unfamiliar surfaces all generated in the same manner (identifying some edges). Reflecting on what is the same and what different about the 1D and the various 2D contexts could reinforce awarenesses that students have begun to educate in themselves.

Reflection

Focusing on the use of applets by a teacher as stimulus to activity by students, and using the framework of six modes of interaction, combined with distinguishing various human powers which can be used and developed mathematically, and with distinctions drawn concerning different ways of attending, it emerges that even these apparently simple ways of using software with students are both complex and demanding. The complexity arises from recognition of the need to vary the modes of interaction so as to keep the whole of the psyche involved, and to prompt a reflexive stance in order to learn from experience. The demanding nature of these pressures arises from the need to have come-to-mind appropriate pedagogic strategies in order to maximise the learning potential for students.

The three studies are representative of only a restricted range of stimuli to mathematical thinking afforded by software. The stance taken here is that even taking one mode of interaction as the initial activity, different modes of interaction between stimulus (teacher-applet), student and mathematics are possible and desirable. It is not so much the stimulus that is 'rich' but the ways of working with that stimulus that can be pedagogically rich or impoverished. The narratives offered based on the case studies suggest general observations about what applets can provide:

A means of initiating enquiry and exploration (producing a phenomenon to be explained as in the case of *Secret, Rolling Polygons*);

An environment in which to work (at least some of the time, as in all three studies);

A means of stimulating continued study of a topic;

A means of testing conceptual grasp and manipulative proficiency (as in the case of *Secret Places*) or of reinforcing and clarifying techniques and concept images (as in the case of *Screencasts*);

An environment in which to make use of what has been learned about a topic in further exploration (as in the case of *Secret Places, Rolling Polygons*).

It seems clear that a screencast can, if well constructed, initiate and support conceptual understanding and appreciation, and the details of techniques or procedures. A screencast can even activate desire to master a technique or appreciate a concept. However, screencasts are not well placed to provoke the kind of activity that leads to effective integration, educated awareness that can initiate an action in the future when required. If used in conjunction with routine exercises, then the integration will be only as effective as the structure of the exercises (Mason and Watson 2005; Watson and Mason 2006).

The addition of software into the educational milieu affords both potential and complexity:

Pedagogical complexity arises from the need to develop fresh ways of working effectively, both when students work for themselves or in a small group to make sense of a screencast, and when activity is directed by a teacher using an applet as the focus;
Mathematical complexity arises from the greater scope for a mismatch between the mathematical potential and the teacher's grasp of the topic or concepts; and,
Learning complexity arises from the demands made on students' commitment to learning deeply and effectively.
It may be that within this complexity lay some of the obstacles to greater use within the mathematics classroom.

References

Ainley, J., & Pratt, D. (2002). Purpose and utility in pedagogic task design. In A. Cockburn & E. Nardi (Eds.), *Proceedings of the 26th annual conference of the international group for the psychology of mathematics education Vol. 2* (pp. 17–24). Norwich: PME.
Ascari, M. (2011). *Networking different theoretical lenses to analyze students' reasoning and teacher's actions in the mathematics classroom.* Unpublished Ph.D. Thesis, Universita' Degli Studi Di Torino.
Atkinson, R., Derry, S., Renkl, A., & Wortham, D. (2000). Learning from examples: Instructional principles from the worked examples research. *Review of Educational Research, 70*(2), 181–214.
Baird, J., & Northfield, F. (1992). *Learning from the peel experience.* Melbourne: Monash University.
Bennett, J. (1966). *The dramatic universe* (Vol. 4). London: Routledge.
Bennett, J. (1993). *Elementary systematics: A tool for understanding wholes.* Santa Fe: Bennett Books.
Brousseau, G. (1997). *Theory of Didactical Situations in Mathematics: didactiques des mathématiques* (1970–1990). (trans: Balacheff N., Cooper M., Sutherland R., Warfield V.). Dordrecht: Kluwer.
Brown, S., Collins, A., & Duguid, P. (1989). Situated cognition and the culture of learning. *Educational Researcher, 18*(1), 32–41.
Bruner, J. (1966). *Towards a theory of instruction.* Cambridge: Harvard University Press.
Bruner, J. (1990). *Acts of meaning.* Cambridge: Harvard University Press.
Claxton, G. (1984). *Live and learn: An introduction to the psychology of growth and change in everyday life.* London: Harper and Row.
Davis, B. (1996). *Teaching mathematics: Towards a sound alternative.* New York: Ablex.

Davis, B., & Simmt, E. (2006). Mathematics-for-teaching: An ongoing investigation of the mathematics that teachers (need) to know. *Educational Studies in Mathematics, 61*(3), 293–319.

Dawes, L., Mercer, N., & Wegerif, R. (2004). *Thinking together: A programme of activities for developing speaking, listening and thinking skills* (2nd ed.). Birmingham: Imaginative Minds Ltd.

Festinger, L. (1957). *A theory of cognitive dissonance*. Stanford: Stanford University Press.

Gardiner, A. (1992). Recurring themes in school mathematics: Part 1 direct and inverse operations. *Mathematics in School, 21*(5), 5–7.

Gardiner, A. (1993a). Recurring themes in school mathematics: Part 2 reasons and reasoning. *Mathematics in School, 23*(1), 20–21.

Gardiner, A. (1993b). Recurring themes in school mathematics: Part 3 generalised arithmetic. *Mathematics in School, 22*(2), 20–21.

Gardiner, A. (1993c). Recurring themes in school mathematics, part 4 infinity. *Mathematics in School, 22*(4), 19–21.

Gattegno, C. (1970). *What we owe children: The subordination of teaching to learning*. London: Routledge & Kegan Paul.

Gattegno, C. (1987). *The science of education part I: Theoretical considerations*. New York: Educational Solutions.

Gibson, J. (1979). *The ecological approach to visual perception*. London: Houghton Mifflin.

Habermas, J. (1998). In M. Cooke (Ed.), *On the pragmatics of communication* (pp. 307–342). Cambridge: The MIT Press.

Handa, Y. (2011). *What does understanding mathematics mean for teachers? Relationship as a metaphor for knowing* (Studies in curriculum theory series). London: Routledge.

Heidegger, M. (1962). *Being and time*. New York: Harper & Row.

Hein, P. (1966). *T.T.T. in Grooks*. Copenhagen: Borgen.

Hewitt, D. (1994). *The principle of economy in the learning and teaching of mathematics*, unpublished Ph.D. dissertation, Open University, Milton Keynes.

Hewitt, D. (1996). Mathematical fluency: The nature of practice and the role of subordination. *For the Learning of Mathematics, 16*(2), 28–35.

James, W. (1890 reprinted 1950). *Principles of psychology* (Vol. 1). New York: Dover.

Jaworski, B. (1994). *Investigating mathematics teaching: A constructivist enquiry*. London: Falmer Press.

Jordan, C., Loch, B., Lowe, T., Mestel, B., & Wilkins, C. (2011). Do short screencasts improve student learning of mathematics? *MSOR Connections, 12*(1), 11–14.

Kang, W., & Kilpatrick, J. (1992). Didactic transposition in mathematics textbooks. *For the Learning of Mathematics., 12*(1), 2–7.

Leapfrogs. (1982). *Geometric images*. Derby: Association of Teachers of Mathematics.

Marton, F., & Booth, S. (1997). *Learning and awareness*. Hillsdale: Lawrence Erlbaum.

Marton, F., & Säljö, R. (1976). On qualitative differences in learning—1: Outcome and process. *British Journal of Educational Psychology, 46*, 4–11.

Marton, F., & Tsui, A. (Eds.). (2004). *Classroom discourse and the space for learning*. Marwah: Erlbaum.

Mason, J. (1979). Which medium, Which message, *Visual Education*, February, 29–33.

Mason, J. (1985). What do you do when you turn off the machine? Preparatory paper for ICMI conference: *The influence of computers and informatics on mathematics and its teaching* (pp. 251–256). Strasburg: Inst. de Recherche Sur L'Enseignement des Mathematiques.

Mason, J. (1994a). Learning from experience. In D. Boud & J. Walker (Eds.), *Using experience for learning*. Buckingham: Open University Press.

Mason, J. (1994b). Professional development and practitioner research. *Chreods, 7*, 3–12.

Mason, J. (1998). Enabling teachers to be real teachers: Necessary levels of awareness and structure of attention. *Journal of Mathematics Teacher Education, 1*(3), 243–267.

Mason, J. (2003). Structure of attention in the learning of mathematics. In J. Novotná (Ed.), *Proceedings, international symposium on elementary mathematics teaching* (pp. 9–16). Prague: Charles University.

Mason, J. (2004). *A phenomenal approach to mathematics*. Paper presented at Working Group 16, ICME, Copenhagen.

Mason, J. (2007). Hyper-learning from hyper-teaching: What might the future hold for learning mathematics from & with electronic screens? *Interactive Educational Multimedia* 14, 19–39. (refereed e-journal) 14 (April). Accessed May 2007.

Mason, J. (2008). Phenomenal mathematics. Plenary presentation. *Proceedings of the 11th RUME Conference*. San Diego.

Mason, J. (2009). From assenting to asserting. In O. Skvovemose, P. Valero, & O. Christensen (Eds.), *University science and mathematics education in transition* (pp. 17–40). Berlin: Springer.

Mason, J., & Klymchuk, S. (2009). *Counter examples in calculus*. London: Imperial College Press.

Mason, J., & Watson, A. (2005). *Mathematical Exercises: What is exercised, what is attended to, and how does the structure of the exercises influence these? Invited Presentation to SIG on Variation and Attention*. Nicosia: EARLI.

Mason, J., Burton, L., & Stacey, K. (1982/2010). *Thinking mathematically* (Second Extended Edition). Harlow: Prentice Hall (Pearson).

Mason, J., Drury, H., & Bills, E. (2007). Explorations in the zone of proximal awareness. In J. Watson & K. Beswick (Eds.), *Mathematics: Essential research, essential practice: Proceedings of the 30th annual conference of the Mathematics Education Research Group of Australasia Vol. 1* (pp. 42–58). Adelaide: MERGA.

Norretranders, T. (1998). *The user illusion: Cutting consciousness down to size* (trans: Sydenham, J.). London: Allen Lane.

O'Brien, T. (2006). What is fifth grade? *Phi Beta Kappan*, January, 373–376.

Piaget, J. (1970). *Genetic epistemology*. New York: Norton.

Piaget, J. (1971). *Biology and knowledge*. Chicago: University of Chicago Press.

Poincaré, H. (1956 reprinted 1960). Mathematical creation: Lecture to the psychology society of Paris. In J. Newman (Ed.), *The world of mathematics* (pp. 2041–2050). London: George Allen & Unwin.

Pólya, G. (1962) *Mathematical discovery: On understanding, learning, and teaching problem solving* (combined edition). New York: Wiley.

Ravindra, R. (2009). *The wisdom of Patañjali's Yoga sutras*. Sandpoint: Morning Light Press.

Redmond, K. (2012). Professors without borders. *Prospect*, July, 42–46.

Renkl, A. (1997). Learning from worked-out examples: A study on individual differences. *Cognitive Science, 21*, 1–29.

Renkl, A. (2002). Worked-out examples: Instructional explanations support learning by self-explanations. *Learning and Instruction, 12*, 529–556.

Rhadakrishnan, S. (1953). *The principal Upanishads*. London: George Allen & Unwin.

Salomon, G. (1979). *Interaction of media, cognition and learning*. London: Jossey-Bass.

Sangwin, C. (2005). On Building Polynomials. *The Mathematical Gazette*, November, 441–450. See http://web.mat.bham.ac.uk/C.J.Sangwin/Publications/BuildPoly.pdf

Shah, I. (1978). *Learning how to learn*. London: Octagon.

Shulman, L. (1986). Those who understand: Knowledge and growth in teaching. *Educational Researcher, 15*(2), 4–14.

Skinner, B. F. (1954). The science of learning and the art of teaching. *Harvard Educational Review, 24*(2), 86–97.

Swift, J. (1726). In H. Davis (Ed.), *Gulliver's travels Vol. XI* (p. 267). Oxford: Blackwell.

Tahta, D. (1981). Some thoughts arising from the new Nicolet films. *Mathematics Teaching, 94*, 25–29.

van der Veer, R., & Valsiner, J. (1991). *Understanding Vygotsky*. London: Blackwell.

van Hiele, P. (1986). *Structure and insight: A theory of mathematics education* (Developmental psychology series). London: Academic Press.

Varela, F., Thompson, E., & Rosch, E. (1991). *The embodied mind: Cognitive science and human experience*. Cambridge: MIT Press.

Watson, A., & Mason, J. (2005). *Mathematics as a constructive activity: Learners generating examples*. Mahwah: Erlbaum.

Watson, A., & Mason, J. (2006). Seeing an exercise as a single mathematical object: Using variation to structure sense-making. *Mathematical Thinking and Learning, 8*(2), 91–111.

Part I
Current Practices and Opportunities for Professional Development

Exploring the Quantitative and Qualitative Gap Between Expectation and Implementation: A Survey of English Mathematics Teachers' Uses of ICT

Nicola Bretscher

Abstract This chapter reports the results of a survey of English secondary school mathematics teachers' technology use ($n=188$). Set within the context of a broader study aiming to develop a deeper understanding of how and why mathematics teachers use technology in their classroom practice, the survey findings are used to explore the widely perceived quantitative gap and qualitative gap between the reality of teachers' use of ICT and the potential for ICT suggested by research and policy. Teachers were asked about their access to hardware and software; their perception of the impact of hardware on students' learning; the frequency of their use of ICT resources; their pedagogic practices in relation to ICT; and school and individual-level factors which may influence their use of ICT. This survey suggests that given the right conditions, at least those currently existing in England, ICT might contribute as a lever for change; however, the direction of this change might be construed as an incremental shift towards more teacher-centred practices rather than encouraging more student-centred practices.

Keywords Technology integration • Mathematics education • Teachers' ICT practices • Hardware and software use

Introduction

This chapter reports the findings of a survey of English mathematics teachers' use of Information and Communication Technologies (ICT) in secondary schools. The survey forms part of a broader research study aiming to develop a deeper understanding of how and why mathematics teachers use technology in their classroom practice.

N. Bretscher (✉)
Kings College, University of London, London, UK
e-mail: nicola.bretscher@kcl.ac.uk

Lagrange and Erdogan (2008) record both a quantitative and a qualitative gap between institutional expectations and teachers' use of digital technologies in classroom practice. The apparent gulf between institutional expectations and classroom reality is particularly significant in the context of unprecedented spending by governments around the world on initiatives to develop educational technology (Selwyn 2000), the emphasis placed on using ICT in the UK National Curriculum for mathematics and the inclusion of technology in mathematics curricula more globally (Wong 2003).

The survey findings are used to explore the widely perceived quantitative gap and more subtle qualitative gap between the reality of teachers' use of ICT in the classroom compared with the legacy of the UK Labour government's vision (1997–2010) and the potential of ICT use highlighted by educational research. Teachers were asked about their access to hardware and software; their perception of the impact of hardware on students' learning; the frequency of their use of ICT resources; their pedagogic practices in relation to ICT; and school and individual-level factors which may influence their use of ICT. Previous surveys have tended to be confused by a lack of differentiation between hardware and software use. In contrast, this survey aims to provide insight into the types of software mathematics teachers choose to use in conjunction with particular types of hardware. More specifically, questions were posed separately regarding teachers' use of software with interactive whiteboards (IWBs) or data projectors in a whole class context and teachers' use of software in the context of a computer suite or using laptops, where students work individually, in pairs or in small groups. In addition, the reasons underlying the gap between expectations and classroom implementation are probed using the data collected in relation to school and individual-level factors.

The Quantitative and Qualitative Gap in Mathematics Teachers' ICT Use

The evidence for a quantitative gap seems fairly unequivocal. The TIMSS 2007 study (Mullis et al. 2008) reports that it was rare for computers to be used for any activity as often as in half the mathematics lessons, even in countries with relatively high availability. In the UK, the ImpaCT2 report (2003) stated that 67 % of pupils at Key Stage 3 never or hardly ever used ICT in their mathematics lessons. In addition, Ofsted (2008) reported that opportunities for pupils to use ICT to solve or explore mathematical problems had markedly decreased, despite the previous years of unprecedented investment by the then Labour government, directing over £5 billion of funding towards educational ICT during the 1997–2007 period (Selwyn 2008). On the other hand, Moss et al.'s (2007) survey on the introduction of IWBs in London schools reports that many teachers are using IWBs in most or every lesson, especially in mathematics and science, and that mathematics teachers made the most use of externally produced subject-specific software.

Citing Ruthven and Hennessy's (2002) study of mathematics teachers in England as evidence, Lagrange and Erdogan (2008, p. 66) define a qualitative gap between the expectation and implementation of ICT as the tendency of teachers to view the benefits of technology in terms of enabling "general 'pedagogical' aspirations rather than for its 'didactical' contribution to mathematics learning". That is, mathematics teachers articulated the benefits of technology as indirectly enhancing students' learning through increased *pace and productivity* and improved engagement (Ruthven and Hennessy 2002) rather than providing a direct means of enhancing mathematics pedagogy. Evidence for a qualitative gap may also be inferred from survey reports of mathematics teachers' typical software use. For example in the US, Becker, Ravitz and Wong (1999) found that drill and practice software was most often used by mathematics teachers. Although this inference is problematic, the use of presentation-oriented software might suggest an additional obstacle to more student-centred practices. Despite this, *The Geometer's Sketchpad* (Key Curriculum Press 2003) was the most favoured mathematical software amongst teachers in Becker et al.'s (1999) study. However, as will be discussed in the paragraph below, surveys tend to give an overview of technology use and are not detailed enough to provide a picture of the different types of software teachers use in conjunction with particular types of hardware. Investigating the choices teachers make about the software and hardware they use in their classrooms is therefore important in order to understand the apparent failure of ICT to make an impression on school mathematics.

Mathematics Teachers' Choices: Hardware and Software

The type of hardware and its deployment appears to be an important factor in structuring teachers' choices about technology use in their classroom practice. In particular, the hardware available affects the types of classroom organisation possible and the nature of pupil interactions with any software used in conjunction with the hardware. It seems reasonable then that the available hardware might also affect teachers' choice of software and how they choose to integrate the use of such software into their classroom practice. For example, investigating teachers' use of technology in the US, Becker et al. (1999) found that teachers with computers in their classrooms were three times more likely to use them compared to teachers who have access to larger numbers of computers but only available in shared computer rooms. One of five key factors in structuring teachers' classroom practice that Ruthven (2009) describes is the *working environment*, which is the physical location and layout of the classroom, the classroom organisation and procedures of a lesson. Indeed, the reported popularity of IWBs amongst teachers in the UK appears due to their ease of use in a whole class context, making this hardware seem a more teacher-oriented form of technology (Moss et al. 2007).

Currently, little is known about what types of software teachers choose to use in conjunction with particular types of hardware (Clark-Wilson 2008). International comparisons of educational technology use such as the TIMSS (Mullis et al. 2008), PISA (OECD 2005) or SITES (Law et al. 2008) surveys can only give a broad overview of technology use and are not fine-grained enough to consider usage of different types of software or hardware at the school level. In terms of hardware, the UK represents a special case since it became the first school-level system to invest heavily in IWBs (Moss et al. 2007). However, large-scale surveys of technology use within the UK tend not to report in detail on technology use within subject areas, such as mathematics, nor to differentiate sufficiently between hardware and software use. Thus whilst large-scale surveys can provide a broad picture of technology use, they cannot provide much insight into the nature of the specific uses by teachers in general or by mathematics teachers in particular. For example, the annual Becta schools survey *Harnessing Technology* reported that 53 % of mathematics teachers use subject-specific software in half or more lessons (Kitchen et al. 2007). However, no further detail is given on what type of subject-specific software is used, nor any indication of the hardware involved. Surveys focusing on mathematics teachers' use of technology, such as the survey conducted by the Fischer Family Trust (2003) or Hyde's (2004) small-scale survey, give a more detailed picture of the types of software used by mathematics teachers; however, this picture is again confused by the lack of distinction between hardware and software use. Similarly, Forgasz's (2002) survey of mathematics teachers' use of technology in Victoria, Australia, gives a detailed picture of the types of software used by mathematics teachers with computers. However, it is not clear whether other types of hardware were available or used by teachers, nor how frequently specific software was used. Miller and Glover's (2006) study of UK mathematics teachers' use of IWBs reports that fewer than 5 % of lessons observed used 'Other ICT' such as geometry packages, spreadsheet or graphing programs; however, they note their lack of use may simply be a consequence of the topics being taught at the time of observation.

The survey reported in this chapter builds on previous surveys by providing an insight into the types of software mathematics teachers choose to use in conjunction with particular types of hardware. In particular, teachers were asked to report their frequency of use of a list of software types in a whole-class context with IWBs or data projectors and their use of the software in the context of a computer suite or when using laptops, where students work individually or in pairs. Teachers were also asked to give an indication of their pedagogic practices using ICT in each of these contexts. Responding to more teacher-centred statements like "I use ICT for presentation purposes" (IWB context) and "Students use ICT to practice mathematical skills" (computer suite context) alongside more student-centred statements like "I use ICT to follow up and explore students' ideas" (IWB context) and "I let students 'get a feel' for the software" (computer suite context), teachers indicated how often these practices occurred in their classroom teaching using ICT. Thus the data from this survey provides a basis for an exploration of the nature of both the quantitative and qualitative gap between expectation and implementation of ICT in English mathematics classrooms.

Factors Influencing Mathematics Teachers' Use of ICT

Summarising previous surveys, Assude et al. (2010) note the similarity of factors encouraging or discouraging mathematics teachers' use of ICT across a range of national and international settings spanning more than a decade. At school level in particular, they raise as issues: access to hardware and software; professional development needs; and technical support and resources as factors which appear to outweigh individual-level factors such as confidence, in preventing teachers from integrating technology into their mathematics teaching (Assude et al. 2010, p. 416). Based on the findings of previous surveys, the survey reported in this chapter also asked teachers about school and individual-level factors which may influence their use of ICT. In the school context, teachers were asked for the level of their agreement with statements addressing factors such as access to hardware, software issues, collegial and technical support, provision of professional development and ICT integration in schemes of work. Again in contrast to previous surveys, questions regarding individual-level factors were posed separately in relation to teachers' use of IWBs or data projectors in a whole class context and teachers' use of computer suites or a class set of laptops, where students work individually or in pairs. This data should provide for a more nuanced discussion of the factors underlying classroom use of ICT, in particular the apparent popularity of IWBs in comparison to other forms of hardware, beyond a common-sense statement that IWBs are a more teacher-oriented form of technology.

Understanding Teachers' Use of Technology from a Socio-Cultural Perspective

The broader aim of this study is to develop a deeper understanding of both how and why mathematics teachers use technology in their classroom practice. As with any curriculum resource, how and why teachers make use of the resource in their teaching is a central research question. In this sense, I view digital technologies simply as a particular type of resource amongst a wider range of curriculum resources and not as something special or unique. This approach is similar to that adopted by Ruthven (2009) and Gueudet and Trouche (2009), and contrasts to some extent with Zbiek et al.'s (2007) approach of singling out and focusing on certain digital technologies as *cognitive tools*. Addressing the broader aim of this study, I assume a socio-cultural perspective on teachers' use of resources in accordance with that described by Remillard (2005) as "*curriculum use as participation with the text*". Remillard's (2005) perspective was developed in relation to 'curriculum materials', specifically referring to printed, often published resources designed for use by teachers and students during instruction. Nevertheless, this perspective is appropriate in the light of my stance towards technology as simply one amongst a range of resources, essentially as a particular type of 'text'. In addition, similar perspectives have been applied to a much wider range of resources and in particular to digital technologies (Gueudet and Trouche 2009; Ruthven et al. 2008; Ruthven 2009).

Applying Remillard's Perspective to Teachers' Use of Technology

Remillard's (2005) perspective of *"curriculum use as participation with the text"* views teachers as sense-makers (Spillane 2006), actively interpreting curriculum materials through a process of dynamic interaction. Underlying this perspective are Vygotskian notions of tool use, wherein tools both shape and are shaped by human action through their constraints and affordances (Remillard 2005, p. 221). Applying Remillard's perspective to technology implies that, although the constraints and affordances inherent in digital technologies may help to shape its end use in the classroom, inevitably, the end user, in this case individual teachers, will also work to shape the technology. Thus the design and nature of hardware or software is an ingredient in, but does not determine, the way individual teachers interpret and make use of particular technologies in their classroom practice. For example, Ruthven's (2008, 2009) research on mathematics teachers' use of technology, in his notion of *interpretative flexibility* and claims of interaction between teachers' *curriculum scripts* and *resource systems* coincide with the perspective described by Remillard. Similarly, Gueudet and Trouche's (2009) outline of the documentational approach, extending the widely influential instrumental approach to teachers' appropriation of technology, shares the same Vygotskian roots as Remillard's perspective. Put more simply, there is no guarantee that teachers will use mathematical software designated *cognitive tools* (Zbiek et al. 2007), such as dynamic geometry, graphing or spreadsheet software, if they use them at all, in ways approaching those envisaged by their designers or advocated in policy literature or mathematics education research. Significantly, Ruthven and Hennessy (2002) provide empirical evidence of teachers using such technology to indirectly enhance students' learning through increased *pace and productivity* and improved engagement rather than providing a direct means of enhancing mathematics pedagogy.

Remillard's (2005) perspective also recognises the impact of contextual features in enabling or constraining teachers' interpretations of technology. Stein et al. (2007) identify *context* as one of the factors influencing the participatory relationship between teachers and curriculum materials. In particular, they highlight contextual features, such as *time* available for planning and instruction, *locale* (school and departmental) *cultures* and *teacher support* through professional development, that can constrain or enable teachers' interpretations of curriculum materials. Similarly, Ruthven (2009) describes *working environment* and *time economy* as two of five structuring factors of classroom practice in relation to technology and Gueudet and Trouche (2009) include institutional influences as part of their model of the documentational approach.

Theoretical Issues in Using Self-Report Data to Understand Technology Use

This sub-section outlines a theoretical approach to understanding the possibilities and limitation in using self-report data to gain insight into teachers' use of technology. To capture elements of the ways mathematics teachers in England interpret and use digital technologies in their classrooms, items relating to teachers' pedagogic practices using ICT in a whole-class context with interactive whiteboards or data projectors and their use of ICT in the context of a computer suite or using laptops, were included in the survey instrument. These self-report pedagogic practice items attempt to access the ways teachers interpret and use these types of hardware, to explore the qualitative gap in technology use, however they cannot provide an indication of how teachers interpret specific software packages within these contexts. Further, a distinction must be acknowledged between what we say we do and what we do, relating to Argyris and Schon's (1974, pp. 6–10) definition of 'espoused theory' (theory to which we give our allegiance) and 'theory-in-use' (theory which governs actions). Thus, teachers' self-reports must be considered "as being their account for us of what they do, refracting their espoused theory of teaching practice, through the items in the instrument that refer them to their concrete, practical actions" (Pampaka et al. 2012). In this sense, teachers' self-reports of pedagogic practice cannot be assumed to correspond exactly with what they do in the actuality of the classroom. Nevertheless, Adler (2001) argues there is some relation or overlap between espoused theories and theories-in-use, although one cannot be reduced to the other. Hence, in the absence of direct observation data, teachers' self-reports may be taken to give some insight into their use of hardware in classroom practice, whilst acknowledging the imperfections of the measure. Viewed as espoused theories, these self-reports of pedagogic practice may also provide insight into teachers' conceptions (Thompson 1992; Zbiek et al. 2007) of mathematics teaching with regard to technology, mediated by the items in the instrument.

Consequently, my broader study is directed not simply at documenting the extent of teachers' use of technology and the degree to which the quantitative and qualitative gap exists, but also at highlighting ways in which teachers, as sense-makers, interpret and shape the technology within the constraining or enabling features of their local school and departmental contexts. This chapter focuses primarily on detailing the types of hardware and software teachers use in their classroom practice, together with indications of how technology is being used, thus any conclusions with regard to why teachers use technology in their classroom practice are necessarily tentative.

The Survey: Instrument, Sample and Data Analyses

The survey instrument has been progressively developed over the course of various phases of piloting. The initial questionnaire design was informed by previous surveys of mathematics teachers' use of ICT, primarily Hyde's (2004) survey of mathematics teachers in Southampton and Forgasz's (2002) survey of mathematics teachers in Victoria, Australia. This questionnaire was trialled with students on the Post-Graduate Certificate of Education[1] (PGCE) mathematics course at King's College London, before being piloted with 27 schools working in partnership with King's College London to offer initial teacher education in secondary mathematics. The results of the pilot survey are reported in Bretscher (2011). As a result of this piloting, the questionnaire was re-developed to include items relating to teachers' pedagogic practices with ICT and to highlight more clearly the division of questions between using ICT in a whole-class context and using ICT in the context of a computer suite or using laptops. Items relating to school and individual factors affecting teachers' use of ICT were also re-written to aid clarity. The re-designed questionnaire was trialled in two further think-alouds[2] with PGCE students and with three experienced in-service teachers, who completed the questionnaire and then gave verbal feedback. The theoretical perspective outlined above implies that survey respondents engage in a participatory relationship with the text of the questionnaire, actively interpreting questionnaire items in the light of their own circumstances, whilst the questionnaire items may also shape respondents' perception of these circumstances. Indeed, one of the three experienced in-service teachers, with whom the questionnaire was trialled, commented with surprise on how she perceived shifts in her own conception of what 'ICT use' meant as she progressed through different sections of the questionnaire.

The final survey instrument contained mainly closed Likert-type response formats grouped under the following sections:

A. *ICT in your school* – items on access to hardware/software and school/departmental level factors effecting ICT use;
B. *ICT use in your own mathematics teaching*

   i. *Your use of hardware* – perceived impact and frequency of use of hardware;
   ii. *Using an interactive whiteboard or data projector in maths lessons* – items on frequency of software use, individual factors effecting ICT use and pedagogic practices with an IWB or data projector in a whole-class context;
   iii. *Maths lessons in a computer suite or using laptops* – similarly, items on frequency of software use, individual factors effecting ICT use and pedagogic practices with ICT in the context of a computer suite or using laptops;

C. *Your own mathematics teaching in general* – Pampaka et al.'s (2012) items relating to pedagogic practices in general (not specific to ICT use); and
D. *About You* – personal background details.

---

[1] The Post-Graduate Certificate of Education is a 1-year initial teacher-training course.

[2] A think-aloud is an interview where the survey respondent offers a verbal explanation of their responses as they progress through the questionnaire.

In addition, two open-ended response questions were included so that teachers could comment more widely on issues relating to access to hardware or software and on using ICT in general in maths lessons. The list of software was derived mainly from Hyde's (2004) list, checked against a survey of software use by the Fischer Family Trust (2003), with the notable inclusion of IWB software and the MyMaths.co.uk website (Oxford University Press 2012). IWB software refers to presentation-type software that is designed specifically for use with IWB hardware, for example SMART Notebook or Promethean ActivInspire. The growing presence of IWBs in mathematics lessons in England, indicated by the pilot study and other reports (e.g. Moss et al. 2007), suggests that IWB software may be used regularly by mathematics teachers and it was therefore included in the list of software for this survey. The *MyMaths* website was included since this site was known anecdotally to be widely used in UK schools (see for example, the school case studies reported in Clark-Wilson 2008, pp. 103–104). It is a subscription site offering teachers pre-planned lessons, on-line homework and many other resources. The lessons and homework are linked to an 'Assessment Management system', allowing teachers to track individual student's progress.

Questionnaires were sent to 87 secondary schools selected through contacts with mathematics educators in three English universities. The schools were thus situated mainly within three rough geographic areas: Greater London, West Yorkshire and the South of England (taken as comprising the counties of Hampshire, West Sussex and Dorset). Nine questionnaires were sent to each school and 50 schools agreed to take part. A total of 188 completed individual teacher questionnaires returned, an average of 3.8 questionnaires per school. Twelve schools returned only one completed questionnaire, whilst one returned all nine. The sample cannot be said to be statistically representative, nevertheless, the participating schools cover a range of characteristics including a wide range of attainment in national tests; most were state schools but some were private schools; some have speciality status and some do not; some are single sex and some are selective. The participating teachers (101 F; 86 M; 1 unspecified) had an average age of 38.5 years and an average length of service of 10.5 years. The majority of respondents (96) described their main responsibility as classroom teacher. The sample also included 24 heads of department, 18 deputy heads of department and 24 Key Stage[3] coordinators. There may be a potential bias in the sample towards teachers who are relatively well-disposed towards ICT or those wishing to be seen as frequent users of ICT. Comparing themselves to their colleagues in the maths department, only 9.0 % of survey respondents thought they use ICT less or much less frequently whereas 33.5 % thought they use ICT more or much more frequently.

Data that could be analysed statistically were manually entered into PASW Statistics 18.0. This package was used to generate descriptive statistics (i.e. frequency

---

[3] A Key Stage coordinator is a teacher with responsibility for overseeing the delivery of the mathematics curriculum to certain year groups. For example, Key Stage 3 refers to the first three years of secondary school, whilst Key Stage 4 refers to the remaining two years of compulsory secondary schooling.

distributions and means) and inferential statistics (t-tests and χ^2 tests) were calculated as appropriate. An independent data coding check, based on a 10 % sample of questionnaires, gave a coding accuracy of greater than 99.9 %.

In order to investigate the influence of contextual factors relating to teachers' local school and departmental contexts on their technology use, a crude measure of school level support for ICT use was also calculated for each school, as follows. A school score for each of the 'school level' factors relating to ICT use was calculated i.e. the mean of the responses given by the teachers in that school. An overall support score for each school was then calculated as the mean of its scores for the 'school level' factors (negatively worded items were reverse-coded). Schools were labeled 1 if their overall support score was higher than the school sample mean and 0 if their mean support score was lower than or equal to the school sample mean.

Results

Access to Hardware and Software

All schools in the sample equipped their teachers with either IWBs or data projectors. The near ubiquity of IWBs in English mathematics classrooms can be ascribed in large part to funding initiatives put in place by the previous Labour government, allowing the purchase of this technology by schools on a large scale. Indeed, in only two schools did all the responding teachers say they had no access to IWBs: in school 90, one teacher responded, reporting access to data projectors (but not IWBs). Similarly in school 42, eight teachers responded, seven of these reporting access to data projectors. The eighth teacher in school 42 was the only respondent in the survey to report having access neither to an IWB nor to a data projector, specifically commenting on the questionnaire that s/he never used this hardware in classroom teaching – despite his/her colleagues' access to and frequent use of this technology.

In contrast only 71.8 % of teachers reported having access to a computer suite shared with other departments. This seems surprisingly low, especially when compared with the coverage of IWBs (93.1 % – see Table 1). Although 53 teachers report having no access to a shared computer suite, 21 of these teachers report having access instead to a computer suite dedicated to the maths department. This leaves 32 teachers (17.0 %) saying they have no access to a computer suite at all (either shared or dedicated to the maths department). Looking across schools however, there are only three schools in which none of the teachers report having access to a computer suite of either type. In each of these three schools only one or two teachers completed questionnaires. Furthermore, 55 % of the 53 teachers who report having no access to a shared computer suite, conflictingly report using a shared computer suite with some frequency during their teaching. Based on this measure, the apparent unreliability of reporting access to shared computer suites was far higher than for other types of hardware. 32 % of teachers who reported no access to

Table 1 Number of teachers with access to hardware, $n = 188$

	With access (%)	
Interactive whiteboard	175	(93.1)
Data projector	36	(19.1)
Computer suite (shared)	135	(71.8)
Computer suite (maths only)	39	(20.7)
Laptops	41	(21.8)
Graphic calculators	65	(34.6)

a data projector claimed to use them in their teaching, possibly reflecting confusion between the IWB and data projector categories.[4] For all other types of hardware included in the survey this figure was below 10 %. The lack of consistency in teachers' responses, both across schools and individually, suggests that while some teachers are reporting access to shared computer suites on the basis of their awareness of the existence of hardware, others are responding according to their perception of availability of the hardware for use (and there may be other interpretations too). Difficulties in booking computer rooms mean that, although shared computer suites exist, their availability for actual use is often restricted. 23.1 % of teachers' responses to an open-ended question regarding issues with access to hardware and software commented on difficulties relating to gaining access to computer suites. The quote below gives a sense of these teachers' comments on hardware access and neatly summarises the contrast in accessibility between IWBs and computer rooms:

> It is easy for us to use ICT with the software from the front but difficult to gain access to the computers for an ICT lesson where students use the computers.

In addition, 25.6 % of teachers' responses commented on the unreliability or slowness of ICT facilities: thus even where access was not an impediment, technical issues could make lessons involving ICT highly problematic as illustrated by the following comment:

> Main issue is unreliability of ICT – so that you cannot guarantee that a planned lesson using ICT will run to plan.

A computer suite dedicated to the maths department or class sets of laptops might be a potential solution to difficulties in gaining access. However, these facilities are still fairly rare and increased access does not overcome technical issues – indeed, in the case of laptops at least, they may carry additional technical difficulties, as one frustrated teacher commented:

> I have access to a class set of laptops (one between two) but [I] never use [them] as the batteries do not last a full lesson. There is very limited access to computer rooms as an alternative.

Access to generic software tools such as word-processing, spreadsheet and presentation software is almost universal (above 90 %). The majority of teachers appear

---

[4] In the survey, the IWB category was referred to as 'Interactive whiteboard with a data projector' whereas the data projector category was defined as 'Data projector only, linked to a computer'.

Table 2 Number of teachers with access to software, $n=188$

		With access (%)	
Spreadsheet		176	(93.6)
PowerPoint		171	(91.0)
Word		171	(91.0)
MyMaths.co.uk website		168	(89.4)
Interactive whiteboard software		166	(88.3)
Email		160	(85.1)
Graphing software		148	(78.7)
Other websites		141	(75.0)
Interactive geometry software		124	(66.0)
CD Roms		117	(62.2)
Database		81	(43.1)
Logo		54	(28.7)
SMILE		39	(20.7)

to have access to graphing software (e.g. *Autograph* and *Omnigraph*) and dynamic geometry software (e.g. *Cabri*, *The Geometer's Sketchpad* and *GeoGebra*), however a significant minority report no access to these resources (21.3 % and 34.0 % respectively). Of course, this may reflect teachers' lack of awareness of the existence of the software at their school or, as with hardware, teachers may be responding based on their perception of the ease of accessing software rather than its existence. Nevertheless, the number of teachers reporting no access to these types of software is surprising perhaps, given recommendations in national curricula that pupils be given opportunities to use such software in mathematics, although there is no compulsion to do so through national examinations, for example (Table 2).

Perhaps more surprising is the near ubiquity of the *MyMaths* website, with 89.4 % of teachers reporting access, costing secondary schools around £540 in annual subscription fees. Indeed, in only two schools did all the teachers consistently report not having access to the *MyMaths* website. Using databases in work on data-handling was a statutory requirement of the original National Curriculum in 1989 (DES 1989) and Logo appeared more frequently than any other form of software in algebra and geometry contexts, although references to these software disappeared in later revisions (Andrews 1997). Despite this only 43.1 % of teachers report access to database software and 28.7 % of teachers responded positively for access to Logo. Some teachers complained about restrictions on downloading and installing software, such as *GeoGebra*, and access to some websites being unnecessarily blocked. Although software might exist in a school, teachers expressed uncertainties over whether it had been installed on all computers or whether it was available at any given time, thereby adding complexity to conducting lessons in a computer suite, as the following comments illustrate:

> Migration of software to new network has caused several items of software to be inaccessible.
>
> The school system is sometimes slow which makes accessing the software time-consuming at times. Changes in our school status mean we have lost some software. Updates in SMARTboard have caused squared paper options to disappear.

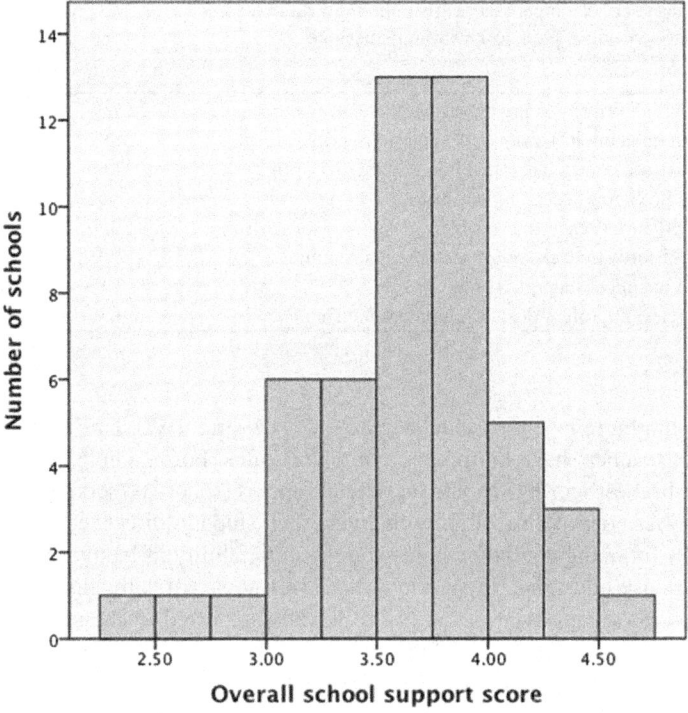

Fig. 1 Distribution of overall school support scores, $n = 50$, mean $= 3.64$, s.d. $= .44$

School Level Factors Relating to ICT Use

Overall, teachers' responses were positive about the factors affecting ICT use, casting their departments and schools in general as supportive communities in which to develop their mathematics teaching using ICT. Indeed Fig. 1 shows that the overall support score for the sample of schools was skewed towards the positive agreement end of the response scale. This could be interpreted as resulting from sample bias – that schools in the sample were more likely to be supportive of ICT use than is the norm – or that teachers simply tend to represent their schools and departments in a positive light. The overall school support score for the three lowest-scoring schools were based upon only one or two respondents from each school; this was not necessarily the case for high scoring schools.

Table 3 shows the mean school score for the school level factors included in the survey instrument. In particular, teachers highlighted their departmental colleagues as supporting their use of ICT. Surprisingly perhaps, given the comments in the previous section, in general teachers tended to disagree that they often had problems accessing hardware. It is important to note here that this question didn't discriminate between access to IWB hardware or computer hardware: thus the positive

Table 3 Mean school support scores for school level factors, $n=50$. Scored on a 5-point Likert-scale where 5 = strongly agree to 1 = strongly disagree

	Mean	(SD)
ICT use is a high priority in my department	3.64	(.72)
I get support on using ICT from colleagues in my department	4.01	(.53)
ICT resources are poorly integrated into schemes of work	2.60	(.78)
I often have problems accessing hardware	2.62	(.77)
Access to software is easy and reliable	3.50	(.76)
The available software lacks relevance to the curriculum	2.03	(.56)
The level of technical support is poor	2.15	(.88)
I have had relevant professional development in using ICT	3.37	(.76)

response might reflect the relative ease of accessing IWB hardware, masking difficulties teachers have in booking computer suites. For example, teachers from two of the highest scoring schools for overall support (school 60 scored 4.18; school 80 scored 4.25) raised difficulties with giving their students direct access to ICT due to problems booking computer suites and the unreliability of laptops via the open-ended response questions. In addition, there was no statistically significant difference in access to any type of hardware or software listed in the survey between schools identified as providing high and low support for using ICT in teaching mathematics (based on χ^2 tests at the 5 % level).

Frequency and Perceived Impact of Hardware Use

The majority of teachers use IWBs and data projectors in almost every lesson, with 85 % of teachers using IWBs in almost every lesson, see Table 4. The ready availability of IWBs and data projectors in normal classrooms makes it unsurprising that they are the most frequently used hardware. IWBs stand out from the other types of hardware as having the highest perceived impact (see Table 5) – this is likely to be linked to their high frequency of use. Interestingly, the perceived impact of data projectors is little different from and actually slightly lower than that of computer suites and laptops in general. Of the 139 teachers reporting impact on student learning for data projectors, only 12 did not have access to IWBs. The relatively low mean impact score for data projectors compared to that of IWBs may reflect a perception that the additional 'interactivity' of IWBs makes them superior for teaching purposes.

Computer rooms shared with other departments have a much lower frequency of use, with 77 % of teachers using them once or twice a term or less. As with IWBs, the frequency of use is to some extent reflected by difficulties in access and in turn reflects the lower impact score of shared computer rooms. Computer suites dedicated to the mathematics department appear to be used slightly more frequently, with a smaller percentage of teachers claiming they never use the resource and a

Table 4 Frequency of hardware use, in %. Note the 'Never' column excludes those who reported having no access to the hardware

		Never	Annually	Once or twice a term	Once a week	Almost every lesson
Interactive whiteboard	n=175	4	1	2	8	**85**
Data projector	n=35	3	0	9	26	**63**
Computer suite (shared)	n=131	6	11	**60**	21	1
Computer suite (maths)	n=37	3	16	**51**	30	0
Laptops	n=41	32	7	**37**	20	5
Graphic calculators	n=63	30	29	**40**	2	0

Table 5 Mean perceived impact score. Scored on a 4-point Likert scale where 4=substantial; 3=significant; 2=some; 1=very little

		Mean impact	(SD)
Interactive whiteboard	n=182	3.16	(.84)
Data projector	n=139	2.43	(.89)
Computer suite (shared)	n=168	2.53	(.80)
Computer suite (maths only)	n=131	2.53	(.99)
Laptops	n=133	2.47	(.97)
Graphic calculators	n=140	2.21	(.90)

somewhat larger percentage saying they use the resource once a week. Nevertheless, the increased frequency of use is marginal since 70 % of teachers still only use a computer suite dedicated to the maths department once or twice a term or less. Similarly there is no apparent difference in teachers' perception of the impact on students' learning of a computer suite dedicated to the maths department compared to one shared by other departments.

The portability of laptops might make it easier to give students direct access to ICT within a 'normal' classroom context; however, this survey suggests that they do not lead to an increase in usage compared to a shared computer suite. Indeed, a greater proportion of teachers with access to laptops report never using them, perhaps due to the kinds of technical difficulties alluded to in previous sections. Whilst the mean impact score of computer suites, laptops and data projectors are fairly similar, the perceived impact of graphic calculators is appreciably lower than this cluster. Likewise, graphic calculators have the lowest profile of frequency of use with 99 % of teachers reporting usage of once or twice a term or less.

Differences in the frequency of use of IWBs between schools with higher and lower support for ICT use were statistically significant ($\chi^2 = 16.67$, df=2, p=.0002). Specifically, teachers in schools with higher support reported higher frequency of use for IWBs in almost every lesson than those in schools with lower support. The difference in the frequency of data projector usage between schools with higher and lower support for ICT was also statistically significant ($\chi^2 = 17.04$, df=3, p=.001). In higher support schools, more teachers claim never to use data projectors than was expected compared to those in lower support schools. Thus teachers in higher support schools used data projectors less frequently than those in lower support schools. There were no statistically significant differences in frequency of use for any of the other hardware listed in the survey, tested at the 5 % level. The differences between higher and lower support schools in terms of teachers' perceptions of the impact of using hardware on students' learning followed a similar pattern. Those in higher support schools were significantly more likely to perceive IWBs as having a substantial impact on students' learning, whereas those in lower support schools thought IWBs only had some impact ($\chi^2 = 22.38$, df = 2, p < .0001). There were also significant differences in the perception of impact of data projectors between teachers in higher and lower support schools ($\chi^2 = 8.61$, df = 3, p = .035). In higher support schools, more teachers than expected thought that data projectors had either very little impact or substantial impact. Although it is not so easy to interpret this result, it could be taken to suggest that teachers in higher support schools have more extreme views about the impact of data projectors. Again there were no statistically significant differences in teachers' perception of impact for any of the other hardware listed in the survey, tested at the 5 % level. These results can be interpreted in at least two ways: when considering school or departmental factors relating to ICT use, teachers appear to equate ICT use with IWB use. An alternative interpretation is that whilst supportive departments can apparently facilitate teachers' use of IWBs, they do little to ameliorate obstacles to giving students direct access to ICT via computer suites or laptops. These results also tend to support the finding noted above that teachers appear to prefer IWBs to data projectors.

Frequency of Software Use

Table 6 compares the mean frequency of software use in lessons with an IWB or data projector to lessons where students are given direct access to the software, i.e. those that take place in a computer room or with laptops. A score of above 2 indicates the software is used more than once or twice a term. Databases, SMILE and Logo scored very low in both contexts, with a score of below 1 indicating less than annual use, so no satisfactory comparison can be made for these software packages. IWB software was the most frequently used piece of software (3.19) in a whole-class context with an IWB. This was followed by PowerPoint, other (unspecified) websites and the *MyMaths* website which also scored above 2. All other types of software including graphing, geometry and spreadsheet software were used on average less than once or twice a term in a whole-class context with an IWB. Thus in general, presentation-oriented software dominates IWB use. Whilst the theoretical stance adopted in this study suggests that making any inferences regarding teachers' actual use of such software is problematic, it is reasonable to note that the design of such presentation-oriented software tends to be more teacher-centred and may therefore present an additional obstacle to the development of more student-centred practices.

Table 6 Mean frequency of software use (a) with IWB or data projector and (b) giving students' direct access via a computer suite or with laptops. Scored on a 5-point Likert-scale where 0 = never, 1 = annually, 2 = once or twice a term, 3 = once a week, 4 = almost every lesson

(a)			(b)		
For IWB/data projectors, $n = 147$	Mean freq.	(SD)	For direct student access, $n = 158$	Mean freq.	(SD)
IWB software	**3.19**	**(1.39)**	IWB software	1.24	(1.54)
PowerPoint	**2.57**	**(1.24)**	PowerPoint	1.46	(.98)
Other websites	**2.56**	**(.95)**	Other websites	1.85	(1.13)
MyMaths.co.uk	**2.41**	**(1.22)**	**MyMaths.co.uk**	**2.03**	**(1.25)**
Word	1.95	(1.23)	Word	1.34	(1.11)
Graphing software	1.89	(1.09)	Graphing software	1.39	(1.01)
Spreadsheet	1.82	(1.04)	Spreadsheet	1.39	(1.30)
Geometry software	1.53	(1.11)	Geometry software	1.20	(.99)
Email	1.37	(1.60)	Email	.66	(1.09)
CD Roms	1.46	(1.29)	CD Roms	.55	(.92)
Database	.84	(1.19)	Database	.58	(.94)
SMILE	.50	(.95)	SMILE	.25	(.62)
Logo	.35	(.67)	Logo	.37	(.72)

The frequency of use in lessons where students were given direct access to the software was low in comparison to lessons with an IWB: only *MyMaths* had a mean frequency score above 2 – a finding supported by the pilot study. This is unsurprising given the frequency of hardware use in mathematics lessons reported above: computer rooms are used much less frequently than IWBs. However the decrease in use is not uniform across all types of software. The mean frequency score of IWB software (−1.95) and PowerPoint (−1.11) dropped the most. Since the main purpose of IWB software and PowerPoint is for presentation, it appears well suited to teacher exposition in lessons with an IWB but not so relevant in lessons where students have direct access to the software. *MyMaths* (−0.38) and geometry software (−0.33) had the smallest drops in frequency use between contexts. Although the *MyMaths* website can be used for teacher presentation, one of its main features are textbook-like exercises and on-line homework, linked to an 'Assessment Management system', allowing teachers to track individual student's progress. Hence it can also be used in lessons where students are given direct access to computers. Similar to geometry software, graphing (−0.50) and spreadsheet (−0.43) software also have relatively low drops in use between contexts, maintaining a mean frequency of use between once or twice a term and annual usage.

In a whole-class context with an IWB, teachers in higher support schools tended to use IWB software ($\chi^2 = 28.93$, df = 3, p < .0001) and email ($\chi^2 = 8.89$, df = 3, p = .031) statistically significantly more than those in lower support schools. Although teachers in higher support schools also used Logo significantly more often than those in lower support schools ($\chi^2 = 7.27$, df = 2, p = .026), this still corresponds to very low levels of use overall. There were no statistically significant differences in frequency of use with an IWB for any of the other software listed in the survey tested at the 5 % level, in particular graphing, geometry and spreadsheet software, PowerPoint and *MyMaths* showed no significant difference in use between higher and lower support schools. Higher support school teachers' more frequent use of IWB software is likely to be a reflection of their more frequent use of IWBs in general, although perhaps of more interest is that they do not use other software significantly more or less often than those in lower support schools. Nevertheless this result might be seen to offer support to the suggestion that when considering school or departmental factors relating to ICT use, teachers appear to equate ICT use with IWB use.

In the context of giving students direct access to ICT in a computer suite, the only software with a significant difference in frequency of use between teachers in higher and lower support schools was Logo ($\chi^2 = 12.15$, df = 2, p = .002). Teachers in higher support schools used Logo more frequently, however again this still corresponds to very low levels of use overall. Again there were no statistically significant differences in frequency of use with a computer suite for any of the other software listed in the survey tested at the 5 % level, in particular graphing, geometry and spreadsheet software and *MyMaths* showed no significant difference in use between higher and lower support schools.

Individual Level Factors Relating to ICT Use

In general, teachers agreed that ICT makes an important contribution to students' learning and helps them to understand mathematics, irrespective of whether students are given direct access to ICT or they experience ICT indirectly through whole-class teaching using an IWB (see Tables 7 and 8). The pattern of response differed little between using IWBs in a whole-class context and giving students direct access to ICT via a computer suite: there were no statistically significant differences between the two contexts, according to a paired t-test at the 5 % level. Similarly, teachers agreed that using ICT improves students' engagement in lessons, with no significant difference between the two classroom contexts. These results suggest mathematics teachers in England generally have a favourable outlook towards using ICT in their teaching and to a similar extent whether students are given direct access to ICT or they experience ICT indirectly through whole-class teaching using an IWB.

Time is highlighted by Stein et al. (2007) as one of many contextual factors impacting on the participatory relationship between curriculum materials and teachers. In terms of time needed for lesson preparation, overall, teachers tended to agree slightly that lessons involving ICT in both classroom contexts took more time to prepare (see Tables 7 and 8); there were no statistically significant differences between the two contexts. However, in both contexts there was a relatively large variation in the perceived time costs across the sample, with 29.0 % ($n = 183$) disagreeing or strongly disagreeing that lessons with an IWB took more time to prepare and similarly 30.1 % ($n = 176$) for lessons in a computer suite. The large variation in perceived time costs for ICT use may reflect that while start-up costs can be high in terms of designing lesson materials, once made, the materials can be

Table 7 Mean score for individual level factors using ICT with an IWB or data projector. Scored on a 5-point Likert-scale where 5 = strongly agree to 1 = strongly disagree

For lessons using an IWB or data projector in a whole-class context		Mean	(SD)
I am confident using ICT in lessons	$n=181$	4.24	(.90)
Lessons using an IWB/data projector take more time to prepare	$n=183$	3.18	(1.14)
ICT makes an important contribution to students' learning of mathematics	$n=184$	3.87	(.82)
Using ICT improves student engagement in lessons	$n=185$	3.97	(.75)
Students' lack of familiarity with software make lessons with ICT difficult	$n=186$	2.68	(1.01)
ICT resources help students to understand mathematics	$n=184$	3.85	(.80)
Classroom management is more difficult when using an IWB/data projector	$n=185$	1.86	(.88)
We cover more ground in lessons with an IWB/data projector	$n=184$	3.57	(.90)

Table 8 Mean score for individual level factors using ICT in a computer suite or with laptops. Scored on a 5-point Likert-scale where 5 = strongly agree to 1 = strongly disagree

Giving students direct access in a computer suite or with laptops		Mean	(SD)
I am confident using ICT in lessons	$n=175$	4.09	(.98)
ICT lessons take more time to prepare	$n=176$	3.18	(1.14)
ICT makes an important contribution to students' learning of mathematics	$n=176$	3.85	(.83)
Using ICT improves student engagement in lessons	$n=176$	3.89	(.81)
Students' lack of familiarity with software make lessons with ICT difficult	$n=175$	2.78	(1.02)
ICT resources help students to understand mathematics	$n=175$	3.75	(.82)
Classroom management is more difficult in ICT lessons	$n=176$	2.64	(1.03)
We cover more ground in ICT lessons	$n=174$	2.92	(.87)

stored electronically, ready to be used in perpetuity. Thus teachers who are new to using a piece of hardware or software or teaching a topic for the first time might agree that preparation time is increased, whereas those who have already accumulated a bank of lesson materials may disagree. Preparation time also depends on how the hardware is used. For example, if an IWB is essentially used as a normal whiteboard then additional time costs may be minimal; however if a PowerPoint presentation is specially prepared for the lesson, the additional time costs could be considerable. Similarly, for lessons in a computer suite, if pupils work through a pre-prepared *MyMaths* lesson and exercises, the teachers' time spent in preparation may be minimal, whereas preparing graphing or dynamic geometry software files for the pupils to interact with could be very time-consuming.

Overall, teachers tended to disagree slightly that students' lack of familiarity with software make ICT lessons more difficult. Still, there was a sizeable minority who either agreed or strongly agreed that students' lack of familiarity with software caused difficulties in ICT lessons: 24.2 % of responding teachers ($n = 186$) in

Table 9 Statistically significant results of paired t-tests at the 5 % level, comparing means for individual level factors between using IWBs in a whole-class context and giving students direct access to ICT in a computer suite

		Mean difference (IWB – CS)	SE	t-score	p-value
Confidence	$n=169$	.154	.054	2.83	.005
Class management	$n=173$	−.786	.085	−9.29	p<.001
Ground covered	$n=170$	.665	.080	8.35	p<.001

the context of using an IWB and 26.3 % of responding teachers ($n = 175$) where students were given direct access to ICT via a computer suite. Surprisingly there were no statistically significant differences in this regard between using IWBs in a whole-class context and giving students direct access to ICT in a computer suite – despite the dominance of teacher control over software when using IWBs, see the following section.

Three individual level factors stood out as showing statistically significant differences between using IWBs in a whole-class context and giving students direct access to ICT in a computer suite: teachers' confidence in using ICT; teachers' perception of the difficulty of classroom management; and the amount of ground (i.e. the amount of curriculum material) covered in ICT lessons – see Table 9 above. Teachers do appear to feel confident in using ICT in lessons both with an IWB and in a computer suite. Although mathematics teachers' confidence has appeared as an obstacle towards using ICT in previous surveys (e.g. Hadley and Sheingold 1993), a more recent survey of mathematics teachers (Forgasz 2006) suggested that teachers' personal confidence and relevant skills were consistently one of the factors most encouraging their use of ICT. However, according to the results shown in Table 9, teachers do appear less confident using ICT in lessons in a computer suite than in lessons with an IWB.

Classroom management is perceived by teachers as being significantly more difficult in ICT lessons taking place in a computer suite than in lessons involving ICT using an IWB. Around 83 % of responding teachers ($n=185$) disagree or strongly disagree that classroom management is more difficult when using an IWB or data projector, suggesting that, in the main, teachers believe that using an IWB facilitates classroom management. Although overall it appears that teachers also slightly disagree that classroom management is more difficult in a computer suite, by comparison this figure is much lower with nearly 50 % of teachers disagreeing or strongly disagreeing that classroom is more difficult when giving students direct access to ICT in a computer suite.

Overall, teachers have the perception that they cover slightly more ground in lessons when using an IWB, with over half agreeing or strongly agreeing with the statement ($n=184$). However, for lessons where students are given direct access to ICT in a computer suite, 50 % of responding teachers ($n=174$) thought it made little difference to the amount of ground covered, whilst slightly over 25 % disagreed or strongly disagreed – presumably suggesting they think less ground is covered in lessons taking place in a computer suite. Thus in general, teachers believe that they

Table 10 Chi-squared tests for differences in individual level factors using an IWB in a whole-class context between high and low support schools

For IWB lessons	χ^2-value	df	p-value
Confidence	21.03	4	$p<.001$[a]
Preparation time	10.57	4	.032[a]
Contribution to learning	25.47	4	$p<.001$[a]
Student engagement	8.71	3	.033[a]
Students' lack of familiarity	11.60	4	.021[a]
Help understanding	22.95	4	$p<.001$[a]
Classroom management	17.30	3	$p<.001$[a]
Ground covered	15.31	3	.002[a]

[a]Indicates a statistically significant result at the 5 % level

Table 11 Chi-squared tests for differences in individual level factors using ICT in a computer suite between high and low support schools

For computer suite lessons	χ^2-value	df	p-value
Confidence	12.33	4	.015[a]
Preparation time	3.08	4	.545
Contribution to learning	16.09	4	.003[a]
Student engagement	4.79	4	.309
Students' lack of familiarity	2.52	4	.641
Help understanding	12.19	4	.016[a]
Classroom management	6.78	4	.148
Ground covered	10.18	4	.038[a]

[a]Indicates a statistically significant result at the 5 % level

cover significantly less ground in lessons where students are given direct access to ICT compared to those conducted in a whole-class context using an IWB. This result is particularly interesting given the pace and productivity rationale for using ICT identified by Ruthven and Hennessy (2002).

Comparing individual level factors in high and low support schools, all showed significant differences with regard to using an IWB in a whole-class context (Table 10).

Overall, teachers in high support schools were more positive in the perceptions of using ICT in a whole-class context with an IWB than those in low support schools. More teachers in high support schools strongly agreed to being confident using IWBs compared to those in low support schools. Similarly, teachers in high support schools agreed more strongly that ICT makes an important contribution to students' learning; that using ICT improves student engagement and helps students to understand mathematics when using an IWB in a whole-class context compared to teachers in low support schools. However, teachers in high support schools disagreed more strongly that ICT lessons take more time to prepare; that classroom management is more difficult and that students' lack of familiarity with software causes difficulties when using an IWB in a whole-class context compared to teachers in low support schools (Table 11).

Table 12 Mean frequency of self-reported pedagogic practices using an IWB in a whole-class context. Scored on a 5-point Likert-scale where 5 = almost always to 1 = almost never

		Mean	(SD)
I use ICT for presentation purposes	$n = 182$	4.12	(1.07)
I use ICT to generate student discussion	$n = 184$	3.28	(1.09)
I control the software on the interactive whiteboard or data projector	$n = 183$	4.03	(.97)
I use ICT to follow up and explore students' ideas	$n = 184$	2.79	(1.12)
I manage software carefully to prevent mathematical discrepancies arising	$n = 173$	3.06	(1.32)
Students control the software on the interactive whiteboard or data projector	$n = 184$	2.03	(.86)
I draw attention to mathematical discrepancies in the software	$n = 176$	2.66	(1.40)
Using ICT, I avoid students making mistakes by explaining things carefully first	$n = 181$	3.15	(1.16)

Comparing individual level factors using ICT in a computer suite between high and low support schools offers a different picture. Teachers in high support schools agreed significantly more strongly that they are confident using ICT in computer suite and that ICT makes an important contribution to students' learning and helps their understanding when students are given direct access to it than those in low support schools. Although there was a significant difference between high and low support schools with regard to teachers' perception of the amount of ground covered in ICT lessons, this result was less easy to interpret. None of the other factors showed statistically significant differences between high and low support schools at the 5 % level.

Teachers' Pedagogic Practices Using ICT with an IWB and in a Computer Suite

Perhaps unsurprisingly given the apparent teacher-centred nature of IWBs, teacher-centred practices such as using ICT for presentation purposes and maintaining teacher-control of the software are the dominant pedagogic practices reported by teachers when using an IWB (see Table 12). Conversely, allowing students to take control of the software on an IWB is reported as the least frequent pedagogic practice. Using ICT to generate student discussion is reported as fairly frequent, though substantially less often than using ICT for presentation purposes. In particular, teachers relatively rarely report using ICT to follow up and explore student ideas, suggesting perhaps that the discussion might be rather one-sided. Interpreting this data, it is important to recall that teachers' self-reports may not accurately reflect classroom practice, since they represent espoused-theories rather than theories-in-action, and that direct observation data is required to validate any assertions made on the basis of the survey data. The lower number of responses for the statements regarding mathematical discrepancies (such as rounding errors) in the software is

Table 13 Mean frequency of self-reported pedagogic practices using giving students direct access to ICT in a computer suite. Scored on a 5-point Likert-scale where 5 = almost always to 1 = almost never

		Mean	(SD)
Students use ICT to practice mathematical skills	$n = 170$	3.41	(1.16)
I encourage students to work collaboratively	$n = 175$	3.35	(1.03)
I let students 'get a feel' for the software	$n = 175$	3.19	(1.15)
Students explore mathematical discrepancies in the software	$n = 167$	2.07	(1.13)
Students work on their own, consulting a neighbour from time to time	$n = 172$	3.15	(1.06)
Students use ICT to investigate mathematical problems and concepts	$n = 175$	2.90	(1.14)
I provide precise instructions for software use	$n = 170$	3.42	(1.10)
I prepare software files in advance to avoid student difficulties using the software	$n = 170$	2.55	(1.36)

due to teachers' confusion over the meaning of 'discrepancies'. It is difficult to offer any interpretation of the data as a result, beyond noting perhaps that it may indicate that teachers are generally not aware of discrepancies between mathematics as modeled by the software and standard mathematics.

Using ICT in a computer suite where students have direct access, the most common pedagogic practices were getting students to use ICT to practice mathematical skills and providing precise instructions for software use. Using ICT for students to investigate mathematical problems and concepts was one of the least frequent self-reported pedagogic practices. Surprisingly, preparing software files in advance was also one of the least frequent reported practices. Due to the dominant use of *MyMaths* in lessons taking place in a computer suite, perhaps it is unnecessary for teachers to prepare software files in advance, alternatively they may download materials from the Internet or from their own pool of resources rather than having to create new resources on a frequent basis. Again teachers found it difficult to understand what was indicated by 'mathematical discrepancies'. The reduced number of responses in comparison to IWB practices is partly due to a number of teachers omitting this question as they felt unable to give reliable responses because they use computer suites so infrequently (Table 13).

Conclusion and Discussion

This study underlines the quantitative gap between institutional expectations and classroom reality in maths teachers' use of both hardware and software. Allowing students direct access to digital technology remains at the margins of teaching practice, with over 75 % of responding teachers ($n = 131$) using computer suites shared with other departments – the most commonly available resource for students' direct access – one or twice a term or less. In contrast, IWBs are used almost every lesson by 85 % of responding teachers ($n = 175$), where control of the technology is rarely devolved to students. The quantitative gap is further emphasised by software use in both classroom contexts. Use of mathematical analysis software (Pierce and Stacey 2010), most commonly associated with theoretical notion of *cognitive tools* (Zbiek et al. 2007) and thus advocated by mathematics education research and government policy, is relatively rare in either classroom context. Presentation-oriented software dominates IWB use, whilst surprisingly the *MyMaths* web-site offering pre-prepared lessons dominates teachers' use of computer suites as well as featuring prominently amongst software used with IWBs. Coming to understand the ways in which software such as the *MyMaths* web-site and IWB software may be viewed as cognitive tools, might help provide insights into why teachers rely on these resources as well as reducing the impression of a deficit in teachers' use of technology.

Difficulty in gaining access to computer suites clearly remains an obstacle to use. Yet even in schools more supportive of ICT use, where conceivably access might be ameliorated by other supporting factors, use of shared computer suites is not significantly higher, although IWB use *is* higher. An alternative interpretation is that teachers judge the support for ICT given by their school based mainly on the ease of use of IWBs, essentially equating ICT use with IWBs. Neither interpretation offers a particularly positive outlook on closing the quantitative gap in ICT use. Similarly, a supportive school context does not improve the use of mathematical analysis software neither does it decrease the reliance on more presentation-oriented software like *MyMaths* and PowerPoint. These findings serve to illustrate how aspects of the *working environment* (Ruthven 2009), such as classroom 'ownership' and organisation, interacting with features of local departmental culture (Stein et al. 2007), both enable and constrain teachers' use of technology and thus curriculum resources more generally.

It has been suggested that the success of IWBs lies in their teacher-centred design, since they allow teachers to incorporate ICT without disturbing well-established teaching practices. This study does nothing to disrupt this viewpoint, however it does offer a slightly more nuanced account. In general, teachers believe that giving students direct access to ICT through computer suites and using IWBs in a whole-class context support students' learning and understanding of mathematics to a similar extent (although the perceived impact of IWBs is higher perhaps due to the increased frequency of use). Likewise, teachers appear to see both classroom contexts as similarly engaging to students. However, teachers are more confident using IWBs than conducting a lesson in a computer suite, perhaps due in part to their teacher-centred design and to their high frequency of use. Again this reflects the influence of the working environment enabling teachers' use of IWBs, whilst limiting their use of computer suites, despite favourable orientations towards both types of hardware. More tellingly, perhaps, teachers perceive IWBs to make general pedagogic aspirations easier to attain: classroom management was seen to be significantly easier with IWBs than a lesson in a computer suite and teachers felt able to cover significantly more ground in lessons with an IWB than those in a computer suite. This finding supports the pace and productivity rationale for using ICT identified by Ruthven and Hennessy (2002) and also suggests the influence of *time economy* (Ruthven 2009) on teachers' use of technology. The positive effect of a supportive departmental culture (Stein et al. 2007) also appears to encourage more favourable individual orientations towards both IWB and computer suites. However in the case of computer suites, this positive effect is limited to enhancing enabling factors and does relatively little to effect what might be regarded as hindering factors to teachers' technology use (Zammit 1992).

This study uses survey data to extend the evidence for a qualitative gap in ICT use amongst mathematics teachers in England beyond that provided by case studies or intermediate studies such as Ruthven and Hennessey (2002). In particular, the survey offers some insight into teachers' pedagogic practices using ICT, although as it is based on self-report data, any inferences must be treated with caution. In the first instance, the dominance of presentation-oriented software in an IWB context and *MyMaths* in computer suite lessons may be taken as evidence for a qualitative gap in ICT use. This inference is, of course, problematic. Just as *interpretative flexibility* (Ruthven et al. 2008) implies that cognitive tools may be used in ways that deviate from those envisaged by their designers or advocated in mathematics education research, digital technologies often associated with replicating 'traditional' or teacher-centred practices, such as IWBs or presentational software like PowerPoint and *MyMaths*, may be interpreted and used by teachers in ways that run counter to this association, to support student-centred practices. Nevertheless, the frequent use of this type of software might suggest an additional obstacle to more student-centred practices. Whilst a strength of the theoretical perspective adopted in this study is the acknowledgement of teachers' sense-making of software, it is also important to temper this with an awareness that the software design is an important factor influencing the participatory relationship between teacher and software, as Stein et al. (2007) point out. Further evidence for a qualitative gap may be inferred from the data on teachers' self-reported pedagogic practices. Of the practices reported in both contexts, the most frequently occurring tended to be more teacher-centred and those with lowest frequency tended to be more student-centred. For example, using ICT for presentation purposes and maintaining teacher-control of the software were highest for using IWBs in a whole-class context, whereas using ICT to follow up and explore students' ideas and allowing students control of the software was lowest. For ICT lessons in computer suites, providing precise instructions for software use and using ICT to practice skills were the practices with the highest reported frequency, whilst using ICT to investigate mathematical problems and concepts was among the lowest. Drawing firm conclusions regarding teachers' pedagogic practices using ICT based on this data is problematic due to the reliance on self-report data, thus these findings should be investigated and validated through further research.

The Second Information Technology in Education Survey (SITES) concluded that, whilst ICT cannot be considered as a catalyst that will necessarily bring about change, given the right conditions, ICT might contribute as a lever for such changes (Law et al. 2008). The direction of this change is implied as a shift in teaching towards a focus on 'twenty-first century skills' associated with more student-centred practices. Roughly half the educational systems included in SITES maintained a similar pattern practices whether or not the teachers used ICT (Law et al. 2008, p. 146). The majority of systems where the pattern of practice was dissimilar showed a stronger 'twenty-first century' orientation in their teacher practices involving ICT. One exception was Hong Kong, which showed a slight increase in the tendency towards teacher-centred practices when using ICT. Cuban (1993) argues that reforms intent on shifting teaching towards student-centred practices tend at best to achieve incremental changes and only marginally reshape existing practices. In particular, Cuban (2001) argues that even if computing technology is taken up on a large-scale it is unlikely to fundamentally change teaching practice. Somewhere between these two viewpoints, Ruthven and Hennessy (2002, p. 85) suggest that as well as providing a 'lever' to make established practices more effective, technology also appears to act as a 'fulcrum' for some degree of reorientation of teachers' practice. The evidence from this survey pointing to a quantitative gap in ICT use, broadly concurs with Cuban's argument: computer use remains low in frequency and therefore at the margins of practice. In the case of IWBs, where technology has been adopted on a large-scale, its use appears to cohere with existing structures of whole-class teaching especially through the predominant use of presentation-oriented software. Coupling the dominance of the IWB in whole-class teaching and presentation-oriented software in both classroom contexts with the evidence of a qualitative gap in teaching practices with technology, this survey suggests that given the right conditions, at least those currently existing in England, ICT might indeed contribute as a lever for change. England did not participate in the SITES study, however in common with Hong Kong and in opposition to the majority of systems in the SITES study, the direction of this change might be construed as an incremental shift towards more teacher-centred practices.

References

Adler, J. (2001). *Teaching mathematics in multilingual classrooms*. Dordrecht: Kluwer.
Andrews, P. (1997). Information technology in the mathematics national curriculum: Policy begets practice? *British Journal of Educational Technology, 28*(4), 244–256.
Argyris, C., & Schoen, D. A. (1974). *Theory in practice: Increasing professional effectiveness*. San Fracnsico: Jossey-Bass.
Assude, T., Buteau, C., & Forgasz, H. (2010). Factors influencing implementation of technology-rich mathematics curriculum and practices. In C. Hoyles & J.-B. Lagrange (Eds.), *The 17th ICMI study: Mathematics education and technology – rethinking the terrain* (pp. 405–419). New York: Springer.
Becker, H. J., Ravitz, J. L., & Wong, Y. T. (1999). Teacher and teacher-directed student use of computers and software. *Technical Report #3: Teaching, learning and computing, 1998 national survey*. Irvine: University of California at Irvine.
Bretscher, N. (2011). A survey of technology use: The rise of interactive whiteboards and the MyMaths website. *Proceedings of the Seventh Congress of the European Society for Research in Mathematics Education CERME 7*. Rzeszow.
Clark-Wilson, A. (2008). *Evaluating TI-Nspire in secondary mathematics classrooms*. Chichester: University of Chichester.
Cuban, L. (1993). *How teachers taught: Constancy and change in American classrooms 1880–1990* (2nd ed.). New York: Teachers College Press.
Cuban, L. (2001). *Oversold and underused: Computers in the classroom*. Cambridge, MA: Harvard University Press.

DES. (1989). *Mathematics in the national curriculum*. London: HMSO.
Fischer Family Trust. (2003). *ICT Surveys & Research: ICT resources used in mathematics*. http://www.fischertrust.org/ict_sec.aspx. Accessed 2009.
Forgasz, H. (2002). Teachers and computers for secondary mathematics. *Education and Information Technologies, 7*(2), 111–125.
Forgasz, H. (2006). Factors that encourage or inhibit computer use for secondary mathematics teaching. *Journal of Computers in Mathematics and Science Teaching, 25*(1), 77–93.
Gueudet, G., & Trouche, L. (2009). Towards new documentation systems for mathematics teachers? *Educational Studies in Mathematics, 71*(3), 199–218.
Hadley, M., & Sheingold, K. (1993). Commonalities and distinctive patterns in Teachers' integration of computers. *American Journal of Education, 101*(3), 261–315.
Harrison, C., Comber, C., Fisher, T., Haw, K., Lewin, C., Lunzer, E., et al. (2003). *ImpaCT2: The impact of information and communication technologies on pupil learning and attainment*. Coventry: Becta.
Hyde, R. (2004). A snapshot of practice: Views of teachers on the use and impact of technology in secondary mathematics classrooms. *International congress on mathematics education*. Copenhagen
Key Curriculum Press. (2003). *The Geometer's Sketchpad v.4*. Emeryville: Key Curriculum Press.
Kitchen, S., Finch, S., & Sinclair, R. (2007). *Harnessing technology: Schools survey 2007*. Coventry: National Centre for Social Research.
Lagrange, J.-B., & Erdogan, E. O. (2008). Teachers' Emergent goals in spreadsheet-based lessons: Analyzing the complexity of technology integration. *Educational Studies in Mathematics, 71*(1), 65–84.
Law, N., Pelgrum, W. J., & Plomp, T. (2008). *Pedagogy and ICT use in schools around the world: Findings from the IEA SITES 2006 study*. Hong Kong: Springer.
Miller, D., & Glover, D. (2006). *Interactive whiteboard evaluation for the secondary national strategy: Developing the use of interactive whiteboards in mathematics*. Keele: Keele University.
Moss, G., Jewitt, C., Levaaic, R., Armstrong, V., Cardini, A., & Castle, F. (2007). *The interactive whiteboards, pedagogy and pupil performance evaluation: An evaluation of the schools whiteboard expansion (SWE) project: London challenge*. London: Institute of Education.
Mullis, I., Martin, M., & Foy, P. (2008). *TIMSS 2007 international mathematics report*. TIMSS & PIRLS International Study Centre, Boston College.
OECD. (2005). *Are students ready for a technology-rich world? What PISA studies tell us*. Paris: OECD.
Ofsted. (2008). *Mathematics – understanding the score*. London: Ofsted.
Oxford University Press. (2012). *MyMaths.co.uk*. Oxford: Oxford University Press. www.mymaths.co.uk. Accessed 15 Oct 2012.
Pampaka, M., Williams, J., Hutcheson, G. D., Wake, G., Black, L., Davis, P., et al. (2012). The association between mathematics pedagogy and learners' dispositions for university study. *British Educational Research Journal, 38*(3), 473–496.
Pierce, R., & Stacey, K. (2010). Mapping pedagogical opportunities provided by mathematics analysis software. *International Journal of Computers for Mathematical Learning, 15*(1), 1–20.
Remillard, J. T. (2005). Examining key concepts in research on teachers' use of mathematics curricula. *Review of Educational Research, 75*(2), 211–246.
Ruthven, K. (2009). Towards a naturalistic conceptualisation of technology integration in classroom practice: The example of school mathematics. *Education and Didactique, 3*(1), 131–152.
Ruthven, K., & Hennessy, S. (2002). A practitioner model of the use of computer-based tools and resources to support mathematics teaching and learning. *Educational Studies in Mathematics, 49*(1), 47–88.
Ruthven, K., Hennessy, S., & Deaney, R. (2008). Constructions of dynamic geometry: A study of the interpretative flexibility of educational software in classroom practice. *Computers in Education, 51*(1), 297–317.
Selwyn, N. (2000). Researching computers and education – glimpses of the wider picture. *Computers in Education, 34*, 93–101.

Selwyn, N. (2008). Realising the potential of new technology? Assessing the legacy of New Labour's ICT agenda 1997–2007. *Oxford Review of Education, 34*(6), 701–712.

Spillane, J. P. (2006). *Standards deviation: How schools misunderstand education policy*. London: Harvard University Press.

Stein, M. K., Remillard, J. T., & Smith, M. (2007). How curriculum influences student learning. In F. K. Lester Jr. (Ed.), *Second handbook of research on mathematics teaching and learning: A project of the National Council of Teachers of Mathematics* (pp. 319–370). Charlotte: Information Age Publishers.

Thompson, A. G. (1992). Teachers' beliefs and conceptions: A synthesis of research. In D. A. Grouws (Ed.), *Handbook of research on mathematics teaching and learning: A project of the National Council of Teachers of Mathematics* (pp. 127–146). Oxford: Macmillan.

Wong, N.-Y. (2003). The influence of technology on the mathematics curriculum. In A. J. Bishop, M. A. Clements, C. Keitel, J. Kilpatrick, & F. K. S. Leung (Eds.), *Second international handbook of mathematics education* (pp. 271–321). Dordrecht: Kluwer.

Zammit, S. A. (1992). Factors facilitating or hindering the use of computers in schools. *Educational Research, 34*(1), 57–66.

Zbiek, R. M., Heid, M. K., & Dick, T. P. (2007). Research on technology in mathematics education: A perspective of constructs. In F. K. Lester Jr. (Ed.), *Second handbook of research on mathematics teaching and learning: A project of the National Council of Teachers of Mathematics*. Charlotte: Information Age Publishers.

Teaching with Digital Technology: Obstacles and Opportunities

Michael O.J. Thomas and Joann M. Palmer

Abstract A key variable in the use of digital technology in the mathematics classroom is the teacher. In this chapter we examine research that identifies some of the obstacles to, and constraints on, secondary teachers' implementation of digital technology. While a lack of physical resources is still a major extrinsic concern we introduce a framework for, and highlight the crucial role of, the intrinsic factor of teachers' Pedagogical Technology Knowledge (PTK). Results from a research study relating confidence in using technology to PTK are then presented. This concludes that confidence may be a critical variable in teacher construction of PTK, leading to suggestions for some ways in which professional development of teachers could be structured to strengthen confidence in technology use.

Keywords Technology • PTK • Instrumental genesis • TPACK

The implementation of digital technology in schools has sometimes been slower than many predicted 20 years ago, with Ruthven and Hennessey (2002) concluding that "Typically then, computer use remains low, and its growth slow" (p. 48). It has also produced variable results in terms of student learning, leading some even to doubt whether it has any real value in schools (Cuban 2001). While some research has demonstrated clear advantages of the technology (Pierce et al. 2010; Zbiek and Heid 2011) other studies report students who are openly opposed to technology use and have a strong belief in the superiority of by-hand work for mathematics (e.g., Stewart et al. 2005). Still other research documents procedural applications,

---

M.O.J. Thomas (✉) • J.M. Palmer
The University of Auckland, Auckland, New Zealand
e-mail: moj.thomas@auckland.ac.nz

such as the checking of answers, as the primary use of technology in secondary mathematics (Thomas and Hong 2004, 2005a).

In this chapter we examine the role of the teacher in using digital technology and present some results from a 10-year longitudinal study examining the pattern of digital technology use in secondary schools in New Zealand. This research describes teacher pedagogical practice and raises the issue of a number of obstacles to technology use. We also suggest that if the construct of pedagogical technology knowledge (PTK) (Hong and Thomas 2006; Thomas and Hong 2005a) is used as a lens for examining crucial variables related to teacher use (and non-use) of technology in mathematics, then these obstacles may be changed into opportunities. Finally, the question of how PTK may be enhanced through suitable professional development is briefly addressed.

Teaching with Digital Technology

Insight into some possible reasons for the slow uptake and variation in terms of student learning outcomes may be afforded by Brousseau's (1997) theory of didactical situations. In his framework the role of the teacher is crucial in orchestrating components of the classroom milieu in such a way that a cognitive epistemological learning situation results. Adding technology to the milieu requires a shift in focus to a broader perspective of the implications of the technology for the learning of the mathematics. Also, constructing a didactical situation involves organisation of an increased number of relationships, necessitating a change in thinking for teachers. A crucial part of the teacher orchestration is the management of affordances and constraints (Gibson 1977), the former describing the potential for action in the situation, while the latter impose the structure for that action. The term *obstacle* is employed by Thomas (2006) and Thomas and Chinnappan (2008) for anything that prevents an affordance-producing entity from being in a classroom situation. Thus in the context of our discussion of digital technology, the physical hardware may be an affordance, the instrumental genesis of the teacher, lesson time available, and curriculum content would be constraints, and a lack of funds and negative teacher attitudes could be obstacles. We will return to some of these below. When we attempt to identify obstacles and constraints that influence implementation of technology in mathematics teaching it is useful to divide them into extrinsic and intrinsic factors. As an example, a recent discussion of CAS technology use by Heid et al. (2013) cites extrinsic factors such as negative attitudes of students, parents and society, and external assessment practice, as well as intrinsic ones such as the attitude and capabilities of teachers and problems inherent in integrating technology into current practice (which could also involve extrinsic factors). Some of these obstacles and constraints were the subject of the longitudinal study described below.

A Survey Revealing Obstacles to Practice

One attempt to understand the practice of teachers and the reasons behind it was the first author's 10-year longitudinal survey study (1995–2005) of secondary mathematics teachers in New Zealand. Replies from the postal surveys were received from 90 of the 336 secondary schools (26.8 %) and 339 teachers in 1995, along with 193 of the 336 schools (57.4 %) and 465 teachers in 2005 (see Thomas 1996, 2006). Both closed and open questions were used to elicit feedback on: the amount of technology in schools; the level of access to the technology; the pattern of technology use in mathematics teaching; and teachers' perceived obstacles to technology use. This data enables us to draw some conclusions about the changing nature of technology use in the learning of mathematics in New Zealand secondary schools.

For example, the teachers were asked in which areas of the secondary curriculum (along with specific topics of graphs, trigonometry and calculus) they used the computer. The figures were very similar in the two years with the exception of an increase in the use of computers for the learning of statistics (from 38 % to 59.5 % citing it as the most common use of computers). This is not surprising since Statistics is a separate subject from Mathematics in New Zealand schools and there is a strong emphasis on learning it.

Over the 10 years there was a change in the kinds of software used in mathematics classrooms away from specific content-oriented graphical, mathematical and statistical packages towards generic software, especially the spreadsheet (from 31.9 % to 62.6 % citing it as the most common software used). One reason for this may be that spreadsheets appear to handle statistical work well enough for secondary schools.

A number of obstacles to increased use of technology were identified by the teachers. With regard to computer use, the extrinsic obstacle of availability of computers remained a major issue, mentioned by 58 % of teachers (see Table 1). While the number of computers in schools is increasing, giving potential affordances, a major constraint is that they are primarily located in large ICT rooms and access by mathematics teachers is often difficult due to competing demands from other curriculum areas. In addition, in 2005 18.4 % mentioned other constraints including the time and effort needed by both students and teachers in order to become familiar with the technology, and the impact on time available for learning mathematics.

Table 1 Constraints and obstacles preventing teachers using computers more in their teaching

Obstacle	% of 1995 teachers ($n=229$)		% of 2005 teachers ($n=452$)	
	First mentioned	Mentioned	First mentioned	Mentioned
Available software	17.4	52.5	10.8	39.4
Available computers	43.7	67.8	42.7	58.0
Lack of training	17.4	45.4	7.5	31.9
Lack of confidence	12.7	34.8	5.3	22.4
Government policy	4.1	12.4	–	–
School policy	0.6	8.0	0.4	9.3

Table 2 Constraints and obstacles preventing teachers using calculators more in their teaching

Obstacle	% of 1995 teachers ($n=64$)		% of 2005 teachers ($n=257$)	
	First mentioned	Mentioned	First mentioned	Mentioned
Calculator availability	76.6	81.3	52.5	71.6
Lack of PD	4.6	12.5	19.1	48.2
Lack of confidence	4.7	10.9	13.6	42.4
Government policy	1.6	9.4	1.9	6.2
School policy	3.1	10.9	0	5.1

The situation is similar with regard to calculators. The major obstacle to using them more remains a lack of available calculators, although there has also been an increase in the need for professional development and greater teacher confidence (see Table 2). Given that 86 % or more of students in the survey owned their own calculator the perceived lack of calculators was surprising, and it may be an absence of GC's is what teachers were mentioning, since only 27.1 % owned these.

In another question teachers were asked whether the existence of classroom resources, including good ideas that work in the classroom, was an obstacle to technology use. Over the ten years the number agreeing increased significantly ($\chi^2 = 76.5$, $p<0.0001$) from 41.0 % to 71.1 % in 2005, with a corresponding drop in those who disagreed from 32.2 % to just 11.0 %. Clearly there is still a great need for classroom resources with good ideas for teachers to use when teaching with technology. This would likely explain the high number of teachers citing a lack of training or professional development (PD) as an obstacle to increased use of both computers (31.9 %) and calculators (48.2 %).

These results agree with factors influencing teacher adoption and implementation of technology in mathematics teaching identified by other researchers. These authors also describe constraints or obstacles such as the teacher's previous experience in using technology, lack of time, few opportunities for PD, poor access to technology, limited availability of classroom teaching materials, little support from colleagues, pressures of curriculum and assessment requirements and inadequate technical support (Forgasz 2006a; Goos 2005; Thomas and Chinnappan 2008). Further, Forgasz (2006a) lists access to technology as the most prevalent inhibiting factor, with lack of professional development and technical problems, including lack of technical support next. Thus there is some consensus with regard to implementation of digital technology availability of that technology is a major issue, followed by a lack of resources, training and confidence. In the next section we discuss a construct that may assist in thinking about a way forward.

Addressing the Issues: Pedagogical Technology Knowledge

It seems clear that addressing intrinsic teacher-related issues, such as those mentioned above, is crucial in the successful implementation of technology in mathematics learning, and this process starts with recognition that didactical use of technology requires teachers to have a particular set of skills and attitudes. As we have seen above, there are a number of factors, often extrinsic, that may negatively influence a teacher's decision to try to use technology. However, intrinsic factors are crucial and include the teacher's orientations; their instrumentation and instrumentalisation of the tools (Artigue 2002; Guin and Trouche 1999; Rabardel 1995; Vérillon and Rabardel 1995); their perception of the nature of mathematical knowledge and how it should be learned (Zbiek and Hollebrands 2008); their mathematical content knowledge; and their mathematical knowledge for teaching (MKT – Ball et al. 2005; Hill and Ball 2004; Zbiek et al. 2007), which includes Shulman's pedagogical content knowledge (PCK – Shulman 1986). The idea of MKT covers appropriate structuring of content and relevant classroom discourse and activities to form the didactical situation.

The factors mentioned above help us understand that while many mathematics teachers claim to support the use of technology in their teaching (Forgasz 2006a; Thomas 2006) the degree and type of use in the classroom remains variable (Zbiek and Hollebrands 2008). One further aspect that should not be overlooked is that a sizeable minority of teachers are either not convinced of its value (Forgasz 2006b) or actively oppose its use (Thomas et al. 2008). This latter study reported that 60.5 % of teachers disagreed with the statement that "All types of calculators should be allowed in examinations," with only 21.7 % in favour, and that 27 % of teachers thought that using calculators can be detrimental to student understanding of mathematics.

A consideration of factors influencing teacher use of technology led Thomas (Hong and Thomas 2006; Thomas and Hong 2005b) to propose an emerging framework for *pedagogical technology knowledge* (PTK) as a construct that could be a key indicator of teacher progress in implementation of technology use. A teacher's PTK applied to mathematics incorporates the principles, conventions and techniques required to teach mathematics through the technology. Thus PTK includes the need to be a proficient user of the technology, but more importantly, to understand the principles and techniques required to build didactical situations incorporating it, to enable mathematical learning through the technology. A number of teacher factors combine to produce PTK, including: instrumental genesis; mathematical knowledge for teaching; teacher orientations and goals (Schoenfeld 2011), especially beliefs about the value of technology and the nature of learning mathematical knowledge, and other affective aspects, such as confidence (see Fig. 1).

Some comparisons could be made between PTK and the Technological Pedagogical Content Knowledge (TCPK), later TPACK, framework (Mishra and Koehler 2006; Koehler and Mishra 2009), which appears to have developed independently around the same time. This more generic framework articulates relationships between the pedagogical content knowledge (PCK) of Shulman (1986), technological pedagogical knowledge (TPK) and technological content knowledge (TCK). However, it differs from PTK in several aspects. Firstly although its original formulation could have been seen as generic, PTK has always been focussed specifically on mathematics, which has its own nuances of content knowledge. The use, in the latest version of PTK (see Fig. 1), of Ball and Bass's mathematical knowledge for teaching, which includes, but

Fig. 1 A model of the framework for PTK

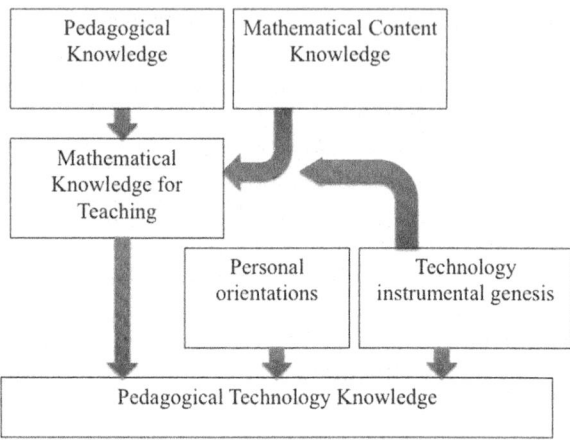

extends and builds on Shulman's generic PCK, emphasises this. Secondly, PTK employs the theoretical base of instrumental genesis (Rabardel 1995), with its explanation of the conversion of a tool into a didactic instrument, while TPACK relates to "knowledge of the existence, components and capabilities of various technologies as they are used in teaching and learning settings, and conversely, knowing how teaching might change as a result of using particular technologies" (Mishra and Koehler 2006, p. 1028), using the Fluency of Information Technology theoretical base (Koehler and Mishra 2009). This appears to have less emphasis on the epistemic value of the technology, that of producing knowledge of the (mathematical) object under study (Artigue 2002; Lagrange 2003; Heid et al. 2013). Thirdly, PTK includes the crucial element of the personal orientations of the teacher who is using the technology and their role in influencing goal setting and decision making, which seems absent in TPACK. However, while there are differences in the frameworks it seems clear that both can provide useful conceptual lenses for analysing classroom practice, with researchers who have used TPACK reporting elsewhere in this volume.

We believe that this latter aspect of teacher orientations and their effect on confidence in using technology has been given less attention in research and development than it deserves. For example, if we look again at Tables 1 and 2 we note that a lack of confidence was mentioned as a constraint or obstacle to further use of computers by 22.4 % and for calculators by 42.4 %, making it a significant factor.

A theoretical framework developed by Schoenfeld (2002, 2008, 2011) helps to explain why we should pay more attention to the role of teacher attitudes and beliefs in teaching practice. Based on the view that teaching is a goal-oriented practice it seeks to identify reasons behind the decisions teachers make during teaching. The framework links these decisions to the *Resources, Orientations, and Goals* (ROG) of teachers, where it is our *orientations*, dispositions, beliefs, values, tastes and preferences, that not only shape the way we see the world, but also, in any given situation, direct the goals we establish to deal with those situations. They also prioritise the marshalling of resources, such as knowledge, that is used to achieve the goals (Schoenfeld 2011). Hence, we should expect teacher orientations, including beliefs about the value of technology and their confidence in using it to be crucial factors in its implementation.

In an individual teacher where PTK is strong there will be good mathematical content knowledge, good understanding of how to use the technology, positive beliefs and attitudes to its use and the confidence to put it into practice. Thus, with this strong PTK such a teacher is more able to promote techniques that have *epistemic value*, capable of producing knowledge of the mathematical object under study, rather than those perceived and evaluated in terms of their 'productive potential' or *pragmatic value* (Artigue 2002, p. 248). Thus with a high level of PTK specific mathematical conceptions are placed firmly at the centre of classroom activity assisting teachers to appropriate structuring of content and classroom discourse and activities for didactical situations. So the question naturally arises, how might we strengthen teacher PTK? In summary, there are three main aspects of PTK that need attention. The first is to increase the mathematical knowledge for teaching of teachers, which is crucial, and includes addressing content knowledge, which may be done through in-service professional development (see below). The second area is the instrumental genesis of teachers with regard to the digital technology. Once again this can be considered in short professional development courses, but their target must be the teaching of mathematics. For example, the process of instrumental genesis involved in changing a car artefact to become an instrument for taking young children to a school in town is quite different from that involved in changing the same car artefact into a rally driving instrument. It's the same artefact but completely different instruments. We would argue that the same is true of digital technology. Using it as an instrument to improve the checking of answers or the speed with which they are acquired is totally different from using it as an instrument to improve conceptual understanding of mathematics, and the gap is probably as wide as the example of the car above. It is instrumental genesis for the second kind of instrument that should be the target of PD. However, it is the third area of PTK that is often not considered at all, and that is the effect of teacher orientations on the use of technology. It is this area, and in particular the role of confidence, that is considered below.

Strengthening the Teacher Orientations Component of PTK

Teacher confidence was a target variable in a study of 22 teachers from Auckland, New Zealand, who were using GCs in teaching, 17 of them experienced in their use and five not. They were given a Likert-style attitude test with five subscales, comprising attitude to: mathematics, technology in general, personal learning, technology and GC in learning mathematics, a number of lessons were observed and finally they were interviewed for about 40 min. The level of confidence of the teachers in GC use in mathematics was inferred following discussion with them and classroom observation of their teaching, and it was deduced that 12 had strong

Table 3 Comparison of subscale means for confident and non-confident GC users

	Attitude to maths	Attitude to technology	Technology in learning maths	GC's in learning maths	Personal learning
Strong Confidence ($N=12$)	3.96	4.15	4.38	4.05	4.19
Weak Confidence ($N=10$)	3.75	3.23	3.98	3.49	3.88
p-value	n.s.	<0.005	<0.05	<0.0005	0.06

and 10 weak confidence, thus forming two groups. Of the 12 teachers with strong confidence in their ability to teach with the GC, 11 were experienced users, 11 had strong school support, 10 had strong HOD support, and 11 had the support of other teachers in their department. In contrast, from the 10 teachers with weak confidence in their GC use, eight were inexperienced users and seven had little head of department or other teacher support in their schools. Interestingly the level of school support for these teachers was split, with five being supportive and five not. This initial analysis suggested that key variables to produce confident users of GC's are the teacher's own experience and the affordance of support from others in the mathematics department. However, there were exceptions to this general situation, with two teachers confident in their use in spite of having no support from other teachers or their department and one who, although an experienced user with strong support from both department and school lacked confidence in his ability to teach with the GC.

The results of a comparison between the attitude scales for the group of teachers with strong confidence and those with weak are seen in Table 3. This shows that, apart from the attitude to mathematics subscale, there is evidence of a significant difference in attitude on all the other subscales, with a higher level for those with greater confidence, although this is weaker on the personal learning scale.

It seemed reasonable to infer from these results that strong confidence in one's ability to teach with the GC is linked to a more positive attitude to technology, to the use of technology in the learning of mathematics and, possibly, to one's attitude to personal learning, which is necessary to learn the new perspectives on teaching required to use technology with a focus on the mathematics. In addition, it was found that those with strong confidence were more likely to have more ideas for using technology to teach mathematics (Mean$_{strong}$=4.3, Mean$_{weak}$=3.6) and to believe that students understand maths better when using a GC (Q25, Mean$_{strong}$=4.0, Mean$_{weak}$=3.0).

Since this initial research (see also Hong and Thomas 2006; Thomas et al. 2008) suggested that improved confidence in classroom use of technology is not only a factor in digital technology use but may be a key driver of the growth of PTK we sought to find further evidence to confirm or refute this hypothesis.

A second study was conducted to test the hypothesis that confidence and PTK are linked (see Palmer 2011). The participants here were 42 female teachers from Auckland, New Zealand, working in a wide range of schools. The teachers' confidence in using GC technology was measured out of 161 using a questionnaire that included targeted questions (31 points – for the number of classes using GCs and

Fig. 2 The instrument for self-evaluation of confidence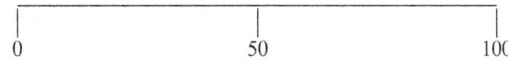

the frequency of use; greater use implying greater confidence), a self-evaluation on a 0–100 scale (100 points – see Fig. 2) and a Likert attitude subscale (30 points). A Likert scale with 100 point continuum response format and semantic anchors was chosen in order to allow teachers a wider range of response than 5-point Likert response format permits, thus better differentiating the teachers. This score was a major component of the total confidence score, with the Likert subscale and open questions providing additional components, along with an opportunity for insight into the confidence levels. Measurement of the variable of teacher confidence was crucial since investigation of any relationships between it and other attributes was a major focus of the study. The questions used in the 5-point response Likert subscale were:

Learning how to use graphic calculators is difficult for me
I have lots of ideas about how I can make use of technology in mathematics
I lack the confidence to use graphic calculators to solve mathematical problems
I often need to ask colleagues when I am not sure about an aspect of using technology
I am more confident when I observe other teachers using graphic calculators with their classes
I lose confidence when another teacher observes my use of graphic calculators with my classes

(scores on negative items were reversed to give a positive measure of confidence).

Indicate your confidence in using graphics calculators in your teaching on the following scale, where 0 is no confidence at all and 100 is totally confident.

Some researchers have criticised the abuse of Likert scales (see e.g., Jamieson 2004), for example, when used as interval scales. However, such concerns have been addressed by Carifio and Perla (2007). They describe the need to distinguish clearly between the use of the words (sub)scale for the measurement of the underlying attitude, etc., and its use for the response format of individual items. As they note (ibid) a measurement (sub)scale constructed with at least 6–8 carefully chosen items is capable of producing required variations in (sub)scale score. Further, if this is coupled with a 5–7 point response format "…then it is perfectly acceptable and correct to analyse the results at the (measurement) **scale** level using parametric analyses techniques…" (ibid, p. 115, original bold). Thus, an aggregated total, say, may be analysed. What is to be avoided is the analysis of data at the item level. We contend that the two measurement (sub)scale requirements above have been met in the study and in the analysis that follows only Likert measurement subscale aggregated data is employed.

As outlined above, PTK is a theoretical construct comprising three major components: mathematical knowledge for teaching; instrumental genesis of the digital technology; and teacher orientations relating to the use of technology, that is especially their beliefs about the value of the technology for the learning of mathematics and their confidence in its use. All of the teachers in the study taught mathematics in Years 9–13 (age 14–18 years) and so the content from these years, primarily elementary algebra, graphs, statistics and beginning calculus, was of interest. This content knowledge of the teachers was not assessed in this study but their pedagogical understanding was considered through some of the interview questions. Hence, in order to see if the rest of the measures correlated with confidence, a single measure for PTK was produced as a composite of equally weighted scores for belief in the value of technology in teaching mathematics, level of personal instrumentation and pedagogical approach, in terms of benefits for student learning.

Teacher belief in the value of technology in teaching mathematics was measured with the following three questions in the questionnaire and also a 26 question Likert subscale, given below. The three questions were:

Do most of your students own graphic calculators? (each year asked for)
Would you like to use graphic calculators more often in your mathematics lessons?
Do your students use calculators in their mathematics lessons only when directed by you?

The Likert subscale questions were:

Learning mathematics is mostly memorising a set of facts and rules
Students understand mathematics better if they solve problems using paper and pencil
Students should not be allowed to use technology during mathematics tests or examinations
Students would understand maths better if they had a graphics calculator
Technology can be used as a tool to solve problems students could not solve without it
Students would be more confident in maths if they had a graphic calculator
Technology can make mathematics more fun
Students should use technology less often in mathematics
Using a graphic calculator will cause students to lose basic computational skills
Students rely on graphic calculators too much when solving problems
Technology should only be used to check work once the problem has been worked out on paper
Mathematics students need to know how to use technology
Students should not be allowed to use technology until they have mastered the idea or the method
Mathematics is easier if technology is used to solve problems
Using graphic calculators makes students better problem solvers
More interesting mathematics problems can be done when students have access to technology
When doing mathematics it is more important to know how to do a process than to understand why it works

Learning mathematics means exploring problems to discover patterns and make generalisations
Students would be better motivated in maths if they could use a graphic calculator
Using a graphic calculator removes some learning opportunities for students
I think technology is a very important tool for learning mathematics
Graphic calculators are only a tool for doing calculations more quickly
Since students can use a graphic calculator, they do not need to learn to draw graphs by hand
I feel that computer algebra system calculators should be allowed in mathematics tests and examinations
Using a graphic calculator to solve statistics makes the problems easier to understand

The level of instrumental genesis of the teachers was assessed using two open questions from the questionnaire and the Likert subscale:

When I use a graphic calculator I often have problems finding the right keys
I find the menus on graphic calculators and computers easy to navigate
I often need to ask students how to do specific things on the graphic calculator
I usually know how to set up the graphic calculator to find answers I want
Many of the graphic calculator's functions are a mystery to me

The open questions were: Please give the main advantage or benefit you have found, or feel to be true, of using graphic calculators in mathematics lessons; and What is the main criterion by which you would identify a good mathematics lesson using graphic calculators? In these the highest score was given for responses that represented a focus on the calculator being used to address mathematical concepts or generalisations, the next for a focus on time-saving and getting correct answers, then for a focus on particular functions of the calculator as a tool or the generating of interest and, finally, the lowest score for a focus on pushing particular buttons on the calculator. The aim here was to capture the *primary* area that the teacher responses focussed on. While some teachers made reference to one or more specific uses of the calculators in each case it was easy to determine for each teacher what their primary criterion for its use, or the main focus, was.

The pedagogical score was an attempt to measure the value for mathematical learning that the teachers assigned to the GCs. For the same two open questions above reducing scores were assigned for recording the graphics calculator as providing some benefit: addressing mathematical concepts or generalisations, theory or investigations; in the ability to move between representations and make links; for looking at graphs or comparing data; and as a tool to get an answer. In addition, the following ten question Likert subscale was used:

Technology can be used as a tool to solve problems students could not solve without it
Students rely on graphic calculators too much when solving problems
Technology should only be used to check work once the problem has been worked out on paper
Using a graphic calculator removes some learning opportunities for students

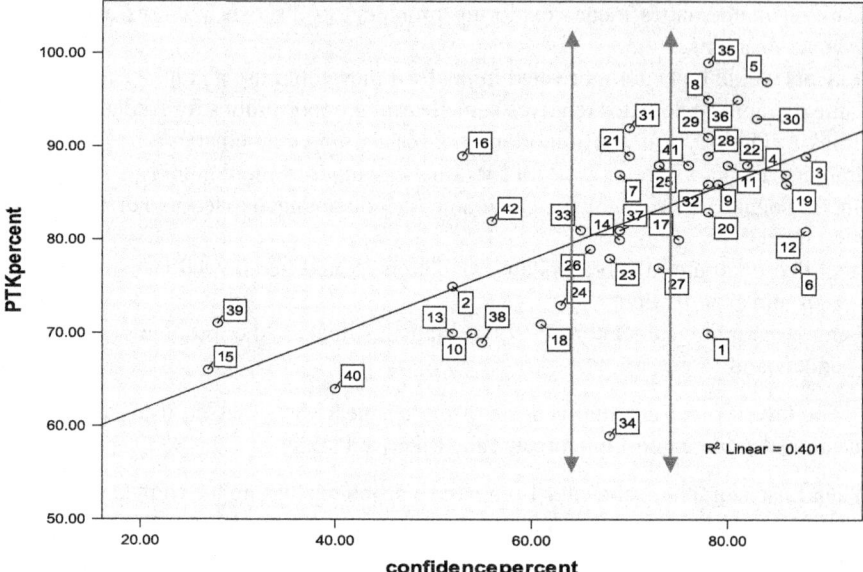

Fig. 3 Correlation between the measures of confidence and PTK

Graphic calculators are only a tool for doing calculations more quickly
Since students can use a graphic calculator, they do not need to learn to draw graphs by hand
I think that the focus should be on the maths not on the graphic calculator
I would prefer students to have a graphic calculator when I teach with a graphic calculator
Teaching with technology is harder than teaching without technology
When I have a good idea for using graphic calculators I try to fit it in with the teaching I have always done

Using the measure of confidence described above, three distinct groups among the teachers emerged, with statistically significant differences in mean levels of confidence: a low confidence group in the range 27–63 %; a medium group in the range 65–76 %; and a high group with confidence scores in the range 78–88 % (Palmer 2011).

Examining the correlation between the measures of confidence and PTK (see Fig. 3, where the blue vertical lines divide up the members of the three groups), the research found that "There was also a highly significant correlation between confidence scores and the PTK score of teachers" (Pearson correlation coefficient 0.633, $p < 0.001$) (Palmer 2011, p. 53). Although this correlation does not demonstrate that either variable is a cause of the other, it does suggest that the two were closely related.

The mean of the aggregated attribute scores for each of the three confidence groups are given in Table 4, and Table 5 shows that there were highly significant differences between the mean confidence scores of the three groups. It is especially

Table 4 Mean (%) attribute scores by confidence group

Mean Score %	Confidence	Belief in value of technology	Open to personal learning	Instrumentation	Pedagogy	PTK
Low group	49	67	75	62	73	73
Mid group	70	70	74	77	77	81
High group	82	75	75	84	81	88

Table 5 t-Tests for the difference between means of attribute scores in confidence grouping

Attribute	Low and medium confidence	Medium and high confidence	Low and high confidence
Confidence	0.0002**	1.35×10^{-9}**	
PTK	0.0129**	0.0373**	
Belief in the value of technology	0.403	0.081*	0.016**
Instrumentation	0.017**	0.153	0.00015**
Pedagogical practice	0.275	0.128	0.037**

*Weak evidence of a difference
**Evidence of a significant difference

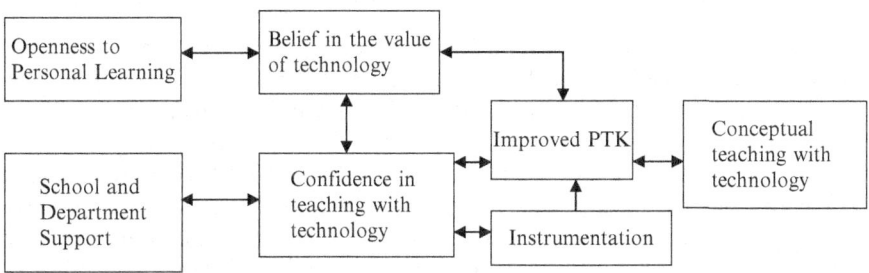

Fig. 4 A tentative model of factors affecting conceptual teaching with technology

worth noting that there was a significant difference in the PTK score between all pairs of groups. Not only that but in all other cases except two, namely Belief in the Value of Technology and Pedagogical Practice between the low and medium groups, the attributes were significantly different from one group to the next. In the two cases mentioned where this was not the case there was a significant difference between the medium and high confidence level groups.

These results seem to indicate that there is a strong correlation between confidence in using technology in the mathematics classroom and PTK, which promotes use of digital technology in a pedagogical manner that facilitates learning of mathematical concepts, as well as procedures. Further we may suggest that often teachers are at identifiably different levels of confidence and PTK. A tentative model of the relationships between some of the critical variables discussed above is presented in Fig. 4, which suggests teacher confidence as the pivotal attribute.

Reporting on a national US survey of over 4,000 teachers Becker (2000a) concluded that "...computers have *not* transformed the teaching practices of a majority of teachers" (p. 29). However, he noted that for certain teachers, namely those with a more student-centred philosophy, who had sufficient resources in their classroom (five or more computers), and who had a reasonable background experience of using computers, a majority of them made 'active and regular use of computers' in teaching. Becker (2000b) has added a description of some characteristics of such an 'exemplary' computer-using teacher, but concludes that extending these to other teachers would be expensive. With specific reference to mathematics teaching, Ruthven and Hennessey (2002) have outlined a model, comprising 12 themes, that "...highlights key processes and critical states which require active—and reactive—planning and management on the part of the teacher for ICT use to successfully support teaching and learning" (p. 83) in the hope this might assist teachers to make more effective use of technology in the classroom. Hence, a crucial question to address is, how can we use PD to assist teachers with lower PTK to move forward?

The Role of Professional Development

The teachers in the study above were asked how they had learned about using the technology and what kinds of PD they would like to have. Two of the teachers in the lower confidence level group said that they learned to use the calculators from a manual or website and two from a workshop. Each member of the group mentioned learning from students and most had referred to notes in mathematics workbooks and textbooks. They were motivated by the fact that using the GCs was fun, was fast, and also had advantages in terms of student learning in particular topics. This was the only group where members mentioned that finding the time to play or 'fiddle around' with the calculators was an issue. They all commented that they would like to learn more about teaching with the calculator with a typical goal of "incorporating it constructively in lessons" expressed in her interview by one teacher (T40).

In contrast, the teachers in the medium confidence level group gained knowledge of how to use the calculators from other people, either at training college or within their school mathematics departments. Some of this learning took place in formal professional development sessions within mathematics departments, but informal interaction with other colleagues was described as the most valuable learning experience for this group. In their interviews, in response to the question "How did you learn to use a graphics calculator yourself?" two of the teachers said "Just learning from each other, incidental informal learning" (T31) and "I find my colleagues are always keen to share their knowledge" (T27). As a result this group seemed to have more time to practice with others. In terms of future professional development they were interested to learn from other teachers to "see how someone else uses it in a different way" (T31) and to "find out about specific things to work in the best interest of the kids" (T34). Similar responses came from the high confidence group who had all learned to use the calculator from their colleagues in the mathematics department. Interestingly it was this group who had more specific topics that they would like for future professional development such as learning how to use CAS calculators or examining the variation in instrumentation between different brands of calculator. Only one teacher, who had been instrumental in training many of her colleagues, did not express a desire for further professional development in the use of calculators.

There seem to be several implications of these findings for both pre- and in-service professional development of teachers with regard to technology. It appears that it is very beneficial to teacher confidence to be part of a group that shares and reflects on their knowledge of instrumentation, practical classroom activities and ideas about the calculator use, especially in the initial stages of learning about the calculators. In this way the medium and higher confidence level groups seem to have emerged from the period of frustration mentioned by the lower confidence level group, and this has helped them to persevere with graphic calculator use with their classes. In contrast, learning from a manual, workbook or from students did not help teachers reach a point where they became confident users of the technology. What kind of activities could form part of the professional development the teachers want, and how might the sessions be structured?

To answer we note that teachers with lower levels of PTK and confidence see technology benefits as a function of visualisation, speed and accuracy of calculation, saving of time and student motivation (see also Thomas and Hong 2005b). They are still coming to grips themselves with basic operational aspects of the technology, such as key presses and menu operations. Their practice is often characterised by an over-emphasis on teaching operational procedures, such as key presses and menu operations, to the detriment of mathematical ideas. Furthermore, with the emphasis on technology rather than mathematics, student work tends to be process-oriented; based on procedures and calculating specific answers to standard problems. They find it difficult to engineer didactic situations. There is little or no freedom given to students to explore and generalise using the technology, which can tend to be seen as an add-on to the lesson rather than an integral part of it. These features then become part of the teacher-initiated expectations in the didactic contract (Brousseau 1997).

In contrast, teachers with high PTK and confidence tend to relate the technology to linking multiple representations of constructs, understanding of ideas, generalisation and moving from step-by-step processes to an overview. They have advanced to the point where they are competent in instrumentation of the technology and are able to focus on other important aspects, such as the linking of graphical, tabular, algebraic, ordered pair and other representations. With high PTK they see digital technology as having a wider application than simply calculation. They feel free to loosen control and encourage students to engage with conceptual ideas of mathematics through individual and group exploration, investigation of mathematical ideas, and the use of methods, such as prediction and testing. For these teachers the mathematics rather than the technology has come to the foreground, and technology has been integrated into lessons and the didactic contract as a way to improve mathematical understanding. The medium level group may be seen as moving towards the position of the high level group. If we think that the approach of the group with high PTK and confidence is preferable, then we need professional development that will assist teachers to progress towards it. This must have relevant resources, a classroom focus and good lines of communication.

The first issue identified through the PTK framework is that of teachers' mathematical knowledge. First, it seems desirable that initial teacher training courses specify minimum requirements of content knowledge for primary and secondary teachers. Second, universities could also be providing PD opportunities for deepening and widening teachers' mathematical knowledge. Experience suggests that this is something that many teachers would be very pleased be involved in. The second issue arising from PTK is to improve instrumental genesis of teachers. This is a huge task since every new artefact requires individual attention. This needs to be done through supportive in-service professional development programmes that promote mentoring of individual teachers by others, preferably in the same school, who have already reached good instrumental genesis. This might be achieved by what is often called a cascade approach. The crucial step though is to put the newly acquired instrument into practice in the mathematics classroom. There is a continuing need for high-quality classroom-based resources that will assist teachers to build didactical situations, and for the corresponding PD to implement them.

While we don't have strong empirical evidence, the comments of the teachers in the study do appear to agree with Jaworski's (2003) argument that an individual's development of mathematics teaching practice "is most effective when it takes place in a supportive community through which knowledge can develop and be evaluated critically" (p. 252) – a community in which all participants are co-learners (Jaworski 2001), engaged in action and reflection she terms a *community of inquiry* (Jaworski 2003). A key aspect of such a community is a critical alignment in which all participants align themselves with agreed aspects of practice but are still able to engage in "critically questioning roles and purposes as a part of their participation for ongoing regeneration of the practice" (Jaworski 2006, p. 190). Our contention is that teaching practice PD is best constructed around such a supportive community of inquiry in a manner that gives teachers the opportunity to observe, practice and reflect on the use of digital technology in a classroom environment. This last factor is usually missing from current PD. How could it be achieved? There may be few places where there is the luxury of a school attached to institutes that train teachers (although these do exist and can be effective), and there may be few schools that would be willing to 'loan' classes for PD, so other means may be required. One suggestion would be to organise a small heterogeneous cluster of teachers on a PD course so that they form a classroom audience, each of whom, in turn, presents a (shortened) prepared lesson incorporating technology to the others, which is then the centre of community discussion and reflection. Doing this in the presence of some teachers with high PTK the resulting supportive discussion would likely be highly valuable. Such group discussions have already proven beneficially in other environments, such as university lecturing (Paterson et al. 2011). Using technology in this way may lead to a positive change in the perception of its value, and in turn to increased confidence, and improved classroom practice based on the epistemic value of the technology.

References

Artigue, M. (2002). Learning mathematics in a CAS environment: The genesis of a reflection about instrumentation and the dialectics between technical and conceptual work. *International Journal of Computers for Mathematical Learning, 7*, 245–274. doi:10.1023/A:1022103903080.
Ball, D. L., Hill, H. C., & Bass, H. (2005). Who knows mathematics well enough to teach third grade, and how can we decide? *American Educator, Fall*, 14–22–43–46.
Becker, H. J. (2000a). Findings from the teaching, learning and computing survey: Is Larry Cuban right? Paper presented at the *2000 school technology leadership conference of the council of chief state officers*, Washington, DC.
Becker, H. J. (2000b). How exemplary computer-using teachers differ from other teachers: Implications for realizing the potential of computers in schools. *Contemporary Issues in Technology and Teacher Education, 1*(2), 274–293.
Brousseau, G. (1997). Theory of didactical situations in mathematics: Didactique des mathematiques, 1970–1990. (trans & Eds: Balacheff, N., Cooper, M., Sutherland, R., & Warfield, V.). Dordrecht: Kluwer Academic Publishers.
Carifio, J., & Perla, R. J. (2007). Ten common misunderstandings, misconceptions, persistent myths and urban legends about Likert scales and Likert response formats and their antidotes. *Journal of Social Sciences, 3*(3), 106–116.
Cuban, L. (2001). *Oversold and underused: Computers in the classroom*. Cambridge, MA: Harvard University Press.
Forgasz, H. (2006a). Factors that encourage and inhibit computer use for secondary mathematics teaching. *Journal of Computers in Mathematics and Science Teaching, 25*(1), 77–93.
Forgasz, H. J. (2006b). Teachers, equity, and computers for secondary mathematics learning. *Journal of Mathematics Teacher Education, 9*(5), 437–469.
Gibson, J. J. (1977). The theory of affordances. In R. Shaw & J. Bransford (Eds.), *Perceiving, acting and knowing: Towards an ecological psychology* (pp. 67–82). Hillsdale: Erlbaum.
Goos, M. (2005). A sociocultural analysis of the development of pre-service and beginning teachers' pedagogical identities as users of technology. *Journal of Mathematics Teacher Education, 8*, 35–59.
Guin, D., & Trouche, L. (1999). The complex process of converting tools into mathematical instruments: The case of calculators. *International Journal of Computers for Mathematical Learning, 3*, 195–227.
Heid, M. K., Thomas, M. O. J., & Zbiek, R. M. (2013). How might computer algebra systems change the role of algebra in the school curriculum? In A. J. Bishop, M. A. Clements, C. Keitel, J. Kilpatrick, & F. K. S. Leung (Eds.), *Third international handbook of mathematics education* (pp. 597–642). Dordrecht: Springer.
Hill, H., & Ball, D. L. (2004). Learning mathematics for teaching: Results from California's mathematics professional development institutes. *Journal for Research in Mathematics Education, 35*(5), 330–351. doi:10.2307/30034819.
Hong, Y. Y., & Thomas, M. O. J. (2006). Factors influencing teacher integration of graphic calculators in teaching. *Proceedings of the 11th Asian technology conference in mathematics* (pp. 234–243). Hong Kong.
Jamieson, S. (2004). Likert scales: How to (ab)use them. *Medical Education, 38*, 1212–1218.
Jaworski, B. (2001). Developing mathematics teaching: Teachers, teacher-educators and researchers as co-learners. In F.-L. Lin & T. J. Cooney (Eds.), *Making sense of mathematics teacher education*. Dordrecht: Kluwer.

Jaworski, B. (2003). Research practice into/influencing mathematics teaching and learning development: Towards a theoretical framework based on co-learning partnerships. *Educational Studies in Mathematics, 54,* 249–282.

Jaworski, B. (2006). Theory and practice in mathematics teaching development: Critical inquiry as a mode of learning in teaching. *Journal of Mathematics Teacher Education, 9,* 187–211.

Koehler, M. J., & Mishra, P. (2009). What is technological pedagogical content knowledge? *Contemporary Issues in Technology and Teacher Education, 9*(1), 60–70.

Lagrange, J.-B. (2003). Learning techniques and concepts using CAS: A practical and theoretical reflection. In J. T. Fey (Ed.), *Computer algebra systems in secondary school mathematics education* (pp. 269–283). Reston: National Council of Teachers of Mathematics.

Mishra, P., & Koehler, M. J. (2006). Technological pedagogical content knowledge: A framework for teacher knowledge. *Teachers College Record, 108*(6), 1017–1054.

Palmer, J. M. (2011). *Examining relationship of teacher confidence to other attributes in mathematics teaching with graphics calculators*. Unpublished M.Sc. Thesis, The University of Auckland.

Paterson, J., Thomas, M. O. J., & Taylor, S. (2011). Decisions, decisions, decisions: What determines the path taken in lectures? *International Journal of Mathematical Education in Science and Technology, 42*(7), 985–996.

Pierce, R., Stacey, K., & Wander, R. (2010). Examining the didactic contract when handheld technology is permitted in the mathematics classroom. *ZDM International Journal of Mathematics Education, 42,* 683–695. doi:10.1007/s11858-010-0271-8.

Rabardel, P. (1995). *Les hommes et les technologies, approche cognitive des instruments contemporains*. Paris: Armand Colin.

Ruthven, K., & Hennessy, S. (2002). A practitioner model of the use of computer-based tools and resources to support mathematics learning and teaching. *Educational Studies in Mathematics, 49,* 47–88.

Schoenfeld, A. H. (2002). A highly interactive discourse structure. In J. Brophy (Ed.), *Social constructivist teaching: Its affordances and constraints* (Volume 9 of the series advances in research on teaching, pp. 131–169). Amsterdam: JAI Press.

Schoenfeld, A. H. (2008). On modeling teachers' in-the-moment decision-making. In A. H. Schoenfeld (Ed.), *A study of teaching: Multiple lenses, multiple views* (Journal for Research in Mathematics Education Monograph No. 14, pp. 45–96). Reston: National Council of Teachers of Mathematics.

Schoenfeld, A. H. (2011). *How we think. A theory of goal-oriented decision making and its educational applications*. New York: Routledge.

Shulman, L. C. (1986). Those who understand: Knowledge growth in teaching. *Educational Researcher, 15,* 4–41.

Stewart, S., Thomas, M. O. J., & Hannah, J. (2005). Towards student instrumentation of computer-based algebra systems in university courses. *International Journal of Mathematical Education in Science and Technology, 36*(7), 741–750. doi:10.1080/00207390500271651.

Thomas, M. O. J. (1996). Computers in the mathematics classroom: A survey. In P. C. Clarkson (Ed.), *Technology in mathematics education* (Proceedings of the 19th mathematics education research group of Australasia conference, pp. 556–563). Melbourne: MERGA.

Thomas, M. O. J. (2006). Teachers using computers in the mathematics classroom: A longitudinal study. *Proceedings of the 30th conference of the international group for the psychology of mathematics education, Prague, 5,* 265–272.

Thomas, M. O. J., & Chinnappan, M. (2008). Teaching and learning with technology: Realising the potential. In H. Forgasz, A. Barkatsas, A. Bishop, B. Clarke, S. Keast, W.-T. Seah, P. Sullivan, & S. Willis (Eds.), *Research in mathematics education in Australasia 2004–2007* (pp. 167–194). Sydney: Sense Publishers.

Thomas, M. O. J., & Hong, Y. Y. (2004). Integrating CAS calculators into mathematics learning: Issues of partnership. In M. J. Høines & A. B. Fuglestad (Eds.), *Proceedings of the 28th annual conference for the Psychology of Mathematics Education* (Vol. 4, pp. 297–304). Bergen, Norway: Bergen University College.

Thomas, M. O. J., & Hong, Y. Y. (2005). Learning mathematics with CAS calculators: Integration and partnership issues. *The Journal of Educational Research in Mathematics, 15*(2), 215–232.

Thomas, M. O. J., & Hong, Y. Y. (2005b). Teacher factors in integration of graphic calculators into mathematics learning. In H. L. Chick & J. L. Vincent (Eds.), *Proceedings of the 29th conference of the international group for the psychology of mathematics education* (Vol. 4, pp. 257–264). Melbourne: University of Melbourne.

Thomas, M. O. J., Hong, Y. Y., Bosley, J., & delos Santos, A. (2008). Use of calculators in the mathematics classroom. *The Electronic Journal of Mathematics and Technology (eJMT)*, [On-line Serial] 2(2). Available at. https://php.radford.edu/~ejmt/ContentIndex.php and http://www.radford.edu/ejmt

Vérillon, P., & Rabardel, P. (1995). Cognition and artifacts: A contribution to the study of thought in relation to instrumented activity. *European Journal of Psychology of Education, 10*(1), 77–101.

Zbiek, R. M., & Heid, M. K. (2011). Using technology to make sense of symbols and graphs and to reason about general cases. In T. Dick & K. Hollebrands (Eds.), *Focus on reasoning and sense making: Technology to support reasoning and sense making* (pp. 19–31). Reston: National Council of Teachers of Mathematics.

Zbiek, R. M., Heid, M. K., Blume, G. W., & Dick, T. P. (2007). Research on technology in mathematics education: A perspective of constructs. In F. Lester Jr. (Ed.), *Second handbook of research on mathematics teaching and learning* (pp. 1169–1207). Charlotte, NC: Information Age Publishing.

Zbiek, R. M., & Hollebrands, K. (2008). A research-informed view of the process of incorporating mathematics technology into classroom practice by inservice and prospective teachers. In M. K. Heid & G. W. Blume (Eds.), *Research on technology and the teaching and learning of mathematics: Volume 1* (pp. 287–344). Charlotte: Information Age.

A Developmental Model for Adaptive and Differentiated Instruction Using Classroom Networking Technology

Allan Bellman, Wellesley R. Foshay, and Danny Gremillion

Abstract This paper presents a detailed explanatory model for adaptive and differentiated instruction. The model combines current practices for mathematics instruction with recommended practices for formative assessment. The model can best be implemented using classroom network technologies (such as TI-Nspire Navigator with TI handhelds), but it can also be used with manual data collection means such as personal whiteboards for each student. The model is presented for mathematics, but could be easily extended to science instruction or other subjects. Experience with adaptive and differentiated instruction suggests that teachers grow to full master level proficiency over time, often over a period of years, and that some teachers never reach that level. Accordingly, two transitional models are presented, an immediate (entry-level) model and an expert model for adaptive instruction. Fully differentiated instruction is incorporated in the 'Master' model. Growth from immediate, to expert, to master level requires development of skill with the technology, but more important are critical changes we infer in the teacher's beliefs, as well as growth in their pedagogical content knowledge (PCK).

Keywords Adaptive learning • Classroom networking • Connected classroom • Differentiated instruction • Mathematics education • STEM education

---

A. Bellman (✉)
University of Mississippi, Oxford, MS, USA
e-mail: aebellman@olemiss.edu

W.R. Foshay • D. Gremillion
Walden University, Minneapolis, MS, USA

Texas Instruments Education Technology Group, Dallas, TX, USA
e-mail: rfoshay@foshay.org

Introduction

There is a great deal of current interest in formative assessment, adaptive and differentiated instruction. Teaching with formative assessment for adaptive and differentiated instruction has been identified as a powerful strategy for closing the achievement gap (Black and Wiliam 1998; Heritage and Stigler 2010; Popham 2008). However, these terms do not have uniform operational definitions. Popham (2008) defines formative assessment as:

> a planned process in which assessment-elicited evidence of students' status is used by teachers to adjust their ongoing instructional procedures or by students to adjust their current learning tactics. (p. 6)

While we agree in principle with this definition, we use the terms slightly differently. In this chapter we will use the following definitions[1] of these common terms:

- *Formative assessment* is the assessment (data collection and interpretation process) used to support instructional decision-making by teachers and students, whether *synchronous* (embedded in an instruction sequence) or *asynchronous* (immediately following it) and whether planned or spontaneous. We emphasise that any kind of information collection, not just quizzes and tests, can be used by teachers and students for this kind of guidance of the teaching-learning process,
- *Adaptive* instruction is the process of selection of the next teaching activity based on formative assessment data, typically in a large group instructional context.
- *Differentiated* instruction involves the organisation of the class into multiple parallel learning activities intended to meet the needs of individual students, based on formative assessment of individual students. This typically is accomplished using a variety of small group/collaborative activities and/or individual study.

Implicit in this definition of formative assessment are a number of requirements for an effective system for supporting adaptive or differentiated instruction:

- *Planning* a curriculum structure based on a *learning progression*.
- *Evidence of students' status* comes from assessments (not limited to conventional tests), but note that what is important here is the information on (learning) status, and its timeliness, not the assessment. Stated differently, teachers (and students) need diagnostically useful and actionable information, in near-real time, not just classification data such as numeric test scores. The assessment's purpose is to suggest adjustments in instruction, in a timely fashion. As Popham (2008) argues, this is substantially different from the purpose of benchmark tests used to predict performance on summative tests, or of high-stakes (summative, standardised) tests meant to show attainment of a proficiency level relative to a population of students. Since these tests provide asynchronous data, often weeks or months after test administration, they can at best help a teacher to build a broad picture of a student. The teacher can then use these data by examining how this longer-term

---

[1] Thanks to Jeremy Roschelle of SRI International for his contribution to these definitions.

(asynchronous) picture is consistent with the more timely (short-term asynchronous, or within-lesson synchronous) assessments based on daily classroom data.
- *Adjustment of ongoing instructional procedures* can be invoked by teachers in response to the assessments. For example, teachers may increase instruction or practice time for some students; change the feedback strategy to provide diagnostic guidance or to correct misconceptions; raise or lower difficulty of examples, problems and questions; change assumptions about students' prior knowledge in explanations; use a wider or more contextually meaningful range of examples and visualisations; invoke an appropriate collaborative learning activity to maximise peer teaching, and so on.
- *Current learning tactics* requires that the actionable information from the formative assessment must be current, that is, closely attached in time to the current instructional activity. This is at a level of granularity that is fine enough to speak to each specific learning tactic which requires adaptation or differentiation. This is probably easiest when synchronous formative assessment strategies are used, but it is also possible with asynchronous strategies if time delays are not too great.

Note how the assumptions underlying this broad interpretation of formative assessment, adaptive and differentiated instruction differ from other practices for modifying instruction according to student characteristics. For example, the long-term organisation of pupils by ability ('ability grouping' or 'setting'), as often implemented, implicitly assumes that students within each group are relatively stable and uniform in their aptitude profile, thus perpetuating the myth of 'teaching to the middle of the class'. By contrast, our definition of differentiated instruction assumes that a student's aptitudes, interests, readiness and learning paths vary by topic and by learning task. In another example, supplemental educational services (SES) that remediate without differentiation are vulnerable to the same weakness, by implicitly assuming that all low-achieving students have the same learning gaps, and all that is needed is repeat teaching (repeated explanations, more practice and review). Mastery learning models that use a remediation strategy have this same limitation. Furthermore, grouping and SES strategies typically make their instructional modification decisions only a few times per year, at most, implicitly assuming that each student's learning progression is linear and uniform.

A 3-Level Explanatory Model for Differentiation

Fully differentiating instruction often requires major changes in teacher practices, supported by their beliefs about teaching and learning, classroom culture and the teacher's role, and about mathematics. We have observed that the full transition can take three years or more and that some teachers never complete the transition to the highest level of expertise. To model our observations of pre-service and in-service teachers, we propose three stages of practice of adaptive and differentiated learning, that is: immediate, expert and master levels. Whilst we have observed these teachers practicing at all three levels, we are not yet ready to propose these levels as a developmental progression of teaching practice.

Origins of the Model

The model has emerged from the work of one of us (Bellman) with over 70 pre- and in-service secondary mathematics teachers over more than a decade, in a programme that includes ongoing participation in a community by programme graduates. Thus, it has been possible to gather longitudinal experience reports, observations and comments from these teachers over a period of up to a decade, to grow and evolve the methods initially taught in pre-service training, and to follow the teachers' further growth in the years after qualifying.

The model presented here is a synthesis of that period of over a decade of data. We characterise the model as an explanatory product of reflective practice and, whilst this synthesis of the model is prepared retrospectively in the spirit of Grounded Theory (Charmaz 2000), we do not mean to imply systematic use of that methodology in its development. The model evolved in three phases:

1. *Phase 1* (first 6 years): the model began as an adaptive instructional approach based on the use of questions for openers, monitoring progress and closing (a check for understanding at the end of the lesson or lesson segment). The technology used included personal white boards for each public school pupil in the classroom and an early version of networked calculators. This technology permitted some class-level synchronous feedback to the student teacher on learning progress, but it did not support detailed insights to the student teachers. Student teachers were trained to make daily reflective journal entries about their teaching practice using six questions:

   1. How did the lesson go? Who was successful and who wasn't? How did you judge that?
   2. What did you do and why did you change any part of your lesson plan or why did you implement the route that you did?
   3. What will you do in the future for those who were not successful?
   4. What did you learn about any of your students today?
   5. What will you do for the class as a whole to follow today's lesson and why?
   6. What will you change next time you teach this lesson?

   As an additional reflective exercise, the teachers were asked to "self-check on how well you knew your class". In this activity, each test and occasional quiz was 'graded' by the student teacher before the assessment was given or while the students were taking the test or quiz. Then the teachers compared their predicted grade with the actual grades and then reflected on the differences by asking themselves these questions:

   1. How accurate were you?
   2. How did you misjudge those for whom you were incorrect, either by predicting too low or too high of a grade?
   3. What will you do to correct your misunderstanding of their ability?

As the year progressed, the teachers tracked the outcomes of these 'self-checks' to see if they were getting better at understanding their students. In this phase, roughly half of the student teachers mastered this teaching technique. Longitudinal data, including test scores, the self-checks from above, classroom observations (using a contact charting protocol which maps teacher-student interactions, their frequency and cognitive level), daily reflective journaling as listed above and self-assessment of lesson plans, confirmed continued use of the method by these teachers, with a trend towards improvement in proficiency.

2. *Phase 2* (next 4 years): Greater precision was added to the adaptive instruction method and experimentation began on differentiated instruction. A culture of data-driven decision-making was made possible by the TI-Navigator 2.x classroom networking system. The pre-service and in-service teachers, mentor teachers and programme faculty collaboratively developed the adaptive model based on quiz and poll data gathered from the technology, together with videotaped classroom observations (multiple times per week), observation logs, lesson plan audits, daily reflective journaling, formal assessment analysis and student homework samples. As understanding of the student teaching practice developed, experimentation began among student teachers on extensions for differentiated instruction. Based on additional data from benchmark work after the opener question, the student teachers began to incorporate more one-on-one remediation, and they moved pupils into small groups based on branched lesson plans. These extensions were further validated and refined using longitudinal feedback from the teacher training programme graduates, collected during summer workshops or (later) weekly Wednesday afternoon workshops. The longitudinal data that were gathered included quarterly reports, lesson plan audits, daily reflective journals and bi-weekly videotaped observations using the contact sheet protocol. These data showed that some (but not all) teachers continued to develop proficiency with the model in their first years of teaching. The data also confirmed the efficacy of TI-Navigator as a tool for gathering student-specific data for differentiation.

3. *Phase 3*: (most recent 4 years) The programme now incorporates a differentiated instruction model using TI-Navigator 3.x. and TI-Nspire Navigator 3.x. Faculty and pre-service teaching students continue to develop the practices of data-driven decision-making, branched lesson plans, graduated difficulty problem sets and various forms of differentiated worksheets used to scaffold small-group collaborative learning. The dynamic mathematics and digital image capabilities of TI-Nspire were incorporated into conceptual teaching for both the whole class instructional mode and for differentiated instructional approaches. The classroom teaching data that was gathered continued to include classroom videos (discussed in weekly workshops) and bi-weekly observations. To support the validation and further evolution of the programme, data collection has continued from the range of sources used in phase 2, including quiz and poll data gathered from the technology, together with videotaped classroom observations (multiple times per week), observation logs, lesson plan audits, daily reflective journaling, formal assessment analysis and student homework samples.

The longitudinal contact with teacher training programme graduates suggests that many continue to use parts of the model, often with student achievement results among the best in their school, even when the context at their school does not allow a full implementation due to issues such as required pacing and classroom tests and high-stakes test preparation. Significantly, some teachers have adopted newer versions of the differentiated instruction model developed after they graduated, but others have not, even when the school context allows. It is this uneven pattern of teacher adoption of the model that has led us to characterise the model in three levels, with potential developmental implications for teachers.

The Three-Level Model

The model that evolved from this programme shows that teachers (whether pre-service students or full-time teachers) work at three levels of practice. They may grow into full differentiation, as they develop their understanding of their own pedagogy, mathematics content and the effective use of classroom networking technology. The use of a developmental progression in technology adoption and use has long been a common feature of innovation integration models such as Levels of Use (LoU) in the Concerns Based Adoption Model (CBAM), and was proposed for technology integration as early as the Apple Classroom of Tomorrow (ACOT) project. The current model thus can be seen as an application of this principle to create a framework for adaptive and differentiated instruction in mathematics using classroom network technology.

The developmental progression implied by the model also includes growing sophistication in the use of some kind of technology for teaching mathematics. The technology most commonly used is the handheld graphing calculator (such as TI-84 or TI-Nspire). In addition, classroom networks (such as TI-Nspire Navigator) may be used to implement any of the three levels of adaptive or differentiated instruction. However, the model treats this technology as optional: some teachers are comfortable using only personal white boards, coloured cards or hand signals for students to provide feedback to the teacher. Considerations in the use of technology for our model are discussed later in this chapter.

Overview of the Three Levels

The developmental continuum of teachers' skills with adaptive and differentiated instruction includes these three levels of definition[2]:

The **Immediate level** describes many teachers' practice in their first year with the model. The nature of familiar large-group instruction changes little, though decisions about "what to do next" from day to day may be informed by feedback from students obtained simply by manually checking quizzes or homework, or by using technology to obtain class summary data. Typically, these data are

---

[2] Thanks to Jeremy Roschelle for his contribution to these definitions.

examined after class. Note also that while teachers at every level often use some form of formative opener/warm-up question or activity, teachers at the Immediate level often perceive the warm-up simply as a classroom management tactic, to get students quiet and on task quickly. Reported benefits at this level usually include better classroom management and higher student engagement. In addition, teachers at this level who use networked technology also often report time efficiency: valuable minutes saved in class and two or more hours per week saved out of class. Teachers at this level receive post-class feedback on their students' learning, which creates the opportunity for self-assessment and professional growth, as data from their classes may lead the teachers to challenge some of their beliefs, and confirm others. Our experience is that some teachers – but not all – engage in this kind of reflective practice, and grow to the next level.

At the **Expert level**, a teacher feels comfortable with the mechanics of obtaining frequent student data and is ready to take ownership of the opportunity to adapt their teaching to better fit student needs and to enrich opportunities for deep student learning. Teachers at this level likely make choices about the time required and the benefits of adapting instruction. They are likely to be open to the use of a wide range of instructional resources, including pre-defined learning activities and lessons, if these resources appear to have a high probability of providing a successful teaching experience in their classroom, with their students. We have seen teachers at the Expert level use classroom data to make 'real time' decisions about the need to re-teach computational procedural skill steps or concepts and they may identify misconceptions needing correction. Their teaching routines (Ruthven's (2009) 'activity scripts') include a mix of large-group instruction and seated work activities such as worksheets and working in pairs or small groups with brief one-on-one tutorial dialogues as the teacher walks around the class (see Ruthven in this volume).

Reported benefits added by the Expert model include the ability to use time efficiently by re-teaching or correcting misconceptions only when most students have the need. Topics that present no difficulties for most students can be skipped. Teachers at this level also like the assurance that no student is 'lost'. Whilst the teachers may make conscious decisions to go on to the next topic and 'leave' some students behind temporarily, they spontaneously develop or may already have a plan to remediate those students either during the same class period or the next time a topic is needed. Gradually at the Expert level, a new classroom culture may emerge, in which helping everyone in the class to learn becomes the shared responsibility of the whole class. Eventually, teacher beliefs can change to support the new culture. In observations, an early indicator of this change in beliefs is a tolerance of a 'constructive buzz' in the classroom as students explain skills and concepts to each other. We often hear teachers who have made the transition to this level comment that they cannot imagine going back to their 'old' way of teaching. They find this model to be much more personally rewarding, because they see who benefits from their teaching.

At the **Master level**, a teacher welcomes the full range of advanced interactive capabilities that their preferred technology offers, and makes innovative use of them in classroom teaching to deepen student understanding and to differentiate instruction to support all learners. It is likely that teachers at this level have firmly established new, more interactive routines for differentiating instruction with their students. Large-group presentations are greatly reduced in length and frequency. The teachers routinely make real-time decisions about how to adapt large-group instruction, or how to group students and differentiate small-group instruction (as often as every 20–25 min), based on information collected throughout the learning activity using various techniques. By using this constant questioning (e.g., using the technology to poll students) the teachers get frequent synchronous checks to confirm that the students are progressing as planned. This kind of differentiation requires sophisticated pattern recognition skills when interpreting classroom data in 'real time'. In addition, the students frequently engage in self-guided learning, by picking their own problems at what they think is the appropriate level for them.

Teachers' comments suggest that their beliefs and practices at this level represent a substantial change from the Expert level. They view their role in the classroom as a guide and instructional manager. They are interested in seeking out knowledge gaps and misconceptions, which they view as the main barrier to learning. Consequently, their lesson planning includes multiple paths that are contingent on what students do. They make sophisticated decisions about which paths to take, based on balancing of the time required versus the importance of an immediate mastering of the concepts being taught. Consequently, they are willing to diverge from the school's prescribed daily pacing guide: they know when they can invest more time now, and gain time efficiencies later. This is a sophisticated strategy for managing what Ruthven (2009) calls the 'time economy' of the classroom (see Ruthven in this volume). Teachers report benefits added at this level include development of their pupils' self-guided learning skills, deeper conceptual understanding by pupils, and more efficient use of time and resources to help the students who need it most. Teachers at this level have reported to us that their classrooms are much more relaxed and rewarding for them and for their pupils at every level of proficiency.

Development of Teaching Expertise

The development of teaching expertise we have observed at the three levels of the model probably depends on growth of three dimensions of knowledge, as they are related to teaching of mathematics:

- Pedagogy, including classroom routines for adapting and differentiating instruction, probably supported by changes in beliefs about teaching and learning.
- Pedagogical mathematics content knowledge (PCK), especially an understanding of students' concepts and common misconceptions of mathematics.

- Technological pedagogical knowledge (TPACK), including the simple operation of the technology and its use in teaching and learning in mathematics.

We will discuss how teachers' knowledge supports each of the three levels of this model. The first section will address the first two knowledge dimensions: pedagogy and PCK at each of the three levels of the model. In the next section, we will discuss TPACK growth.

Immediate Level

We have observed that teachers at the Immediate level show a high degree of teacher-centredness: they tend to focus on improving their explanations and managing the class. Typically, this is the entry-level experience with adapting instruction.

Pedagogy: The comments of Immediate level teachers suggest that these beliefs are the most common about pedagogy:

- The teacher's focus is on finding the 'best' explanation for his/her students. The teacher believes that "if I can answer everyone's questions, and have good examples, everyone will get it".
- The teacher wants to encourage student questions and discussions, but questions often require only recall or low-level convergent reasoning (limited to finding the right answer, computational procedure steps and mathematics facts).
- The teacher implicitly believes in a linear learning path: "I taught it – I tested it – they passed it – they know it – on to the next topic!" without further review, practice, or integration and generalisation of knowledge.
- The teacher believes that if a student doesn't understand, the preferred strategy is to repeat the 'best' explanation ("louder and slower") and to keep practicing. Thus, adaptive learning is primarily through review and modification of the amount of assigned practice (the most simplistic definition of Mastery Learning). However, these teachers may criticise this kind of adaptation as "slowing down the class" and thus preventing them from "covering the content".

PCK: We infer from their comments that teachers at the Immediate level have a relatively simplistic understanding of the subtleties of mathematical content knowledge and how it develops through the learning process (PCK). Most commonly:

- The teacher's main focus in the enacted curriculum is on computational procedures and factual knowledge of mathematics.
- The teacher attributes wrong answers to carelessness or failure to accurately recall and follow procedures. Typically the teacher does not believe most errors are caused by misconceptions, and does not anticipate likely misconceptions as they teach.
- The teacher draws examples from the textbook; the teacher does not construct them to meet a specific learning need.
- The teacher has not considered the distinction between ability to do mathematics problems in the abstract as opposed to application to new and unfamiliar tasks

or contexts (far transfer). The teacher thinks of the curriculum as a linear sequence of topics, with the sequence provided by the textbook. The teacher does not recognise, or minimises, the importance of learning tasks such as conceptual change, integration, inference, generalisation and far transfer. The teacher tends to think of (and teach) each segment of the curriculum in isolation from others.

Expert Level (Adaptive Instruction)

The Expert-level teacher adapts instruction by focusing on diagnosis and systematic correction of errors and misconceptions. This eventually can lead to a culture change in the class, with teachers and students sharing responsibility for learning. In our experience, development from the Immediate to the Expert level often takes at least a year, unless extensive coaching and professional development are available, or the teacher already has developed the pedagogical insights that support this level.

Pedagogy: The comments of Expert level teachers lead us to infer these practices and beliefs as the most common about pedagogy:

- The teacher allocates instructional time to conceptual understanding, as well as computational procedures. Typically, they prefer to build conceptual understanding first, and then introduce the procedural knowledge based on the concepts.
- The teacher attends to pupils' errors in order to diagnose gaps in prerequisite knowledge and misconceptions, based on the belief that most errors are caused by these flaws in understanding.
- The teacher commonly asks questions with a single correct answer (convergent reasoning). However, they may occasionally use open-ended questions requiring divergent reasoning.
- The teacher most commonly does teacher-centred lecture, but with improved questioning and dialogue surrounding errors and misconceptions. The teacher also uses routines which allow students to talk and work in fluid small groups (for example, 'playing dumb' and thus requiring students to correct the teacher's apparent misconception). The teacher tolerates 'constructive buzz' in the classroom.
- The teacher encourages students to take responsibility for building their understanding and helping to diagnose causes of errors and misconceptions they or other students have made.
- The teacher's focus is on what the students are doing, because of the belief that learning is about what the students do, not what the teacher does.
- The teacher's preferred routine is to enlist the class's aid in identifying the cause of a lack of understanding. Once the probable cause is established, teaching strategies used often include construction of better examples, peer 'teachback' or other discussion techniques (often large-group). The teacher believes this kind of discussion of errors and use of peer teaching builds deeper conceptual understanding for the whole class.
- The teacher takes extra time to address gaps and misconceptions, with the expectation that the time can be made up by identifying other topics that need

no further explanation. They use frequent data collection to guide decisions about pacing.

PCK: Expert-level teachers typically show deeper conceptual understanding of the mathematics they are teaching, as well as the ability to anticipate common misconceptions and gaps in mathematical understanding.

- The teacher anticipates likely misconceptions and plans 'multipath' lesson plans with contingencies to use when these misconceptions are encountered as student errors.
- Often, the teacher will prefer to "start where the students are" and defer the use of more difficult problems, which they know are beyond their students. Later, they will return to the more difficult problems when students have built new knowledge or knowledge integration and are ready for far transfer or higher cognitive complexity.
- The teacher spontaneously constructs examples to meet a specific learning need, rather than drawing examples exclusively from the textbook.
- The teacher is more likely to include contextualised examples and problems, to facilitate knowledge integration and far transfer.

Master Level (Fully Differentiated Instruction)

At the Master level, the teacher is generating and interpreting in real time a continuous data stream, generated by embedding questions frequently in the lesson. The teacher then interprets the data and recognises patterns of student understanding, which lead to real-time decisions on how best to group students for small-group learning tasks. The groups are fluid and task-based, and may be homogeneous or heterogeneous, depending on the pedagogical intent. The teacher uses a variety of differentiation strategies. Learning goals include not just 'the content' of mathematics, but also the building of a sense of mathematical curiosity, as well as dealing with the students' anxiety and confidence issues surrounding mathematics. In our experience, the few teachers who reach this level do so typically after two or more years of experience, typically with substantial support, coaching and professional development.

Pedagogy: The comments of Master level teachers lead us to infer these practices and beliefs as the most common about pedagogy:

- The teacher may begin a new topic with a short introductory lecture, but most time is spent in small-group work.
- The teacher routinely uses open-ended questions requiring divergent reasoning, with proportionately fewer questions with a single correct answer (convergent reasoning).
- The teacher's role includes the frequent monitoring of group activity (both by using technology and by walking around). Whenever possible the teacher encourages the small groups of students to engage in peer tutoring to solve problems

independently, before intervening. If a problem occurs frequently, the teacher may convene the class for short 5–10 min lecture/explanation, or may include the explanation in a summary discussion at the end of the learning activity.
- The teacher makes sure that students take responsibility for working collaboratively on assigned learning tasks. Students use feedback from questions to monitor their own understanding and to help diagnose causes of errors and misconceptions they or other students have made.
- The teacher exhibits the role of instructional manager ('guide on the side').
- The teacher can usually use their pedagogical content knowledge to anticipate and prevent learning problems. The teacher uses this knowledge to construct examples and worksheets, and as input to grouping strategy.
- The teacher views a diversity of answers as a teaching opportunity, and may use teaching routines designed to elicit a diverse range of answers.
- The teacher's preferred routine (script) is to enlist the class's aid in identifying the cause of the problem and correcting it through better examples, and through a use of small-group techniques such as peer teaching. The teacher shows a preference for exploratory learning tasks to deeper conceptual understanding for everyone.
- The teacher differentiates learning objectives based on difficulty level (low, average, and high or enrichment) and encourages a range of different problem-solving strategies, rather than a single solution algorithm.
- The teacher devotes considerable attention in class on questions to promote insights on 'big ideas' (see below), the integration of knowledge and far transfer, and justifies this practice with the belief that learning progressions resemble a spiral, not a linear sequence.
- The teacher differentiates learning objectives and problem difficulty, depending on a judgement of what each student needs next.
- The teacher differentiates using both homogeneous and heterogeneous small groups. Typically, when peer teaching to correct omissions, errors and misconceptions is the goal, heterogeneous groups are preferred; when students have a basic understanding but need different levels of problem difficulty, homogeneous groups are preferred. Groups are task-specific and fluid: in a 100-min extended class, up to four different groups per period may be used.
- The teacher makes informed trade-offs on time, often by spending time on initial learning and then on addressing misconceptions, but making up for this by identifying and skipping other topics that need no further explanation, identified through use of frequent data collection. For students who need more time than class allows, out-of-class supplemental strategies are employed.
- The teacher shows high expectations for every student's success in mathematics.
- The teacher uses differentiation to address issues of mathematics anxiety and confidence often within the first 6 weeks of class. However, even these Master teachers may not recognise or have the expertise to deal with long-term (developmental) issues of the sort often encountered with special needs students.

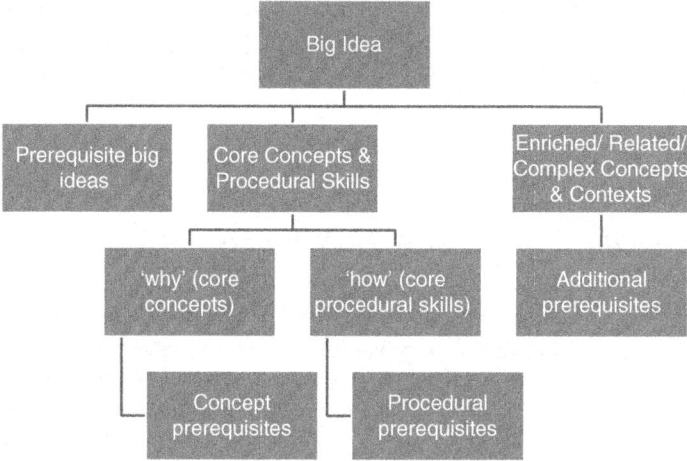

Fig. 1 The Master teacher's hierarchy of knowledge types

- The teacher's lesson plans are not linear. They are planned as a series of learning activities, data collection actions and contingent branch points with decisions made based on how students respond.

PCK: Master-level teachers typically show a hierarchical knowledge structure of the mathematics they are teaching. They use this knowledge structure to structure their enacted curriculum and to anticipate common misconceptions and gaps in mathematical understanding.

- The teachers often think of their curriculum content as differentiated by knowledge type, as shown in Fig. 1.

In this structure, the core concepts and procedural skills are defined by the state test, the textbook and the school or district curriculum leaders. This is the knowledge and skill level that every student must learn. By contrast, the elaborated/related or complex concepts and contexts provide additional depth of understanding and ability to engage in far transfer of the Big Ideas. They go beyond the core, and represent a richer knowledge structure which students can develop as time permits, after having mastered the core.

In a typical year-long curriculum, there are fewer than 100 components to an instantiation of this structure, including fewer than a dozen 'big ideas' with supporting connected concepts and skills. Navigating this structure defines the options for *learning progressions* within a course. While it is rare even for Master teachers to actually draw this structure, their comments about the knowledge structure they are teaching often reveals a structure with these properties. At the Master level, the teacher's insights are constantly evolving, thus the teacher is implicitly 'editing' (adding, deleting, rearranging) this structure, in order to make better decisions on what to do next. In this way, the teacher gradually develops an understanding of alternative *learning progressions* for the curriculum. While the teaching practices of the Master are superficially the same as those of the Expert, this knowledge structure sets the focus for their construction of explanations, worksheets, homework and assessment tasks.

TPACK and the Use of Classroom Networks

One of the biggest barriers to the adoption of adaptive and differentiated instruction on a wide scale is the all but overwhelming data management challenge that these strategies present. A 'connected classroom' within-class network that links every student's personal device can be designed to overcome these challenges and provide an ideal platform for collaborative learning, automated formative assessment, and adaptive or differentiated instruction. High quality, strong evidence shows classroom networking can be used to create a learning environment that is substantially more effective than conventional classroom practice (Center for Technology in Learning 2009; Penuel and Singleton 2010; Penuel et al. submitted; Roschelle 2011).

Our experience is with the Texas Instruments TI-Navigator and TI-Nspire Navigator classroom networks, so the discussion that follows will refer to that system. However, we do not mean to imply that adapting and differentiating instruction can be done only with TI-Navigator. Whilst we have observed that this technology saves time, is less cumbersome and improves the power of teacher and student decision-making when compared to non-technology techniques, it is quite feasible to adapt instruction, at least to some degree, without technology or by using alternative technologies. It might be that learning these techniques without technology can provide a useful starting point for teachers who are not yet confident in the use of the technology. We have observed subjectively, however, that teachers who attempt the Expert or Master levels of the model without use of technology often complain of the amount of preparation time required, and ultimately they are at risk of burnout.

Recent qualitative studies have demonstrated that Master-level teaching with fully differentiated instruction represents a major and challenging change for teachers (experienced or pre-service) who are used to large-group instruction. The previous section discussed the changes in pedagogical and pedagogical content (PCK) knowledge and beliefs as teachers progress to the Expert and Master levels. In this section, we will address the changes in knowledge and beliefs surrounding the use of technology in teaching (TPACK), with particular focus on mathematics teaching (Graham 2011; Hill et al. 2008; Niess et al. 2009). We used the following definition of TPACK, which find useful because of its focus on the teaching/learning process supported by technology:

> TPACK is a knowledge of the dynamic, transactional negotiation among technology, pedagogy, and content and how that negotiation impacts student learning in a classroom context. (Cox 2008, p.77)

The following discussion attempts to paint a developmental picture of how teachers' thinking about the role of technology in their teaching process evolves, with particular attention to use of classroom networking as an important modality for sustaining the 'transactional negotiation' surrounding our three levels of sophistication in adaptive and differentiated instruction.

Immediate Level

If mathematics teachers use technology at the Immediate level, typically it is used only in the most basic ways. Use of a handheld (such as TI-Nspire) is often limited to calculating and checking answers. If a classroom network links the handhelds, students can enter answers (e.g., via TI-Navigator's *Question* tool), but this is seen primarily as a time-saver. Teachers often experience increased student engagement, which they often report as improved classroom management (e.g., "keeping everyone on task"). On the other hand, teachers at this level will likely see time spent with technology as potentially competing with time they need to address their required scope and sequence (e.g., "I'd use the gadgets more, but then I wouldn't have time to cover the content"), so they often limit technology use to one-for-one substitutions with familiar routine activities, such as taking attendance, distributing, collecting and automatically marking homework and worksheets, a warm-up exercise, or an end of unit quiz (automatically marked).

Immediate-Level Teaching Practices Using Technology

While generalisations are always dangerous, we offer these observations about Immediate-level teachers' practice, synthesised from our observations:

- Teachers use the classroom network to save time (e.g., through automated document distribution and collection, automated grading of quizzes).
- Teachers use the TI-Navigator screen capture (which displays one, some or all student handheld screens on a projector) to spot who is off task.
- Teachers use the TI-Navigator projected emulator (which displays a functioning handheld device to the class) and Live Presenter (projection of student screens) to help students operate the technology.
- Teachers commonly use the TI-Navigator Question subsystem to give a warm-up problem to start the lesson.
- Teachers may also use the technology to do the wrap-up post-quiz or 'exit problem.' If the end-of-class activity shows a high error rate, this leads to subsequent re-teaching (a mini-lesson review) and more assigned practice, but not to further diagnostics or differentiated instruction. If only a few students have a problem, the teacher may follow up individually with them, but often this is not a result.
- Teachers often delay interpreting data until after class, rather than using the data to make decisions as the class is working. This delays re-teaching to the next day's class or later.

Expert Level

Expert level teachers assume availability of technology in their working environment and are willing to take chances with trying out new technology tools and teaching routines, because they feel confident they can handle any unexpected results. They also have a good sense of the time required to use technology and feel more freedom to allocate time when they identify opportunities to pursue important mathematics in depth and to correct gaps and misconceptions.

Expert-Level Practices Using Technology

Expert-level teachers still use the classroom network to save time (through automated document distribution and collection, automated grading of quizzes). In addition:

- Teachers examine student work 'on the fly' (e.g., through TI-Navigator's screen capture) to spot 'interesting' errors (which show gaps and misconceptions). They then engage the class in analysing the reason for the error.
- Teachers encourage student explanations and examples presented to the whole class (e.g., using TI-Navigator's *Live Presenter* mode, which displays any student's handheld screen to the whole class). This mode is also used to show students how to operate the technology, as at the Immediate level.
- Teachers administer the warm-up (pre-) and wrap-up (post- or 'exit') problem using the classroom network (e.g., TI-Navigator's *Question* subsystem). A high error rate leads to further diagnostics or peer teaching, as problem topics are identified in real time.
- Teachers may occasionally embed individual questions in an instructional presentation as a 'checkpoint' or 'exploration' to generate a wide range of positive and negative examples for class discussion.
- Teachers may add additional instructional resources to the technology environment, such as examples, problem sets and questions taken from the textbook or their other resources.

Master Level

Master-level teachers often take pride in inventing new teaching routines (activity scripts) that fully exploit the affordances of the technology. They are comfortable with taking risks with the technology, and are 'fault tolerant' when the technology behaves in unexpected ways in class, because they can quickly recover by adapting or switching to a more familiar alternative teaching routine. These teachers are comfortable with importing learning activities from a wide range of sources, inventing their own learning activities and customising or localising the activities provided with the technology, in a constant quest to close learning gaps, prevent and correct misconceptions, stimulate curiosity, improve engagement and build student self-confidence to counteract 'mathematics anxiety'. They often share these activities with their peers within or outside their school and district. Consequently, they are heavy users of tools to import and export content and data, and they often experiment with authoring tools to create examples, questions, simulations and games.

Master-level teachers may play a leadership role among their peers in their school and district, as well as their region and state. However, paradoxically, master-level teachers sometimes report that they feel isolated within their own school and they may have a stronger community of practice with Master teachers in other districts (often through technology user groups) than with their own local peers.

Master-Level Practices Using Technology

- As with teachers at every level, Master teachers still use the classroom network technology to save time (through automated document distribution and collection, automated grading of quizzes and homework). They also use the technology to administer the warm-up (pre-) and wrap-up (post- or exit) problems.
- Teachers are skilled at using real-time data displays (such as TI-Nspire Navigator's screen capture and collaboration tools) to spot 'interesting' errors (which show misconceptions) and engage the class in analysing the reason for the error, and to see multiple correct solutions to a problem.
- Teachers encourage students to monitor their own learning through data summary reports and by comparing their work with others in the class (e.g., through the TI-Navigator screen capture tool).
- Teachers frequently use open-ended exploratory questions (and TI-Navigator's collaboration tools, or just a personal white board for each student) designed to elicit and display a wide range of alternative answers, which then stimulate a rich discussion of the merits of each, alternative reasoning paths, and misconceptions. Students are encouraged to share their 'Aha!' insights verbally and through display of their work (using TI-Nspire Navigator's *Presenter* mode).
- The teachers will use the classroom network technology to identify students with something 'interesting' to discuss, who are not the first to raise their hand or to participate. They see one benefit as confidence-building.
- Teachers continuously embed questions in lessons, to get immediate feedback for both teacher and students to use in deciding on what learning issues/tasks to address next. They often interpret the data display 'on the fly' to spot problem topics and make the first 'path' decisions in their lesson plan.

Concluding Remarks

Figure 2 is a graphical summary of the three levels of our model represented as a potential developmental continuum.

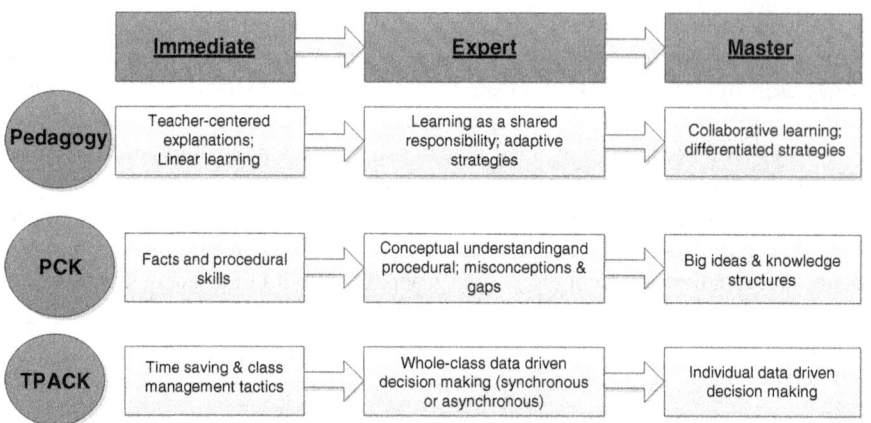

Fig. 2 Three-level model of adaptive and differentiated teaching and learning

Most models of teaching with formative assessment (Heritage and Stigler 2010; Popham 2008) do not deal explicitly with the development of teacher capacity; the implicit expectation is that most teachers can immediately adopt the formative assessment model and differentiate instruction. We believe this is practically unrealistic, and inconsistent with general models of innovation and technology integration (Hall et al. 1975; Dwyer et al. 1992). Our observation of practising and pre-service teachers suggests that the three levels of sophistication in the use of formative assessment presented here constitute three points on what may be a realistic developmental continuum (possibly a learning progression) of teaching skills, knowledge and beliefs. Because progression through the levels of the model often requires major changes in pedagogy knowledge and PCK, it is common for teachers to take years to reach the Master level of the model, even with effective professional development and coaching. However, teachers we have worked with report meaningful benefits to themselves and their students at every level of the continuum.

We also believe that the model can be viewed as consistent with the conclusions of Ruthven, who found that teachers most commonly integrated technology into their current practice, rather than attempting transformative change. It is also consistent with the research of Clark-Wilson (2009, 2010a, 2010b), who found that the most common pattern of technology integration was to substitute a technology-based teaching activity for the same or a similar non-technology activity. Only a few teachers attempted something that was truly 'new' to them.

As is common with other models of teaching with formative assessment, nothing about our model specifically requires use of technology. However, it is difficult to imagine this differentiated instruction as described here to be practical for daily use at scale without the support of a classroom networking system such as TI-Nspire Navigator, and teachers must master this technology as they develop their teaching skills at each level. The technology makes possible real time embedded data gathering and reporting, interaction with students individually or in groups, real time embedded formative assessment with tracking of individual student progress.

Further, it makes student problem-solving work visible by showing the entire solution process, rather than simply gathering an answer. That said, learning the technology is the least challenging part of becoming an Expert-level or Master-level teacher as this explanatory model defines it. The most important learning tasks for teachers who wish to advance through the levels of the model may well be to build (or change) their knowledge and beliefs about mathematics, teaching and learning, classroom culture and their role. The speed and level of implementation of the model depends primarily on these critical changes in teacher beliefs. For that reason, TI-Nspire Navigator technology, and the feedback it provides to teachers, is as valuable to teachers as is the functionality and feedback provided to students. In addition, implementing the model requires carefully planned professional development, coaching and peer support, as well as supportive school policies and well-designed curricula, which stay in place over a sustained period of time, as teachers grow.

Our current research is using this model in two ways: first, one of us (Bellman) is working with colleagues on scaling the teaching of the model to other pre-service and in-service teacher training programmes. This will help us understand if the developmental patterns of the model are stable and reproducible. Second, we have begun to use the model to synthesise and interpret findings on teacher use of TI-Navigator and TI-Nspire Navigator. Roschelle (2009) provides a recent overview of this research. Ultimately, we believe the model has the potential to guide development of improved plans for introducing technology-based formative assessment for differentiated instruction.

References

Black, P., & Wiliam, D. (1998). Assessment and classroom learning. *Assessment in Education: Principles, Policy & Practice, 5*, 1.

Center for Technology in Learning, S. I. (2009). Does teacher knowledge of students' thinking in a network-connected classroom improve mathematics achievement? *TI EdTech Research Note #14*. Dallas: Texas Instruments.

Charmaz, K. (2000). Grounded theory: Constructivist and objectivist methods. In N. Denzin & Y. Lincoln (Eds.), *Handbook of qualitative research* (2nd ed., pp. 509–535). Thousand Oaks: Sage.

Clark-Wilson, A. (2009). *Connecting mathematics in the connected classroom: TI-Nspire™ Navigator™*. Chichester: The Mathematics Centre- University of Chichester.

Clark-Wilson, A. (2010a). Emergent pedagogies and the changing role of the teacher in the TI-Nspire Navigator-networked mathematics classroom. *ZDM*, 1–15. doi:10.1007/s11858-010-0279-0

Clark-Wilson, A. (2010b). Emergent pedagogies and the changing role of the teacher in the TI-Nspire Navigator-networked mathematics classroom. *ZDM, 42*(7), 747–761. doi:10.1007/s11858-010-0279-0.

Cox, S. (2008). *A conceptual analysis of technological pedagogical content knowledge*. Ph.D. dissertation, Brigham Young University, Provo.

Dwyer, D. C., Ringstaff, C., & Sandholtz, J. H. (1992). *The evolution of teachers' instructional beliefs and practices in high-access-to-technology classrooms first-fourth year findings ACOT*. Cupertino: Apple Computer, Inc.

Graham, C. R. (2011). Theoretical considerations for understanding technological pedagogical content knowledge (TPACK). *Computers in Education, 57*(3), 1953–1960. doi:10.1016/j.compedu.2011.04.010.

Hall, G., Loucks, S., Rutherford, W., & Newlove, B. (1975). Levels of use of the innovation: A framework for analyzing innovation adoption. *Journal of Teacher Education, 26*(1), 52–56.

Heritage, M., & Stigler, J. W. F. R. W. (2010). *Formative assessment: Making it happen in the classroom*. Thousand Oaks: Sage.

Hill, H. C., Ball, D. L., & Schilling, S. G. (2008). Unpacking pedagogical content knowledge: Conceptualizing and measuring teachers' topic-specific knowledge of students. *Journal for Research in Mathematics Education, 39*(4), 372–400.

Niess, M. L., Ronau, R. N., Shafer, K. G., Driskell, S. O., R., H. S., Johnston, C.,... Kersaint, G. (2009). Mathematics teacher TPACK standards and development model. *Contemporary Issues in Technology and Teacher Education, 9*(1), 4–24.

Penuel, W. R., Beatty, I., Remold, J., Harris, C. J., Bienkowski, M., & DeBarger, A. H. (submitted). Pedagogical patterns to support interactive formative assessment with classroom response systems. *Journal of Technology, Learning, and Assessment*.

Penuel, W., & Singleton, C. (2010). Classroom network technology as a support for systemic mathematics reform: Examining the effects of Texas instruments' MathForward Program on student achievement in a large, diverse district. *Journal of Computers in Mathematics and Science Teaching (JCMST), 30*(2), 179–202.

Popham, W. J. (2008). *Transformative assessment*. Alexandria: ASCD.

Roschelle, J. (2009). *Towards highly interactive classrooms: Improving mathematics teaching and learning with TI-Nspire Navigator*. Menlo Park: SRI Center for Technology and Learning, SRI International.

Roschelle, J. (2011). Improving student achievement by systematically integrating effective technology. *Journal of Mathematics Education Leadership, 13*, 3–11.

Ruthven, K. (2009). Towards a naturalistic concepualisation of technology integration in classroom practice: The example of school mathematics. *Education & Didactique, 3*(1), 131–149.

Integrating Technology in the Primary School Mathematics Classroom: The Role of the Teacher

María Trigueros, María-Dolores Lozano, and Ivonne Sandoval

Abstract In this chapter, we analyse the role of the teacher when using digital resources in the primary school mathematics classroom in Mexico and its relation to students' mathematical learning. We carry out this analysis through the use of an instrument that we developed in which we relate five different aspects of the role of the teacher we consider important with the three different uses of technology classified by Hughes (*Journal of Technology and Teacher Education, 13*(2), 277–302, 2005) namely ***replacement, amplification*** and ***transformation***. We use an enactivist perspective that considers learning as effective action in a given context (Maturana, H., & Varela, F. *The tree of knowledge: The biological roots of human understanding*, Revised Edition, Boston: Shambhala, 1992) in order to describe the way in which differences both in the uses of technology and in the role the teacher assumes in the classroom contribute to creating classroom contexts in which mathematical learning is promoted to different degrees.

Keywords Digital technology • Enciclomedia • Role of the teacher • Enactivism • Mathematics learning • Uses of technology

---

M. Trigueros (✉)
Instituto Tecnológico Autónomo de México, Mexico D.F., Mexico
e-mail: trigue@itam.mx

M.-D. Lozano
Universidad de las Américas Puebla, Puebla, Mexico

I. Sandoval
Universidad Pedagógica Nacional, Mexico D.F., Mexico

Introduction

The use of digital technology in the classroom has occupied the attention of researchers in mathematics education for several decades. Most of the studies focusing on the teaching of mathematics using specific software have been concerned with middle or high school students and teachers (e.g. Ruthven 2007; Drijvers et al. 2009). Since the teaching of mathematics in primary school has particularities and restrictions that make it very different from the work done at higher levels of education, it is important to carry out more research in order to investigate how the integration of technology occurs at this level. Through the work in this chapter, we intend to contribute in this direction by analysing the role of the teacher when using digital resources in the primary school mathematics classroom in Mexico and its relation to students' mathematical learning.

Background

The use of digital technologies in the primary school in Mexico has been influenced greatly by the introduction of the national teaching programme Enciclomedia, which was created in 2004 with the intention of complementing already existing materials in primary school classrooms – such as the mandatory textbooks – with computer programs and teaching resources designed to be used with an interactive whiteboard. It is a large-scale project, involving more than 7,000 schools and 170,000 classrooms, and is meant to support the teaching and learning of all subjects in grades 5 and 6 of primary school by working with one computer. Enciclomedia's programs were designed with the intention of motivating students to engage in mathematical problems by inviting them to take part in games and other activities and by providing them with interesting contexts. The use of interactive whiteboards was intended to promote classroom interaction and to enhance interactivity with computer programs. Evaluations of the Enciclomedia project have shown positive results in terms of resources' usability and interactivity, a high potential for promoting meaningful and high order operations learning, as well as high motivation of students (Holland et al. 2006; Díaz de C. et al. 2006; Trigueros et al. 2007). Infrastructure and teacher training were, however, found to be problematic (Loredo et al. 2010). Furthermore, difficulties were identified in terms of the integration of Enciclomedia's resources by the teachers in their everyday lessons. Some studies (e.g. Díaz de C. et al. 2006; Sagástegui 2007) report that resources have the potential to change teachers' practice, but more research is needed as there are still few studies focusing on the ways teachers use these resources in their classrooms (e.g. Trigueros and Lozano 2012).

In this chapter we will analyze primary school teachers' work with technology in their classroom. We are interested in a detailed description of when and how they use digital resources in their teaching practices and how this relates to different roles they can undertake.

Theoretical Framework

We are interested in investigating what happens in relation to the role of the teacher when technology is introduced in primary school mathematics classrooms. In order to do this, we use an enactivist perspective (Maturana and Varela 1992), which considers that, in the process of living, individuals carry out those actions which are effective or adequate in a given context and that it is this continuous process of successful interaction with the environment that we call 'learning'.

In this way, learning occurs when individuals interact with each other, changing their behaviour in a similar way. In a particular context or location, the participants create together the conditions that will allow actions to be adequate. As members of a particular community interact with each other, patterns of behaviour are created constituting a classroom culture (Maturana and Varela 1992). With these ideas in mind, we are interested in investigating how patterns of effective behaviour emerge in mathematics classrooms as teachers and students use Enciclomedia. In particular, we focus on those patterns related to the role of the teacher.

Learning Mathematics with Computer Tools

From an enactivist perspective, the use of computer tools is part of human living experience since "such technologies are entwined in the practices used by humans to represent and negotiate cultural experience" (Davis et al. 2000, p. 170). Tools, as material devices and/or symbolic systems, constitute an important part of learning, because their use shapes the processes of knowledge construction and of conceptualisation (Rabardel 1999, 2011). When tools are incorporated into learners' activities they become instruments which are mixed entities that include both tools and the ways these are used. Instruments are not merely auxiliary components or neutral elements in the teaching of mathematics, they shape students' and teachers' actions. Every tool generates a space for action, while at the same time imposing on users certain restrictions. This makes possible the emergence of new kinds of actions. When using the tool, teachers' history and context will then determine which actions are undertaken among the ones made possible by the programs.

In this way, we consider teachers as learners who are modifying their actions in an environment that includes specific characteristics and certain tools. The actions teachers undertake define their different roles in the classroom, and we look at them through an instrument of analysis that considers different aspects of the role of the teacher.

The Role of the Teacher and the Different Uses of Technology: An Instrument of Analysis

We propose five different aspects of the teachers' role that we investigate through this work. We are aware that the different aspects overlap and cannot be clearly separated. This classification is used for purposes of theoretical analysis only.

1. *Role in terms of communication of mathematics*. In mathematics classrooms, students and teachers are in contact with mathematical concepts that are defined by a larger community of mathematicians. Sometimes, these concepts and processes are made available to the students exclusively by the teacher and through textbooks. The inclusion of technology in the classroom often implies that the computer programs become another source of mathematical information. Technology might therefore influence the teachers' role regarding mathematics concepts and procedures by providing a complementary source that the teacher and the students can both comment on and work with. Effective behaviours then might include several forms of interaction with the mathematical content included in the digital programs.
2. *Role in terms of interaction with students*. The role of the teacher in the context of this study refers to the way in which teachers interact with students and how they manage and regulate what happens in the classroom. Sometimes teachers listen attentively to students and respond accordingly, often modifying what they had planned for a lesson. On other occasions, teachers tend to follow closely a determined path, and respond scarcely to students' questions or interests. The inclusion of technology can influence the way in which the teacher regulates interactions by presenting unexpected situations that might have not occurred without the use of particular programs. Effective behaviours might include allowing students to explore the use of the program and discussing unplanned mathematical problems that might arise while using technology
3. *Role in terms of validation of mathematical knowledge*. The teacher as a source of validation for correct mathematical procedures and answers is also an important aspect to be examined when technology enters the classroom. Several interactive programs give feedback to the student when an answer is entered. Teachers might discuss answers with students before the program validates them or they might allow students to use the program as a means for validating their own answers.
4. *Role in terms of the source of mathematical problems*. In mathematics classrooms, the teacher is often the main source of mathematical problems. Even if problems from textbooks are solved, it is the teacher who decides which problems or exercises are to be worked on. Mathematical problems, however, can also emerge from activities in the classroom itself. When technology is used in the classroom, it becomes another possible source of mathematical problems and it can also influence the way in which mathematical problems are selected. Teachers' behaviours might include encouraging students to solve those problems posed by the programs (directly or indirectly), even in those cases

when unexpected uses of technological devices lead to mathematical problems that had not been addressed before and that might not be included in the lesson plan or in the curriculum.
5. ***Role in terms of actions and autonomy of students***. Actions on mathematical objects and tools can be carried out both by teachers and students in the classroom. Sometimes the teacher assumes a more active role, while students mainly listen or copy. At other times students have more autonomy to decide what to do and how to do it. Again, technology may change the dynamics of who does the mathematical actions in the classroom. Teachers' actions might include allowing the students to work with the program and the mathematical problems without much intervention. In other cases, teachers might use the computer to show students certain features or uses of the program before the latter are allowed to explore.

After observing mathematics lessons or videos from lessons involving technology, and having read some of the literature regarding the use of digital programs in the classrooms, we decided that we wanted to explore how the different aspects of the teacher's role related to the particular ways in which technology is used. We considered that looking at both the teacher's role and the use of the programs might give a richer account of the way in which technology is integrated within the classroom culture.

In order to consider different ways of using technology, we use the categories developed by Hughes (2005): "The variation in technology supported pedagogy can be captured through three categories: (a) Technology functioning as replacement, (b) amplifications, or (c) transformation" (p. 281). She defines technology as replacement when it is used in a way that does not change "established instructional practices", that is, "the technology serves as a different means to the same instructional end". Technology as amplifier "capitalizes on technology's ability to accomplish tasks more efficiently and effectively, yet the tasks remain the same"; and technology as transformation may change "teacher's instructional practices and roles in the classroom" (p. 281). From an enactivist perspective, it is important to analyse how the different aspects of the role of the teacher are related to her or his use of technology and how this relation changes through the lessons as he or she takes decisions through the lesson. Taken together we can have a clearer picture of the integration of technology in the mathematics classroom and of which aspects need to be considered to help teachers make innovative and effective uses of technology.

The research questions addressed in this study are:

- How can we describe teachers' effective actions in terms of the different aspects of the role they take when technology is introduced in the classroom?
- How are these aspects of the teacher's role related to their use of technology?
- What kinds of classroom cultures are created by teachers' effective behaviours in terms of the role of the teacher and of the use of technology?
- How are different classroom cultures fostered by different teacher behaviours related to students' learning of mathematics?

Methodology

We decided to investigate the teacher's role in the mathematics classroom when using technology by selecting eleven teachers from five schools in three different states in the country. The states and schools were chosen mainly because we, as participants in a larger study that investigates teaching practices with technology, are in contact with teachers and head teachers from those schools. Three of the schools involved in this study are urban and two of them are semi-rural. Additionally, one of the urban schools is a private school, while the remainder are state schools (see Table 1).

Most grade 5 and 6 classrooms in Mexico have the Enciclomedia project equipment, and some of them have also computer labs with an average of 20 computers. In addition, with the exception of one semi-rural school, all of the schools participating in this project have internet access. The teachers that we selected for this study had at least 1 year of experience using technology in their lessons and had an interest in participating in a research project. We selected them because, from our observations, they each represented a particular group of teachers who use technology in a certain way. The three teachers we selected differed in their background, professional training and experience in teaching mathematics and also in their training experiences on the use of technology.

After observing several lessons and analysing video recordings from the eleven teachers who were initially included in the study, we identified three teachers who were representative of the characteristics of the different groups of teachers, with respect to their experience both as teachers and with the use of technology, and according to the way in which they used digital resources in their classrooms. We decided that, in order to deepen our understanding of the role of the teacher when using technology, we would focus on the analysis of a lesson that can include one or more sessions of these three representative teachers who we will refer to as Gabriel, Juan and Susana.

The Teachers

Gabriel is an elementary school teacher with 30 years of experience and who initially trained as a secondary school mathematics teacher but then decided to work at the elementary level. His use of technology in the classroom started when the

Table 1 The teachers

State	Type of school	Teacher
Distrito Federal	Private	1 teacher
	Public 1	5 teachers
Estado de México	Public 2	2 teachers
San Luis Potosí	Public 3	2 teachers
	Public 4	1 teacher

program Enciclomedia was installed in his school in 2004. He initially received some general training, which consisted only of technical instruction on the use of the software, and has continued with the development of his technological abilities independently. He has not had any training in relation to the teaching of mathematics with digital technologies. Gabriel represents a large group of teachers who are interested in teaching with technology, but who have not received pedagogical training.

Juan is a young primary school teacher with 6 years teaching experience. He has trained himself on the use of technologies; he has never taken any training course, but he is a proficient user of computers and owns a personal computer. During an interview, Juan mentioned that he likes technology because it is helpful to keep track of his records and to search for information. When he uses it in the classroom, it is mainly for teaching mathematics. We selected him because he represents another group of teachers who are proficient users of computers and have recently started to use them in their teaching.

Susana is a teacher-researcher who had been teaching for 5 years in a private urban primary school before she started including digital resources in her lessons. She has been involved in mathematics education for several years and is interested in reading the mathematics education literature. She got involved in 'Enciclomedia's' training workshops from its early stages, and has been using it in her classroom for 6 years. She always has the latest version installed in her computer at home. We selected her because she represented a small group of teachers who were particularly successful in integrating resources in their teaching practice.

Research Tools

In order to study teachers' actions when investigating both the uses they make of the resources available and their role in the classrooms, it is necessary to employ a variety of research tools so that different perspectives are addressed. For this study, we analysed classroom observation notes, video-recordings of teachers' lessons and audio-recordings from interviews. The guides for the observation notes were developed as part of the abovementioned larger study and were very general. Observers described the digital resources used during the lesson and particular incidents they considered important in relation to the use of technology in the classroom. The lessons were video-recorded and teachers were free to decide on the mathematical theme they wanted to teach, the resources they wanted to use and how they wanted to use them. Often they would teach the mathematical topic that was meant to be taught on those dates according to the official programme. During the interviews, which were also carried out as part of the larger project, teachers were asked about their background and training and about the ways in which they worked with technology during their mathematics lessons.

The collected data were reviewed, for this particular study, by all three researchers involved. During a first round we looked at the data together with the aim of selecting those episodes that would be analysed. For our analysis, we focused on teaching episodes where digital technologies were integrated. We then developed collectively an instrument of analysis to characterise each teacher's practice that we describe below. The purpose of the analysis is to reflect on the dynamics of the use of technology in the classroom, and of the different aspects of the role played by the teacher in relation to the mathematics being taught. In order to make our results reliable, the data coming from the selected teaching episodes were also analysed independently on a second occasion by the three researchers using the same instrument of analysis. Final decisions were the result of comparison and negotiation of the independent outcomes.

Throughout a teaching session, it was possible to identify different time intervals in which a specific kind of activity involving the technology was used to work with mathematics problems. We call each one of these intervals an episode. A matrix is used to describe the dynamics of the selected episode by means of an arrow. A full description of how to interpret the resulting matrices is included later in the chapter. This instrument is useful to compare the dynamics of the lessons of the same teacher and of different teachers and can also be used to make teachers aware of their actions and to reflect on their lessons. In this chapter we use it to compare one typical lesson of the selected teachers.

The comparison of the teachers who represent the selected classes is used to differentiate them in terms of the mathematical activity that is favoured during each interval and possibilities for students to learn the intended mathematical topics. We also use data to determine if certain uses of technology are favoured or more frequently linked to particular aspects of the teacher's role and how it influences mathematical activity.

Results

We analyse in detail the different roles that teachers can play when they use digital resources in their lessons. This analysis provides useful information to help explain why certain approaches are taken by teachers and an indication of the possible learning outcomes (in terms of students' learning) that are associated with the teacher's different roles. For that purpose we have divided the descriptions of the episodes according to the changes in class activity.

Gabriel

For the selected session, Gabriel worked with a chapter from the students' textbook in which, according to official documents, students are expected to *"deduce equivalences between the units of volume and capacity for liquids"* (SEP 2009, p. 155). Gabriel started the lesson (episode 1) using the electronic whiteboard to write some definitions. He also used manipulatives like a 1 dm³ glass cube and a plastic bottle. During the whole episode, Gabriel followed closely his lesson plan. He gave explanations and asked some questions, but he did not use the students' responses to review or complete his explanations. He was always in charge of the communication and validation of the mathematical knowledge:

G: *How many centimetres are in a decimetre?*
S_1: *Ten!*
S_2: *A hundred!*
G: *And what do you think? A decimetre is equivalent to how many centimetres?*
S_4: *To ten?*
G: *To ten what…I cannot hear your, speak louder!*
S_4: *To 10 cm?* [...]
G: *How much water can I fit into that cube?*
S_1: *One litre!* S_2: *One half!*
S_4: *A cubic decimetre has a capacity of mmm… litres.*
G: *One litre, because exactly, it was exactly what I could pour here, one, do you agree? Very well.*

(He continues explaining, showing equivalences in capacity using water and writing in the whiteboard).

The interaction with students was limited to the asking of rhetorical questions such as those shown above, to pose some problems from the textbook and to explain them. Students' actions were limited to listening to Gabriel and answering some of his questions, without necessarily reflecting. In this episode the technology was being used as a replacement tool, since Gabriel could have done all of his writing on the blackboard.

Later, Gabriel opened a program (episode 2) called 'Capacity measures' from Enciclomedia project (Fig. 1).

This is an interactive program, which uses the context of a milk factory to invite students to fill containers of different sizes by using smaller containers of different capacities, which are then carried by a truck to the warehouse. With this program it is possible for the students to choose one container from the warehouse and calculate how many of another chosen container would be needed to fill the large

Fig. 1 Capacity Measures

container, prompting a possible class discussion of the comparison of different measurements of capacity. The program includes teaching suggestions and interesting questions to ask while working with the program.

In this episode the teacher shared the communication of mathematical knowledge using the interactive resource, which was a source of mathematical problems. Gabriel chose an activity from the program, asked questions that guided the whole class through the activity, and used the program to verify the results. He remained in control of the computer. His interaction with the students was limited to the posing of questions that students continued to answer, with little evidence of any deeper reflection.

> G: *Which of these containers shall we use (referring to the containers in the truck)?*
> S: *The big one.*
> G: *How many containers are needed to fill up the tank in the warehouse?*
> S: *5 1/2* [Gabriel writes the number in the program and they see that when the all the milk is poured the container is not completely full]
> G: *Let's see if it works… How many more containers are needed?*
> S: *6 ¼, 8 teacher, 11* [Shouting]

Even though some students responded to the questions during the episode, their attitude seemed passive since they were not reflecting on their answers. It can be concluded that in this episode the use of the technology was as an amplifier, since the program showed situations that it would not be possible to illustrate in the classroom. However, the teacher did not exploit fully the possibilities offered by the technology and did not use it as the starting point to further explore the students' understanding.

After working with the program for a while, Gabriel opened the animation 'Metric units of volume' (episode 3), where equivalences and comparisons between different units of volume are illustrated through images of objects such as a swimming pool and a football stadium (Fig. 2).

Students watched the whole animation without interruption. When it finished, Gabriel did not make any comment about it or ask any questions. He left the communication of mathematical knowledge and problems, and the validation of results, to the resource. Students did not have any autonomy to question what was

Fig. 2 Metric units of volume

presented to them and there was no possibility to interact with either the program or the teacher. The technology was used in part as replacement, and in part as amplifier, since the animation illustrates situations, which cannot be observed in the classroom, but at the same time it was left completely in charge of the mathematical explanation.

Finally (episode 4), Gabriel went back to the electronic whiteboard. He referred back to the ideas introduced when he started the lesson, but this time he used some of the examples shown in the animation, and used drawings to repeat some of the animation to ask new questions. Again, he expected immediate responses from the students and he did not invite them to think more deeply about the answers:

> G: *A cubic decimetre, it is equivalent to what? To a litre. One cubic metre, how many litres can it hold?*
> S-G: *A thousand.*
> G: *Imagine that you have a cubic decametre. How many litres can it hold?*
> S: *Ten thousand? Eight thousand?*
> G: *How many cubic metres there are?*
> S: *A hundred?*
> G: *10 by 10 is 100. And 100 by 10 is 1000. Then a cubic decametre is 1000 m^3.*

Later, he showed again the same animation as reinforcement of his explanations, but again the students did not make any comment. He finished the lesson by asking students to work on some problems from the textbook. Later, some students went to the whiteboard to write their answers and neither Gabriel nor the other students asked them about their procedures. Gabriel emphasised the formulae that the students needed to know in order to calculate the volume of different containers, the units of volume and the correct responses for the problems. Finally, as an end to the lesson, Gabriel reviewed the relationships between the units of volume and capacity. He asked some questions but, as students did not give the answer he was expecting, he gave the information that he was asking for:

> G: *How can we calculate the capacity of a container?*
> G: *In order to know how much container can fit, first we calculate its...*
> [No answer]
> G: *Volume*
> G: *Here the volume is*
> S1: *30; S2:50; S3: 60*
> G: *40 right?*

Again he was in control of the communication and validation of the mathematical knowledge, while the source of mathematical problems was the animation and the textbook. There was no real interaction with the students in terms of letting them participate and, although the students had some autonomy as they worked on the textbook problems, they were not able to discuss their ideas. We conclude that, in this final episode, the technology is again used as a replacement since the teacher could have taught the same lesson without the use of the technology, and the animation was not actually discussed.

Table 2 describes the dynamics of Gabriel class. It gives a snapshot of how the different aspects of his role as teacher changed through the lesson and their relation

Table 2 Analysis of Gabriel's aspects of his role as a teacher in relation to the use of technology

Teachers' role / Uses of technology		Replacement	Amplification	Transformation
Communication of mathematics	Students, textbooks, technology and teacher		3	
	Some other elements of communication	2	4	
	Teacher communicates exclusively	1 →		
Interaction with students	Modifies plan according to students' participation			
	Listens to students and answers questions but goes back to plan			
	Little interaction with students, follows predetermined plan	1 → 2	3, 4	
Validation of mathematical knowledge	Multiple sources of validation			
	Teacher as only source of validation	1 → 2	3, 4	
Which is the source of mathematical problems	Other sources of problems including those coming from digital programs and students themselves	1 → 2		
	Problems from teachers and textbook		3	
	Unique source of mathematical problems	← 4		
Actions and autonomy of students	Students have autonomy to decide what to do and how to do it.	1 →		
	Teacher assumes active role, while students mainly listen, copy or answer questions.		2, 3, 4	

to the use of technology. Each arrow represents an episode by showing a movement regarding, on the one hand, the aspects of the role of the teacher (as denoted by upwards or downwards movements) and on the other, the use of the technological resources (represented by movements towards the left or right). The use of consecutive arrows is intended to give information about how the role and the use of technology change during the lesson. The position of the beginning and ending of each arrow are important in determining how close a teacher's actions are to each aspect of the role or to each use of the technology. For example, a teacher can use technology as a replacement, but his actions can reveal that this use can be closer or further away from using it as an amplifier. The arrow corresponding to that episode will be nearer to the border of the table between these two uses. The same is true for the aspects of the role of the teacher; if the arrow point is closer to a border, it means that most of the actions of the teacher can mainly be considered related to that aspect.

In examining the different aspects of Gabriel's role as a teacher in relation to how he used the technology, the patterns of his actions in the classroom made it possible to observe his tendencies to maintain control of the class and to follow his teaching plan closely. The use of the whiteboard and the animation did not contribute any substantial change in his actions. It was only when he introduced a more interactive tool that aspects of his role changed slightly as he allowed the program to be the source of mathematical problems and to determine the correctness of answers. Still, during this episode, Gabriel's behaviour did not allow students to explore their ideas further by following unexpected pathways or incorrect answers. For students, effective actions in Gabriel's classroom included guessing and answering by trial and error. Students' answers led to a limited amount of feedback, which came from the interactive program. Often their responses were not followed up as Gabriel would give the correct answer himself. In this context, effective behaviour is unlikely to promote or be conducive to mathematical learning. It seems that even though Gabriel was interested in using technology in the class, he was not able to use it to stimulate students' reflection on the mathematical content he was teaching.

Juan

For the lesson we selected for analysis, Juan taught a chapter from the textbook on "conversion of fractions into decimal numbers and locating them on the number line" (SEP 2009, p. 47). Juan started the lesson (episode 1) by asking his students to draw 5 apples and 3 children, and to share the apples among the children. He interacted with students by means of questions and comments in response to their answers, but he did not ask the students to explain the reasoning behind their answers:

> J: How many parts of each apple will be given to each child?
> S_1: One whole apple and two thirds.
> J: Let's see. Each one will have one apple. I have shared three of the five apples, but I still have 2 more apples. This one and this other apple [signaling his drawings of the apple], how can I divide them?
> Students: In three parts.
> J: In three. I suppose they have to be equal. [He represents the parts in his drawing of the apples]. Each one will have one part of each apple. So, each child receives, how many?
> Students: One whole and two thirds.

He then said:

> J: We will see an animation with the computer so we can revise what has been learnt and that it will show us some examples about how certain things can be shared.

Before showing the animation, he reminded students of the names of the elements of a fraction (numerator and denominator) and their meaning. Then the animation 'Fractions' (Fig. 3) from the Enciclomedia project was introduced. It shows several situations in which familiar objects are divided into a given number of equal parts so that they can be shared amongst different numbers of clowns.

Fig. 3 Fractions

Once the animation ended, the teacher asked students whether they liked it and he posed another sharing activity. He did not make further comments that might have made students reflect on the animation content in terms of the lesson's objectives. Throughout the whole episode, Juan was in control of the communication of mathematical knowledge, the main source of problems, in control of the validation of knowledge and making the decisions about what the students had to do. Students did not have autonomy and their participation was limited to answering his questions, although it is possible to observe that they did give careful consideration to their answers. The technology was used as a replacement since it seems Juan only used the animation as a way to review some contents and to motivate students, although it may be considered that it was also used as amplifier since it shows concepts in a contextualised and attractive way, which cannot be done by the teacher himself. However, no reflection was initiated by Juan on what the technology had demonstrated.

Juan then used a worksheet that he had prepared using a word processor that included two sets of problems that presented situations concerning the sharing of donuts (episode 2). One set were problems with proper fractions, while the other involved improper fractions. The worksheet was the sole source of the mathematical problems. Students were asked to write the result using decimal numbers. Juan started with the first set and asked:

J: If we want to convert all these fractions to decimal numbers, is it possible?
Students: Yes!
J: How can we do it? Who can tell me how?
S_1: By dividing.
J: What should I divide? [Showing the first exercise]

A similar dialogue was repeated for each of the fractions, but a different student was selected to work on the whiteboard. In response to a problem where 7 donuts were shared by 3 children a student responded 2 wholes and one third where Juan had expected him to respond 7/3. Juan solved the division sum 7 divided by 3 on the board and wrote 2.33 (see Fig. 4) and gave the fact that the number continues indefinitely.

Fig. 4 Juan's solution on the board

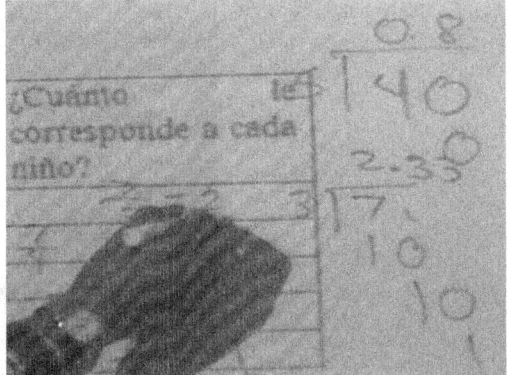

When Juan asked the students for an equivalent decimal expression for 7/3, again they gave an unexpected response:

J: And I can continue like this [referring to the division algorithm]. 7/3 is equivalent to...
Students: Two and three tenths.
J: thirty-three, what?
Students: hundredths.
J: And if a write another three?
Students: Thousandths.
J: Then it would be two and three hundred and thirty three thousandths.

When the students had finished the exercises Juan asked for volunteers to give the answers and he verified the answers with the help of the group. When 7/6 appeared, he asked, *What is this number called as a decimal?* However, none of the students could answer and they did not use repeating decimals in their solution.

In this episode there was interaction with students in the form of questions and answers and the communication and validation of mathematical ideas were shared between the teacher and the students, although the teacher had the last word on the validation of their responses. Students participated actively, but they did not have autonomy. The interactive whiteboard and the word processor was used as replacement since the activity on which the group had worked could have been accomplished in the same way using the blackboard, or paper and pencil.

In episode 3 Juan used the interactive Enciclomedia program 'The Number Line' with the group. In this activity the students are asked to find a number between two given numbers that have been chosen randomly within the context of a game (Fig. 5).

Mixed numbers can appear alongside proper and improper fractions. However, numbers can only be entered in the program as mixed numbers and decimal numbers are only accepted by the program when expressed with two decimal places. At a more advanced level, the students are also asked to approximate the location of the number on the line. Four groups of students play the game and, for each correct answer, they are awarded points. Each team is represented by a different colour token, which moves across a board until one of the tokens crosses the finish line. The program gives automatic feedback to the user by indicating whether the answer

Fig. 5 The Number Line

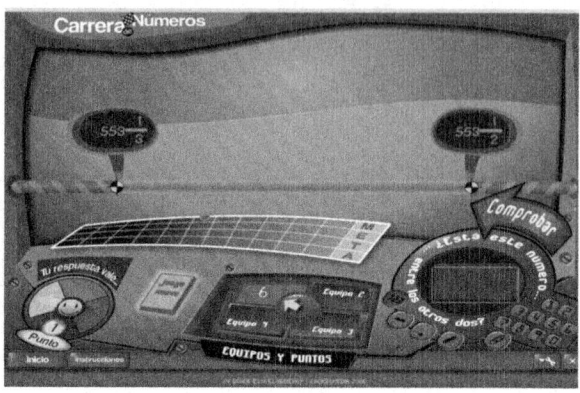

is correct or not. Each time a correct response is given, the program selects the smaller of the two sub-intervals formed on the number line when the number given by the students is introduced. Fractional or decimal numbers soon become a necessary input in order to progress in the game.

Juan explained the goal of the game to the students and chose the intermediate level of the program, which has decimal numbers as the end points of the interval. Hence the technology was the source of the mathematical problems. The teacher selected a student from each team to respond to each activity proposed by the program. There was more interaction between the students in their teams. They participated actively and reflexively in the game. Students compared different numbers and used different operations to make this comparison. Students acted with more autonomy, although the teacher was always in control of the activity as he was the one entering the numbers into the program. However, each time the students chose a wrong number Juan asked the group why it was not correct and helped the students with their explanations. They mainly worked with decimal numbers and hardly used fractions. The validation of knowledge was shared between the interactive program and the teacher, although it was mainly the teacher who justified the feedback given by the program.

At some point they were faced with a situation in which they needed to find a number between 435.36 and 435.37:

J: [After three minutes without a response from the students] Which number have you chosen? What is the whole part?
J: What is the difference between this [.36] and this [.37]?
S_1: One hundredth, isn't it?
J: How do we write it as a fraction?
S_1: 435 and 375 hundredths
S_2: Thousandths

Juan entered a number that lay outside the interval (1/100) so the program marked the answer as incorrect and presented an interval for the next team to play. After all the other teams had participated, the group faced again the interval (435.36, 435.37). This time Juan asked students to use numbers with decimals instead of fractions.

They suggested 435.365 but the program only allowed him to write 435.36. Students commented that the answer was incorrect and Juan tried to use the program to show it was correct but was unable to solve the problem. A student finally found out that the program did not allow the user to write decimal numbers with more than two digits.

J: Here it is again, so your answer is...
S3. Four hundred thirty five point three hundred sixty five.
J: [Enters the number 435.36 into the program] I cannot write such a number.
S3: [After two minutes] If you write that the program will say it is incorrect. If you write point thirty five it will decide it is also incorrect. But there is no more hundredths there [referring to the program]
[Juan continues trying for an other 2 minutes]

Juan finally closed the program. He did not take the opportunity to discuss further any strategies to find a fraction or decimal number between two decimal numbers where the difference between them is one hundredth. Communication of mathematical knowledge was shared between the program, students and the teacher, although in the last part it was the teacher who communicated this knowledge exclusively.

Throughout this episode technology was used as a replacement and as an amplifier. There was a change in the class dynamics as the interactive program was introduced, and students participated more actively and with more autonomy discussing in their teams which could be the answer and the way to present it since the program gives more points if students use fractions than if they use decimals. However, Juan did not use the resource to challenge students' mathematical knowledge, to help them think on new strategies to find fractions and decimal numbers on the number line, or to guide them in their solution of situations, which could not be solved with the program.

Juan ended the lesson (episode 4) by asking students to work on a textbook activity that asked the students to find fractions and decimal numbers on given number lines. He worked with the whole group by reading each problem and asking questions such as *How do you know? How would you turn 4/5 into a decimal number?* He worked on this last problem on the whiteboard, but wrote an incorrect answer (0.4). Students did not notice the mistake and the lesson ended before they finished the activity. In this episode the teacher was again the source of mathematical problems and was in charge of validation. The interaction with students was in terms of questions and answers and students did not have autonomy. Technology was used as replacement. Table 3 shows the analysis tool in this case.

The different aspects of Juan's role as a teacher in relation with his use of technology show that even though Juan's patterns of actions were guided by his teaching plan, he tried to interact with his students by asking questions and listening to their answers. The use of the animation, his prepared worksheet and the whiteboard did not contribute much to change his actions. However, the use of the program, and particularly the need to divide students in four groups, contributed to a change in his actions. He used the program as a source of mathematical problems and to validate the answers; he was open to giving students more time for discussion before making

Table 3 Analysis of Juan's aspects of his role as a teacher in relation to the use of technology

Teachers' role / Uses of technology		Replacement	Amplification	Transformation
Communication of mathematics	Students, textbooks, technology and teacher		3	
	Some other elements of communication	2, 1	4	
	Teacher communicates exclusively			
Interaction with students	Modifies plan according to students' participation		3	
	Listens to students and answers questions but goes back to plan	2		
	Little interaction with students, follows predetermined plan	1	4	
Validation of mathematical knowledge	Multiple sources of validation	2	3	4
	Teacher as only source of validation	1		
Which is the source of mathematical problems	Other sources of problems including those coming from digital programs and students themselves	2	3	
	Problems from teachers and textbook	1	4	
	Unique source of mathematical problems			
Actions and autonomy of students	Students have autonomy to decide what to do and how to do it.		3	
	Teacher assumes active role, while students mainly listen, copy or answer questions.	1, 2	4	

their responses and he made them reflect on the reasons behind their answers. For the students, the effective actions in the classroom included the answering of the teacher's questions and, in episode 3, their reflections on how to respond in order to gain more points in the game, with reasons for their answers. Juan is a young teacher eager to use technology, but he did not use it in an effective way to promote students' learning.

Susana

Susana was particularly interested in the teaching of fractions. She developed a teaching sequence involving two interactive programs that involved the use of fractions in different ways. The first program she used is called 'The Balance' (Fig. 6), which shows a problem situation where scales need to be balanced by using

Fig. 6 The Balance

fractions. The program provides the users with automatic feedback that helps them in identifying which parts of the mobile toy are balanced and which are not.

The lesson took place in the computing room, where every pair of students had a computer with access to the program 'The Balance'. In the first part of the session (episode 1), Susana asked the students to use 'The Balance' to compare different pairs of numbers in order to decide which is greater. She started by using whole numbers and then she introduced simple fractions like 1/2 and 1/3 and decimal numbers like 0.5 and 0.05. Students used the program to compare the numbers thus becoming familiar with how the program functioned. Later, Susana taught the group how to build a balance with the program using two different levels of scales and asked them to work with this kind of scales using fractions. She posed different examples for students to solve. There was a group discussion regarding the meaning of equivalent fractions.

In this case, we consider that Susana, 'The Balance' and the students all participated in the communication of mathematics. The program, by imposing a certain problem, was one source of mathematical problems, while Susana acted as another source by asking students to use specific numbers. The interactive program validated the answers by visually showing whether the scales were balanced or not. Students had little autonomy in this part of the lesson, since they would just enter the numbers the teacher suggested in order to compare them, although they did have discussions in which they talked about why they thought one number was smaller than another one. 'The Balance' was mainly used as an amplifier, as it carried out the calculations automatically and showed if the scales were balanced or not.

In a second episode, Susana asked students to build their own mobile toys using different levels and numbers (episode 2). She emphasised to the students that they should make sure that all of the different levels of the mobiles were balanced. In order to do this, students had to add and subtract fractions, and figure out how to divide a fraction into two different equal parts (see Fig. 7).

Fig. 7 Problem with the balance

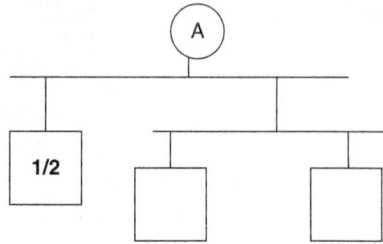

There was a long group discussion in which students talked about how to divide a fraction such as ½ in two equal parts. They used different representations, including drawings of pizzas on the blackboard. They worked with several examples, as after dividing 1/2 by 2 Susana asked about other fractions such as 1/3, 1/4 and later 3/4, 5/8 and so on. In this episode, the students in Susana's class became stronger agents in the communication of mathematics. They even became the source for mathematical problems, since they designed their own mobile toys. The program itself remained the main source for validation, showing every time whether the mobile was balanced or not. Susana adapted the group discussions to the kinds of problems posed by the students and to their explanations. We therefore consider that in the interaction with students she modified her plans according to their participation:

I was not planning on discussing how to divide fractions using graphic representations. I thought maybe they would try different options, like trial and error, or adding two numbers, things like that, because I thought they would work with numbers like 2/4, but I was not expecting them to want to come up with an explanation like "the pizzas". When I saw they wanted to do it that way, I was like, okay, let's go ahead with this! (Susana, 01-07-2011)

The use of technology in this episode became transformational, because, in motivating students to create their own mobile toys and therefore inventing their own mathematical programs, Susana's plans and role were modified:

I did not think that students would create such big mobile toys one their own that quickly! It had never happened before when I worked on that particular chapter from the textbook. They were really engaged when using the program (Susana, 01-07-2011)

We consider that the interactive program alongside the students' ideas and the discussions in the classroom, allowed mathematical learning to happen. In order to act effectively, that is, to balance different kinds of mobile toys that could be constructed using the program, students came up with new explanations and procedures. They had not encountered the partitioning of fractions before, and they developed a procedure that allowed them to solve the problem that was posed by the program as well as the other problems posed by the teacher. During the last episode (episode 3) of the first session, students worked with 'The Balance' in order to solve the problems posed in their textbook (see Fig. 8).

The problems became more difficult. The students finished all of the exercises within the textbook and later they continued to devise their own even larger mobile toys that they had to balance. The discussion on the division of fractions continued.

Fig. 8 Problem with the balance

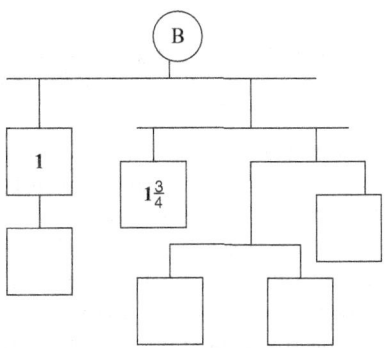

They concluded that, in the case of a division of a fraction by 2, if the numerator was even, it was easy to divide, while if it was odd, more difficult procedures had to be used. They came up with some 'rules' for even and odd numerators, which they wrote on the whiteboard. Later they extended the problem into partitioning fractions in 3, 4 and 5 equal parts and they spent quite a long time discussing different procedures until they came up with a general one (finding equivalent fractions in which the numerator could be divided exactly by the whole number).

In this last episode, the role of the teacher was similar to that of episode 2. The sources of the mathematical problems were the textbook, the program and the students and again Susana allowed the students to formulate their own explanations and procedures, especially in the case in which they had to divide fractions. The use of the technology remained transformational, as students deepened their explanations and justifications.

We consider that, without the visual feedback from the program, students may not have had to conceptualise and develop procedures for the partitioning of fractions. Previously, we had observed that within that particular chapter in the textbook, students would balance mobile toys inadequately, for example by balancing a weight of 1 ½ kg with one of 1 kg and one of ½ kg placed horizontally. This would balance one level of the mobile but not the second level. 'The Balance' motivated students to find ways of dividing a given fraction into equal parts as it was not effective behaviour to have sections of the mobile toy which were not balanced. Susana had asked the students to balance all sections of the mobile. In the end they were exploring the division of fractions by whole numbers in a general manner.

On the one hand, it is interesting to note here that, in this case, the program became a strong influence on the students' effective behaviours. A different type of program might not have had the same effect. On the other hand, it is important to observe that the way in which Susana used this tool also modified the students' behaviour. The restrictions that she posed and the questions that she asked were important influences on the students' learning. The importance of the way in which the teacher uses the digital tools was confirmed by our observation of a second session in which Susana used 'The Number Line' (see Juan above) in order to deepen students' understanding of fractions. During this session, Susana asked questions prior to the interactive program providing feedback. Frequently used questions were: *How did you find the number? How do you know that number is between those two?* After the feedback had been given by the program, she asked additional questions such as: *Why is the answer correct/incorrect? How would you find another number between those two?* Students were invited to come to the board in order to illustrate their explanations and justifications. They were allowed to make drawings and to use concrete materials.

During these discussions the use of the program was also transformational. Susana's practice was modified because, as she mentioned, the interactive programs related to fractions available in Enciclomedia invited her to create a teaching sequence related to fractions that she had not previously considered:

> *When I saw that there were several programs that I could use for the teaching of fractions, I decided to create a longer teaching sequence. The textbook chapters which are related to these topics are not sequential, but I decided it would be interesting to explore fractions from different views in consecutive sessions. (Susana, 01-07-2011)*

In this case, the context that the program created in the classroom, together with the teacher and the students, again promoted the students' mathematical learning. Effective behaviour included using known ideas and procedures with fractions to solve mathematics problems that initially, they did not know how to solve. Table 4 shows the analysis in Susana's case. The thicker arrows represent the work with the Number Line.

Discussion

A first comparison of the Tables 2, 3 and 4 shows that there is a close interaction between the different aspects of the role of the teacher and the use of technological resources.

Regarding *communication* it can be observed that technology often becomes an additional source of mathematical information in the classroom. Interactive programs, animations and worksheets expose students to mathematical concepts and ideas. However, this mathematical information can promote very different kinds of effective behaviours in the classroom. For example, we observed that the use of interactive resources is more likely to promote the emergence of shared mathematical ideas, compared to the use of animations or activity worksheets based on the textbook where the interactive whiteboard is used to replace traditional paper and pencil activities. When using an interactive program such as the 'Number Line' or 'The Balance', it is possible that mathematical information is communicated also by students as they engage with the problems and share their ideas. Of course, not all uses of interactive programs lead to this form of communication, as we saw in the case of Juan's use of 'The Number Line'. The teacher's behaviour can impose restrictions on what it is possible to do with the program.

Technology also plays a role in promoting *interaction* in the classroom. However, it can be deduced from the data and the information we have about the teachers'

Table 4 Analysis of Susana's aspects of her role as a teacher in relation with the use of technology

Teachers' role / Uses of technology		Replacement	Amplification	Transformation
Communication of mathematics	Students, textbooks, technology and teacher		1 2	3
	Some other elements of communication		2	3
	Teacher communicates exclusively	1		
Interaction with students	Modifies plan according to students' participation		2	3
	Listens to students and answers questions but goes back to plan		1	3
	Little interaction with students, follows predetermined plan	1	2	
Validation of mathematical knowledge	Multiple sources of validation		1 2	3
	Teacher as only source of validation		1 2	3
Which is the source of mathematical problems	Other sources of problems including those coming from digital programs and students themselves		2	3
	Problems from teachers and textbook	1	2	
	Unique source of mathematical problems	1		
Actions and autonomy of students	Students have autonomy to decide what to do and how to do it.		2	3
	Teacher assumes active role, while students mainly listen, copy or answer questions.	1 1	2	3

backgrounds, that teacher training, didactical ability and disposition to share the control of the lesson with students and to be flexible regarding the original lesson plan are dominant factors in this aspect. This is exemplified by the way Susana modified her original plan, and managed the use of the resources and the interaction with and between students. Effective behaviours during her lessons included the discussion of mathematical ideas and problems both in small groups and in whole group discussions. Her role as a teacher provided a space in which students could follow their ideas even when they led towards unexpected (to her) places. Gabriel, in contrast, stayed close to his original lesson plan and limited the interaction with students to the use of closed and rhetorical questions which promote effective behaviours such as guessing and trying to please the teacher and do not favour students' mathematical learning.

Validation of mathematical knowledge when using technology appears to be closely related to the design and use of the digital resource, although the teacher can make use of the resource whereby the validation of knowledge is shared with students. Again, a good example of this possibility to share validation among students, technology and the teacher is illustrated by Susana's use of both 'The Number Line' and 'The Balance'. In this case, effective behaviours include giving reasons for why a specific answer is entered into the program and finding procedures that might lead to correct answers before trying them out in the program.

Regarding the role of the teacher in terms of students' *autonomy*, we observed that certain types of resources are more inviting to explore different mathematical ideas and procedures by students themselves than others. Problems like the ones posed by 'The Balance' or 'The Number Line' can motivate students and promote effective behaviours such as working on their own or with friends to find answers and explanations. Other programs such as animations can promote a more passive attitude from students. This again, is strongly influenced by the way in which the teacher uses the digital resource.

In terms of the uses of technology the tables show that the introduction of digital resources alone is not enough to transform the activity in the classroom. During the same lesson different uses of technology can be observed. The creation of an environment that promotes effective actions, which are related to the learning of mathematics depends on the different aspects of the role of the teacher. It is teachers who make decisions that can change both the use of technology and the dynamics in the classroom so that transformation is possible. A combination of the teacher and the use of the technology can create learning contexts where actions are limited or where they are more conducive to mathematics learning.

The analysis of the use of technology by teachers in this study shows a strong tendency for most of the teachers to limit the use of technology as a replacement and amplifier. Most of the teachers have not received training regarding the didactical use of Enciclomedia's mathematical resources and have not experienced, even as spectators, how the programs can be integrated in lessons which provide a context that promotes learning. This is clearly evidenced by Gabriel who is willing to introduce technology in his classes, but does not exploit the potential of the resources to create an environment where mathematical activity is effective and by Juan who does not use it to challenge the mathematical knowledge of his students. Susana, on the other hand, illustrates how teachers who have been trained in the didactical use of technology can benefit from the different possibilities afforded by resources, can combine them and create classroom contexts where students discuss and reflect on their actions and can learn mathematics. In this classroom culture, the teacher herself can learn.

Mathematical knowledge of teachers can also limit the different aspects of their role in the classroom. Observations show how they are not able to use opportunities presented by the technology's potential or limitations in order to discuss students' strategies. For example, when Juan is faced with the need to explain how to find a number between two given numbers when their difference is one hundredth, and there are students who apparently already know the answer, he did not ask these students how they found the number. When both Gabriel and Juan used animations in their classroom, they did not ask any question related to the content. The students acted as spectators and in this environment their learning can be inhibited.

Traditional ways of teaching, in which the teacher intends to transfer mathematical knowledge, can become an obstacle to a use of technology that has transformational potential. When teachers stick to their lesson plans and when they keep the control of what mathematical knowledge is communicated, which problems are worked and the validation of students' answers, technology is naturally used as replacement or amplifier. Again, the class environment is such that mathematical learning can be limited.

From the technological resources used by these teachers it can already be observed that there is a variety among Enciclomedia's resources. Although resources can be classified as animations or interactive resources, some interactive programs demand more interactivity than others. In our examples we can observe that the 'Capacity Measures' program used by Gabriel asks for calculations and approximations from the teacher and students, but as we described before, students can effectively guess or use a trial and error strategy in order to select an answer from the options in the program. The responsibility for questioning and inviting students to modify their actions is on the teacher's side. Other resources, such as 'The Number Line's ask for more involvement from the beginning in terms of the mathematics, since trial and error practices do not help in playing the game and are therefore not effective behaviours in this context. Also, the need to locate the selected number on the line requires reflection on the part of the players. The teacher can take this opportunity to discuss procedures and strategies used by students in terms of their being correct or not, or in terms or their being efficient. This discussion can promote effective behaviours such as mathematical reasoning activities and reflection where learning is possible. In our previous description we could observe that Juan does not profit from this opportunity while Susana does. Finally, 'The Balance' exemplifies a very open resource that can trigger both teachers' and students' creativity in order to design a variety of ways of using it for the learning of concepts related to whole numbers, numbers with decimals and fractions. In the selected episode Susana allows the students to design challenging situations that result in interesting opportunities to discuss different properties and operations with fractions.

The differences in the design characteristics of the technological resources can impact differently in the ways teachers use them. Although it is the teacher who can guide activity with resources in the class, some of them can somehow induce changes in some aspects of the teacher's role.

The results that have been analyzed show that the technological resources used by teachers influenced the learning of their students, as they framed the effective actions of both teacher and students. Although there are differences between the technological resources used, as we previously discussed, each of these create spaces for action and at the same time they impose certain restrictions. It is important for the teacher to be aware of the possibilities for action and of the restrictions so that the kinds of actions that promote the learning of mathematics can be fostered.

Conclusions

The data analysis has enabled us to conclude that the characteristics of digital resources used in the classroom have an important influence on the role the teachers play during their lessons. Some technological resources such as 'The Balance' and 'The Number Line' can, by themselves, help the teacher to create a context where the students have greater possibilities for autonomy and where interaction and discussion of mathematical ideas foster effective behaviour, creating an environment for mathematics learning. However this is strongly dependent on the teachers' effective actions, which are influenced by their mathematical knowledge, the experience they have regarding the use of the technological tools and their didactical strategies and practices. It is important to notice that, even when digital resources can promote active participation in the classroom by opening spaces for mathematical exploration, the actions carried out by the teacher, that is, the different aspects of her or his role can inhibit or enhance such effectiveness.

Enciclomedia's resources were conceived as a tool for the classroom. The didactical idea behind their conception was that teachers would develop didactic strategies where the resources could play an important role in helping to provide a dynamical and participative class environment, which can promote students' learning (SEP 2004). As can be concluded from this study, this goal is possible to achieve when teachers have reviewed and studied the programs, have made an open planning for their lessons and, most importantly, have been trained on how to integrate technological resources to their teaching in an effective way. This is the case with regard to Susana. The dynamics of the different aspects of her role in the two sessions provide a good example of how this integration can promote students' learning. For her, technological resources are instruments she can use to replace, amplify or transform situations involved in teaching specific topics and which help create a classroom context where she and her students can propose and discuss mathematical ideas as well as share and develop them.

However, the data shows that developing this expertise is difficult and needs formal training. In our study there was only one more teacher who worked similarly to Susana. Other teachers, who have only received training on the general use of the software, without a hint of how to introduce them into specific lessons, often develop teaching strategies where technology is used as replacement or amplification. In this case there is no real change in the environment they create in their classroom.

Regarding the programs themselves, it would be important to classify Enciclomedia's programs or any kind of digital resources in terms of their possibility to act as tools that can help teachers to: transform the classroom dynamics; work on making the necessary changes so that they include possibilities for exploration; and enable the students to choose scenarios or discuss situations, with and without the teacher. In this kind of environment, effective actions include the reflection on mathematical ideas and procedures that can be conducive to mathematical learning.

While other kinds of resources are not necessarily open, they can be useful as instruments to validate mathematical knowledge. In this case, the different aspects of the teacher's role such as interaction with students, being the source of mathematical problems, and enabling the shared communication of mathematics and autonomy of the students are fundamental to the creation a productive environment for learning.

Finally, the methodological tool we developed was useful in describing teachers' actions when using technology as it highlights aspects and details of the internal working of classrooms that may otherwise have remained hidden. We believe this instrument might be useful for other researchers and also for teachers who are interested in reflection about their own practice. The matrix also served to highlight the relevant aspects of the role of the teacher with his or her possibility to move towards a transformation use of technology. This does not mean that other uses of technology are not important; each of them can play a part in different moments of the lesson and may complement each other. It is the balance between these uses together with the dynamics of the aspects of the role of the teacher that creates the conditions for effective actions that can be described as mathematical learning.

Acknowledgements This project was funded by Conacyt's Grant No. 145735. We would also want to thank Asociación Mexicana de Cultura A.C. and ITAM for their support.

References

Davis, B., Sumara, D., & Luce-Kapler, R. (2000). *Engaging minds: Learning and teaching in a complex world*. London: Lawrence Erlbaum.

Díaz de, C. R., Guevara, N., Latapí, S., Ramón, B., & Ramón, C. (2006). Enciclomedia en la práctica. Observaciones en veinte aulas 2005–2006. *Centro de investigación educativa y actualización de profesores*. A.C. México.

Drijvers, P., Doorman, M., Boon, P., Gisbergen, S., & Reed, H. (2009). Teachers using technology: Orchestrations and profiles. *Proceedings of the 33rd conference of the international group for the Psychology of Mathematics Education* (Vol. 1, pp. 481–488). Thessaloniki/Grecia.

Holland, I., Honan, J., Garduño, E., & Flores, M. (2006). Informe de evaluación de Enciclomedia en: F. Reimers (Coord.) *Aprender más y mejor. Políticas, programas y oportunidades de aprendizaje en educación básica en México*. México: FCE/ILCE/SEP/Harvard Graduate School of Education.

Hughes, J. (2005). The role of teacher knowledge and learning experiences in forming technology-integrated pedagogy. *Journal of Technology and Teacher Education, 13*(2), 277–302.

Loredo, J., García, B., & Alvarado, F. (2010). Identificación de necesidades de formación docente en el uso pedagógico de Enciclomedia. *Revista Electrónica Sinéctica, 34*, 1–16.

Maturana, H., & Varela, F. (1992). *The tree of knowledge: The biological roots of human understanding*, Revised Edition, Boston: Shambhala.

Rabardel, P. (1999). Eléments pour une approche instrumentale en didactique des mathématiques. *Actes de l'école d'été de didactique des mathématiques*, Houlgate 18–21 Août, IUFM de Caen, pp. 203–213.

Rabardel, P. (2011). *Los hombres y las tecnologías. Visión cognitiva de los instrumentos contemporáneos* (trans: Acosta, M.). Colombia: Ediciones Universidad Industrial de Santander.

Ruthven, K. (2007). Teachers, technologies and the structures of schooling. *Proceedings of the fifth Congress of the European Society for Research in Mathematics Education* [CERME 5] (pp. 52–67).

Sagástegui, D. (2007). Usos y apropiaciones del programa Enciclomedia en las escuelas primarias de Jalisco. *Memorias del Congreso del Consejo Mexicano de Investigación Educativa.*
SEP. (2004). *Programa Encilomedia: Documento Base.* Subsecretaría de Educación Básica y Normal. México: Secretaría de Educación Pública.
SEP. (2009). *Plan de Estudios 2009. Educación básica. Primaria* (2nd ed.). México: Secretaría de Educación Pública.
Trigueros, M., & Lozano, M. D. (2012). Teachers teaching mathematics with Enciclomedia: A study of documentational génesis. In G. Gueudet, B. Pepin, & L. Trouche (Eds.), *From Text to "Lived" Resources*, Chapter 13. (pp. 247–264). New York: Springer.
Trigueros, M., Lozano, M. D., & Lage, A. (2007). Development and use of a computer based interactive resource for teaching and learning probability, in practice of primary classrooms. *International Journal for Technology in Mathematics Education, 4*(13), 205–211.

Technology Integration in Secondary School Mathematics: The Development of Teachers' Professional Identities

Merrilyn Goos

Abstract This chapter reports on research into the impact of digital technologies on Australian mathematics teachers' classroom practice. The aim of the study was to identify and analyse individual and contextual factors influencing secondary mathematics teachers' use of technology, and compare ways in which these factors come together to shape teachers' pedagogical identities. The first section of the chapter examines the teacher's role in terms of their pedagogical identities as users of technology, and introduces two theoretical frameworks for investigating trajectories of identity development. One framework classifies ways in which technology can change teaching and learning roles and mathematical practices. The other is concerned with teacher learning and development, and explains why teachers might embrace or resist technology-related change. The sections that follow provide case studies of two beginning teachers of secondary school mathematics who were integrating digital technologies into their classroom practice. Analysis of these case studies highlights issues related to identity development and demonstrates that identity trajectories are neither random nor fully determined, but instead are constrained by person-environment relationships.

Keywords Identity • Sociocultural perspectives • Technology metaphors • Valsiner • Zone theory • Zone of proximal development • Zone of free movement • Zone of promoted action

---

M. Goos (✉)
School of Education, The University of Queensland, Brisbane, Australia
e-mail: m.goos@uq.edu.au

Introduction

In the twenty-first century, young people live in a world where digital technologies have become essential for managing work and leisure activities. Communication, entertainment, manufacturing, transport, finance, medicine, weather forecasting and many other aspects of life now depend on sophisticated technological systems, much of which is invisible to the user (Confrey et al. 2010). Digital technologies are often personalised and seamlessly integrated into young people's daily lives through use of handheld devices such as mobile phones or tablets that offer a wide range of Web-enabled applications. Mathematics underpins many of these modern-day applications of technology, and yet, despite its ubiquitous presence in the world outside school, technology still plays only a marginal role in many mathematics classrooms (Artigue 2010).

A significant body of research has examined the effects of computer and calculator use on students' mathematical achievement and attitudes and their understanding of mathematical concepts (e.g., see Ellington 2003; Penglase and Arnold 1996 for reviews on calculator use). More recently, research has begun to examine the potential for learning mathematics within digital game environments such as Nintendo and Pokemon (Jorgensen and Lowrie 2011; Lowrie 2005). These studies suggest that engagement with the game, and especially in the repetition of moving back and forth through its different sites, may help players develop complex visualisation and problem solving skills. A second strand of inquiry is focusing on other kinds of technology-immersive environments created via digital mathematical performances. Gadanidis and Borba (2008) introduced this notion to highlight the social and multimodal affordances of new digital media. They noted that the Web offers a medium for sharing mathematical performances using texts, pictures and videos, and suggested that as a result the 'performers' – whether students or teachers – develop new mathematical understandings and new aesthetic appreciation of the power and beauty of mathematics.

In contrast to the longstanding research focus on how students learn mathematics with technology, less attention has been given to teachers' technology-mediated classroom practices and the role of the teacher in technology integration. Internationally there is research evidence that simply improving teachers' access to technology has not, in general, led to increased use of or movement towards more learner-centred teaching practices (Burrill et al. 2003; Cuban et al. 2001; Wallace 2004). Windschitl and Sahl (2002) identified two factors that appear to be crucial to the ways in which teachers adopt (or resist) digital technologies. First, teachers' use of technology is influenced by their beliefs about learners, about what counts as good teaching in their institutional culture, and about the role of technology in learning. Secondly, school structures, especially those related to the organisation of time and resources, often make it difficult for teachers to take up technology-related innovations. These are some of the issues that are considered in this chapter, which draws on the findings from a 3-year research study that sought to identify and analyse individual and contextual factors influencing secondary mathematics teachers' use of technology, and to compare the ways in which these factors come together to shape teachers' pedagogical identities.

The next section of the chapter theorises the teachers' changing roles in technology-enriched learning environments in terms of their pedagogical identities as users of technology, and introduces some theoretical frameworks for investigating the trajectories of their identity development. The sections that follow detail the case studies of two beginning teachers of secondary school mathematics who were integrating digital technologies into their classroom practice. The analysis of these case studies points to the factors that influence the development of their pedagogical identities.

Theorising About Technology-Enriched Mathematics Teaching

Two types of theoretical framework are needed to study implications for teachers of the impact of digital technologies on mathematics education. One type of framework represents ways in which technology can change classroom roles and mathematical practices. The other framework is concerned with teacher learning and identity development, and helps explain why teachers might embrace or resist technology-related change. The research study informing this chapter used both types of framework to investigate implications for technology-enriched mathematics teaching. This research drew on sociocultural theories of learning involving teachers and students in secondary school mathematics classrooms (see Goos 2009a, b). Sociocultural theories view learning as the product of interactions with other people and with material and representational tools offered by the learning environment. Because it acknowledges the complex, dynamic and contextualised nature of learning in social situations, a sociocultural perspective can offer rich insights into conditions affecting innovative use of technology in school mathematics.

Teaching and Learning Roles

In technology-enhanced learning environments, students experience mathematics in new ways that may challenge the traditional role of the teacher as transmitter of knowledge. Technology is not merely an add-on or supplement for pencil and paper work in such environments; instead, it becomes a "conceptual construction kit" that provides access to "new understandings of relations, processes, and purposes" (Olive et al. 2010, p. 138). If digital technologies have the potential to change mathematical knowledge and practices in the classroom, the role of the teacher also changes. The first framework for theorising technology-enhanced mathematics teaching, developed by Goos et al. (2000), classifies ways in which technology can change the teacher's role.

Goos et al. (2000) analysed the effects of digital technologies as cultural tools that both amplify and re-organise mathematical thinking. Mathematics learning is amplified when technology is used to speed up tedious calculations or to verify results obtained first by hand. However, a more profound cognitive re-organisation occurs when students' mathematical thinking is qualitatively transformed through interaction with technology as a new system for meaning-making. Goos et al. developed four metaphors to describe how digital technologies can act as tools that transform teaching and learning roles. Technology can be a *master* if students' and teachers' knowledge and competence are limited to a narrow range of operations. Students may become dependent on the technology if they are unable to evaluate the accuracy of the output it generates. Technology is a *servant* if used by students or teachers only as a fast, reliable replacement for pencil and paper calculations without changing the nature of classroom activities. Technology is a *partner* when it provides access to new kinds of tasks or new ways of approaching existing tasks to develop understanding, explore different perspectives, or mediate mathematical discussion. Technology becomes an *extension of self* when powerful and creative uses are seamlessly integrated into the teacher's mathematical and pedagogical repertoire to support and enhance a teaching program. Although this framework classifies more and less sophisticated uses of technology, it does not imply that only one type will be observed in a lesson or series of lessons (see Goos et al. 2000, for an example of a lesson in which all four technology roles were evident). However, the framework does provide a way of tracing changes in teachers' classroom roles as they appropriate digital technologies into their practice.

Teacher Learning and Development

The second theoretical framework is based on an adaptation of Valsiner's (1997) zone theory of child development to study interactions between teachers, students, technology and the teaching-learning environment. The zone framework extends Vygotsky's concept of the zone of proximal development (ZPD) to incorporate the social setting and the goals and actions of participants. Valsiner describes two additional zones: the zone of free movement (ZFM) and zone of promoted action (ZPA). The ZFM represents constraints that structure the ways in which an individual accesses and interacts with elements of the environment. The ZPA comprises activities, objects, or areas in the environment in respect of which the individual's actions are promoted. The ZFM and ZPA are dynamic and inter-related, forming a ZFM/ZPA complex that is constantly being re-organised by adults in interactions with children. However, children remain active participants in their own development because they can change the environment to achieve their emerging goals. Thus the ZFM/ZPA complex does not fully determine development; instead, development is 'canalised' along a set of possible pathways jointly negotiated by the child in interaction with the environment and other people.

Valsiner (1997) noted that the ZFM/ZPA complex is also observable in educational contexts, and he provided examples of how teachers can set up broad or narrow ZFM/ZPA systems that allow students different choices in completing tasks.

He additionally argued that zone theory is applicable to any human developmental phenomena where the environment is structurally organised, and so it seems reasonable to extend his theory to the study of teacher learning and development in structured educational environments. When considering teachers' professional learning involving technology, the ZPD represents a set of possibilities for developing new knowledge, beliefs, goals and practices. The ZFM is an inhibitory mechanism that structures the teacher's environment, and so could include perceptions of students (their behaviour, social backgrounds, motivation, perceived abilities), access to resources and teaching materials, curriculum and assessment requirements, and organisational structures and cultures. Whereas the ZFM might suggest which teaching actions are *permitted*, the ZPA represents activities, objects, or areas of the environment in respect of which certain teaching approaches are *promoted*. The ZPA could include pre-service teacher education programme formal professional development, and informal interaction with colleagues at school.

Previous research on technology use by mathematics teachers has identified a range of factors influencing uptake and implementation. These include: skill and previous experience in using technology; time and opportunities to learn; access to hardware and software; availability of appropriate teaching materials; technical support; organisational culture; knowledge of how to integrate technology into mathematics teaching; and beliefs about mathematics and how it is learned (Forgasz 2006; Simonsen and Dick 1997; Tharp et al. 1997; Thomas 2006). In terms of the zone framework outlined above, these different types of knowledge and experience represent elements of a teacher's ZPD, ZFM and ZPA, as shown in Table 1. However, in simply listing these factors, previous research has not necessarily considered possible relationships between the teacher's setting, actions and beliefs, and how these might influence the extent to which teachers adopt innovative practices involving technology. In the research discussed in this chapter, zone theory provides a framework for analysing these dynamic relationships.

Taken together, the two theoretical frameworks provide a way of investigating the development of teachers' pedagogical identities as users of digital technologies. From a sociocultural perspective, teachers' learning is understood as changing participation in practices that develop their identities as teachers (Lerman 2001). Wenger (1998) described identity as "a way of talking about how learning changes who we are" (p. 5). He argued that identity has a temporal dimension: because we continually re-negotiate our identities they form trajectories incorporating past, present and future. It is this sense of "learning as becoming" (Wenger 1998, p. 5) that the following analysis attempts to capture.

Research Design and Methods

Participants in the research study were four Australian secondary school mathematics teachers acknowledged by their peers as effective and innovative users of technology. The teachers were selected to represent contrasting combinations of factors known to influence technology integration (see Table 1). They included two

Table 1 Factors affecting teachers' use of technology

Valsiner's zones	Factors influencing teachers' use of digital technologies
Zone of proximal development (Possibilities for developing new teacher knowledge, beliefs, goals, practices)	Mathematical knowledge
	Pedagogical content knowledge
	Skill/experience in working with technology
	General pedagogical beliefs
Zone of free movement (Environmental constraints that limit freedom of action and thought)	Perceptions of students
	Access to hardware, software, teaching materials
	Technical support
	Curriculum and assessment requirements
	Organisational structures and cultures
Zone of promoted action (Activities, objects, or areas of the environment in respect of which teaching actions are promoted)	Pre-service teacher education
	Professional development
	Informal interaction with teaching colleagues

beginning teachers who experienced a technology-rich pre-service programme (described in Goos 2011) and two experienced teachers who developed their technology-related expertise solely through professional development experiences or self-directed learning. This chapter focuses on the professional formation of the two beginning teachers.

The aim of the research was to carry out highly contextualised investigations of how and under what conditions the participating teachers integrated digital technologies into their practice. There were four main sources of data. First, a semi-structured scoping interview invited the teachers to talk about their knowledge and beliefs (which influence their ZPDs), professional contexts (elements of their ZFMs), and professional learning experiences (providing ZPAs) in relation to technology. Thus the structure of the interview was based on the relationship of each zone to factors known to influence technology integration, as outlined in Table 1. For example, teachers were asked about their reasons for using technology in mathematics lessons, their views on how technology influenced student learning and attitudes towards mathematics, their perceptions of any constraints or opportunities in their schools that might affect their use of technology, and their formal and informal experiences in learning to teach mathematics with technology. Interviews were transcribed and teachers' responses were used to 'fill in' the abstract zones of proximal development, free movement, and promoted action with details that were relevant to their own professional histories and contexts.

A second source of data provided additional information about the teachers' general pedagogical beliefs via a Mathematical Beliefs Questionnaire (described in more detail in Goos and Bennison 2002). The questionnaire consisted of 40 statements to which teachers responded using a Likert-type scale based on scores from 1 (Strongly Disagree) to 5 (Strongly Agree).

Third, a snowballing methodology (described by Cobb et al. 2003), involving two rounds of audio-recorded interviews, was used to further probe sources of influence on teaching mathematics. The first round asked participating teachers to identify people who significantly influenced how they taught mathematics, and second round interviews were subsequently conducted with people identified via this process to determine how they attempted to influence how mathematics was taught.

The fourth source of data was lesson cycles comprising observation and video recording of at least three consecutive lessons in which digital technologies were used to teach specific subject matter, together with teacher interviews at the beginning, middle, and end of each cycle. A single video camera was placed on a tripod towards the rear of the classroom and focused on the teacher, the whiteboard, or the data projector screen on which the teacher's computer or calculator output was displayed. Interviews sought information about teachers' plans and rationales for the lessons and their reflections on the factors that influenced their teaching goals and methods.

The next section draws on the data outlined above to present case studies of two beginning teachers, Geoff and Susie (pseudonyms) in order to develop a picture of each teacher's pedagogical identity with respect to technology integration.

Teacher Case Studies

Introducing Geoff

Since graduating from his university pre-service programme in 2000, Geoff had been teaching at an independent girls' school with an enrolment of over 1,000 students in Grades 8–12. This is an academically-oriented school that charges expensive tuition fees, with students who come mainly from upper middle class professional families. Although he was qualified to teach English as well as mathematics, Geoff was assigned only to mathematics classes.

When interviewed, Geoff said that his teaching philosophy was influenced by his love of mathematics, instilled in him as a secondary school student by a mathematics teacher he admired for his 'command of the subject'. Geoff acknowledged that this teacher had been a conservative and a traditionalist at heart, but his 'quirky sense of humour' conveyed a sense of eccentricity that made learning mathematics exciting.

Geoff's passion for mathematics was reflected in responses to the Mathematical Beliefs Questionnaire, where he expressed strong agreement with the statements "Mathematics is an evolving, creative human endeavour in which there is much yet to be known" and "Doing mathematics involves creativity, thinking, and trial-and-error". Questionnaire responses also indicated that Geoff held student-centred views about mathematics teaching and learning; for example, he strongly disagreed that in mathematics something is either right or wrong and that mathematics problems can be solved in only one way, and agreed that teachers should allow time for students to find their own methods for solving problems.

Geoff had been interested in computers since his childhood and was an experienced Excel spreadsheet user when he started his pre-service teacher education programme at university. As a teacher, he enjoyed developing mathematical modeling tasks that were embedded in real life scenarios or stories. (One such task is described in the next section.) In these tasks, he used various types of digital technologies to collect and analyse real data, skilfully blending empirical and analytical approaches to help students develop mathematical models to fit the data.

Illustration of Geoff's Practice

Geoff had participated in an earlier research project that documented the modes of working with technology adopted by pre-service and beginning teachers and investigated personal and contextual factors that shaped their pedagogical identities (see Goos 2005). In his first year after graduation, he taught a Grade 8 mathematics class that was the focus of the research. This was his first experience of using a motion detector in conjunction with a graphics calculator and screen projection unit to teach students how to interpret distance-time graphs. He called on individual students to walk towards or away from the motion detector so as to match a pre-selected distance-time graph displayed on the calculator and view screen. In the following lesson, when he did not have access to the same technology, he devised a simulated graph matching activity in which students 'walked' the graph he had drawn on the whiteboard as he moved his pen along the horizontal (time) axis. In terms of the technology metaphors framework introduced earlier in the chapter, Geoff was using technology as a *partner* because he wanted to provide students with access to a new kind of task that developed their understanding of scale and gradient. This task engaged students in a physical experience that gave instant feedback on the match between the graphical representation and their movement.

When interviewed after the lesson, Geoff explained that he was looking for further challenges in learning to teach mathematics with digital technologies:

> I know what things the graphics calculator can do, and I have a pretty good knowledge of Excel, but really now that teachers know how to include this in their pedagogy, I suppose the emphasis would be now on getting the most out of it. Instead of just knowing what to do, how to really take this technology and explore it to its fullest extent and use all of the resources that [it] has to offer instead of taking bits and pieces that might be good. I suppose unlocking the potential ... of what this technology has to offer.

Geoff's interest in integrating powerful uses of digital technologies into his pedagogical repertoire suggests that his trajectory for development was leading him towards using technology, and especially Excel spreadsheets, as an *extension of self*.

Although Geoff's approach in the Grade 8 lessons was taken directly from the teaching materials accompanying the motion detector, the activity resonated with his more creative interests in using drama, songs and story-telling in teaching mathematics as well as English. This mathematical performance approach was evident in a lesson that was observed during the subsequent research project, 5 years later.

Geoff was teaching an advanced mathematics subject to a Grade 12 class. In a series of lessons on differential equations he planned to introduce Newton's Law of

Cooling via a 'murder mystery' that his students were to solve with the aid of data logging probes and Excel spreadsheets. His approach was aligned with the state-mandated syllabus that emphasised using mathematically-enabled technologies to allow students "to tackle more diverse, life-related problems" (Queensland Board of Senior Secondary School Studies 2000, p. 10). As part of the study of calculus, students were to "appreciate the importance of differential equations in representing problems involving rates of change" (p. 17), through learning experiences such as investigating "life-related situations that can be modeled by simple differential equations such as growth of bacteria, cooling of a substance" (p. 18).

In the first part of the lesson, Geoff introduced differential equations of the type $\frac{dy}{dx} = f(y)$ and reminded students that they had dealt with equations of this type in their earlier studies of mathematics. He noted that there were many instances of such equations in real life and asked students to suggest examples. They remembered that this equation could represent exponential growth or decay, such as in bacterial growth or measuring rates of cooling. One student recalled that the rate of change of temperature was a function of the difference between the object's temperature and room temperature.

Geoff worked through some examples, including one that illustrated exponential decay. He then set the students to work on textbook exercises. During this segment of the lesson another teacher, who had been recruited by Geoff to help set up the modeling scenario, burst into the room and announced that a 'murder' had been committed in a nearby classroom. Not knowing whether to believe the teacher or not, the class followed Geoff to the 'crime scene' where they found an outline of the 'victim' chalked on the floor and two cups of coffee that were still warm. Geoff told the class that police had arrested two suspects who admitted to being in the room making coffee some time earlier but denied committing the crime. According to the 'police report' that Geoff distributed to students, the time of death had been fixed at 11:45 am, 15 min before Geoff's colleague announced the 'murder'. The task for the class was to analyse the cooling rate of the coffee, given the time it was poured and initial temperature, in order to work out whether the suspects could have been in the room at the time the 'murder' was committed.

This task is an application of Newton's Law of Cooling, $T = (T_0 - T_R)e^{kt} + T_R$, where T=temperature of an object undergoing cooling, t=time, k=decay constant, T_0=initial temperature of object, and T_R=room temperature. Geoff set up temperature probes in each coffee cup to collect temperature and time data while he developed the necessary theory with input from the class. This involved eliciting from students the differential equation for the relationship between the rate of cooling and the difference between object temperature and room temperature, $\frac{dT}{dt} = k(T - T_R)$, which was then re-written as $\frac{dt}{dT} = \frac{1}{k}\left(\frac{1}{T - T_R}\right)$. Students integrated this equation to give $t = \frac{1}{k}\log_e(T - T_R) + C$, and found the value of the constant of integration, C, by substituting the initial values $t=0$ and $T=T_0$. This gave $t = \frac{1}{k}\log_e(T - T_R) - \frac{1}{k}\log_e(T_o - T_R)$, a function that the students then expressed in exponential form $T = (T_0 - T_R)e^{kt} + T_R$.

	A	B	C	D	E
1	Time	Actual Temp	Predicted Temp	Difference	Difference Squared
2	0	66.9	66.80	-0.10	0.010000
3	2	64.6	64.61	0.01	0.000137
4	4	62.4	62.58	0.18	0.030788
5	6	60.5	60.68	0.18	0.032639
6	8	58.9	58.92	0.02	0.000306
7	10	57.1	57.28	0.18	0.031260
8	12	55.8	55.75	-0.05	0.002490
9	14	54.5	54.33	-0.17	0.029089
10	16	52.9	53.01	0.11	0.011553
11	18	51.7	51.78	0.08	0.005985
12	20	50.6	50.63	0.03	0.001069
13				SUM	0.155316
14					
15		$T = a \cdot e^{kt} + b$			
16					
17		a	31.5		
18		k	-0.036		
19		b	35.3		

Fig. 1 Newton's Law of Cooling spreadsheet

Geoff transferred the temperature (T) and time (t) data to a spreadsheet (http://extras.springer.com[*]) that plotted the exponential function $T = a \cdot e^{kt} + b$. He had set up the spreadsheet to allow the user to change the values of a, k and b and observe changes in the 'goodness of fit' of the model (square of the difference between actual and predicted temperatures for each data point) and also in the corresponding graph superimposed over the scatterplot of temperature versus time data (Fig. 1). At the end of the lesson Geoff emailed this spreadsheet to the students. They finished the modeling task for homework and emailed Geoff their completed spreadsheets overnight.

Geoff's use of the modeling task allowed students to engage with a practical application of differential equations at the same time as they were developing an understanding of the associated mathematical concepts. His use of multiple technologies – graphics calculator, temperature probes, Excel spreadsheet – allowed

---
[*] Log in with ISBN 978-94-007-4638-1

him to combine empirical and analytical approaches to give meaning to these concepts. Five years into his teaching career, it seemed that Geoff was creatively integrating a range of technologies into his mathematical and pedagogical repertoire as an *extension of self*, as foreshadowed in the interview conducted during his first year of teaching (see above).

While it is not uncommon for teachers to use digital technologies such as spreadsheets and graphics calculators with data probes to illustrate Newton's Law of Cooling, Geoff's approach was distinguished by his creativity in embedding the modeling task in a dramatic 'murder' scenario that aroused his students' curiosity and guaranteed their attention as the underlying theory was developed. Although not a fully digital mathematical performance in the sense described by Gadanidis and Borba (2008), this was still a technology-enriched performance that had the potential to generate an emotional response to the murder and police investigation as well as a cognitive response to the mathematical problem.

Geoff's Developing Pedagogical Identity

According to Valsiner's (1997) zone theory, the zone of proximal development entails a set of possible 'next states' of the developing system's relationship with the environment, given the current state of the ZFM/ZPA complex and the individual's developmental state. Thus the ZPD captures development that lies between the possible and the actual. In Geoff's case, his ZPD as a recently graduated teacher included an appreciation of mathematics as a creative human endeavour, some student-centred understandings of how mathematics is learned, and considerable interest and skill in using digital technologies for learning and teaching mathematics. Thus his ZPD offered possibilities for development as a teacher who uses digital technologies as a 'conceptual construction kit' (Olive et al. 2010, p. 138) rather than only as a replacement for calculations that can be done by hand. Geoff found employment in an apparently well-resourced school that was beginning to implement a policy emphasising technology use in all subjects. In mathematics, this meant that all students from Grade 9 upwards had to buy their own graphics calculator, and the school had invested in data logging peripherals and screen projection units as well as fitting out several mathematics classrooms with data projectors and computers connected to the internet. External curriculum and assessment requirements in senior secondary mathematics included mandatory use of computer software or graphics calculators. On the surface, then, it seemed that Geoff's professional environment offered a zone of free movement with broad boundaries for action that permitted experimentation with digital technologies for teaching mathematics. Similarly, the teaching actions promoted by the school administration – the zone of promoted action – seemed to lie within the ZFM. For example, school-based professional development was provided whenever new technology resources were purchased, and the Head of the Mathematics Department encouraged Geoff to incorporate digital technologies into all of his mathematics teaching. The apparent

ZFM/ZPA complex therefore *promoted* teaching actions that were *permitted* within the school and external curricular environment.

However, the ZFM/ZPA complex that Geoff actually experienced within the school worked to constrain his development in subtle ways. Even though the school employed technical support staff to help teachers integrate technology into their lessons, Geoff said that they responded slowly, if at all, to his frequent requests for new mathematical software to be installed over the school's intranet, and he had been obliged to install programs himself on individual computers in order to teach some lessons. This was the case for the Newton's Law of Cooling lesson described earlier, where temperature and time data had to be manually entered into the modeling spreadsheet because the software that can do this automatically had not yet been installed on the classroom computer. The problem was exacerbated by having limited access to computer laboratories that were regularly booked out to other, non-mathematics, classes. Timetabling practices often allocated mathematics classes to rooms in which the teachers rarely used the available technology, while other mathematics teachers who wished to use these resources could not gain access. Geoff also referred to an organisational culture that was not conducive to risk taking, and especially to the conservative influence of parents who expected mathematics to be taught in traditional ways not involving technology. Despite the support of his Head of Department, Geoff's school-based ZPA was characterised by passive acceptance of technology on the part of the other mathematics teachers. He said he believed that he had brought more ideas to colleagues, in terms of technology, than they had been able to teach him.

Valsiner (1997) pointed out that children can negotiate changes to the ZFM/ZPA complex in order to achieve their emerging goals. Likewise, Geoff was able to find a zone of promoted action outside the school that mapped onto his ZPD in developmentally productive ways. There were three aspects to this external ZPA. The first involved participating in university research projects such as the one described here. Geoff noted that the press for innovation that he felt as a consequence of his participation was beneficial because he was motivated to turn 'a germ of an idea' into a real lesson. Discussing his ideas for the coffee-cup murder mystery some weeks before this lesson, he acknowledged:

> This project is good because it gives me the impetus to do something like that which … otherwise still might just be a happy thought.

The second aspect to the external ZPA saw Geoff looking for formal professional development opportunities, such as the intensive, week-long conference that had recently introduced him to advanced features of Excel. Nevertheless, Geoff was selective about what he took from these professional development experiences:

> The majority of things I see that I'd like to use I don't get to use, probably because I see so much of it. I've got to be a bit choosy about what I plan to do.

The third element of his external ZPA came from his increasing participation in the activities of his local mathematics teacher professional association, and in particular the professional growth he experienced by presenting workshops and seminars on teaching mathematics with digital technologies. Although Geoff had little control over his material circumstances at the school – his ZFM – his decision to access an external ZPA helped him to take charge of his own development and re-interpret the limitations imposed by timetabling rigidity, lack of technical support, and a conservative school culture as not necessarily preventing him from adopting innovative teaching practices. This zone theory analysis provides some sense of Geoff's identity trajectory in 'becoming' a teacher who confidently integrated digital technologies into his practice, and his role in negotiating that trajectory. The other way to observe the development of his pedagogical identity is to recognise that his modes of working with technology became more sophisticated over time, progressing towards *extension of self* as he integrated the range of resources available to him into the mathematical practices of the classroom.

Introducing Susie

At the start of the research study, Susie was in her third year of teaching in an independent secondary school with an enrolment of around 600 students in Grades 8–12. The student population was fairly homogeneous with respect to cultural and socio-economic background, with most students coming from white, Anglo-Australian middle class families. Susie was qualified to teach mathematics and music, but at this school she was assigned to teach only mathematics classes.

Susie's own experience of learning mathematics at school was structured and content based, and this influenced the ideas about mathematics teaching that she brought to the pre-service programme:

> I thought it would be great if I could just put stuff on the board and let them do their work and answer questions if they needed it and write exams, tick, cross and that's my job.

According to Susie, these ideas were first challenged by her mathematics curriculum lecturer at university who opened her eyes to different approaches to teaching mathematics. She was now trying to implement these approaches herself. For example, when interviewed, she explained that in her classroom "we spend more time on discussing things as opposed to just teaching and practising it", and that for students "experiencing it is a whole lot more effective than being told it is so".

Susie's responses to the Mathematical Beliefs Questionnaire were consistent with the student-centred approaches that she was now trying to implement in her teaching. For example, she expressed strong agreement with statements such as "In mathematics there are often several different ways to interpret something", and she disagreed that "Solving a mathematics problem usually involves finding a rule or formula that applies". Other questionnaire responses were strongly supportive of cooperative group work, class discussions, and use of calculators, manipulatives and real life examples.

Aged in her mid-20s, Susie said she felt she was born into the computer age and this contributed to her comfort with using digital technologies in her teaching.

She recognised that technology saved time with calculations and graphing but also saw it as providing opportunities for mathematical exploration:

> You make progress so much quicker than having to do things by hand and you can just do examples like … what does this rule look like? What does this linear function look like? And they can put it into their calculator and check and have a look [...] So it's just quicker to explore things.

Illustration of Susie's Practice

Observations of Susie's Grade 10 mathematics class in the first year of the research study illustrate her preference for using digital technologies to explore mathematical concepts. In one lesson cycle, she introduced quadratic functions via a graphical approach involving real life situations and followed this with algebraic methods to assist in developing students' understanding. Lessons typically engaged students in one or two extended problems rather than a large number of practice exercises. For example, students used the regression function on their graphics calculators to investigated quadratic models for data on the growth of babies, the path of a tennis ball as it is hit over the net, the height of an object dropped from the top of a building, and the cross sectional dimensions of a railway tunnel arch. They then used their models to make predictions that went beyond the data. A characteristic of these tasks was that students were asked to comment on the strengths and limitations of their models in relation to the real life data rather than just accepting the calculator regression output as an indicator of goodness of fit.

The assessment task for this unit of work required students to investigate projectile motion as a practical application of quadratic functions. The task made use of a computer game in which the Sesame Street character Gonzo was shot from a cannon towards a bucket of water some distance away (http://www.funny-games.biz/flying-gonzo.html; see Figs. 2 and 3). The game allows players to vary the angle of projection and the cannon 'voltage' (a proxy for muzzle velocity) and observe the effects on the distance Gonzo travelled as they 'aim' him at the bucket of water.

Susie had discovered this game at a professional development workshop run by the local mathematics teacher association. The presenter was Geoff, the teacher profiled in the previous section of the chapter. Geoff found the game when searching on the internet for applications of projectile motion that he could use with his Grade 12 class. During this Grade 12 lesson, which was observed as part of the research study, Geoff introduced the parametric equations for projectile motion

$$x(t) = Vt\cos\vartheta \text{ and } y(t) = Vt\sin\vartheta - \frac{gt^2}{2}$$

where $x(t)$ is the horizontal displacement, $y(t)$ the vertical displacement, ϑ the angle of projection, V the initial velocity, t the time in flight and g acceleration due

Technology Integration in Secondary School Mathematics... 153

Fig. 2 Opening screen of Flying Gonzo game

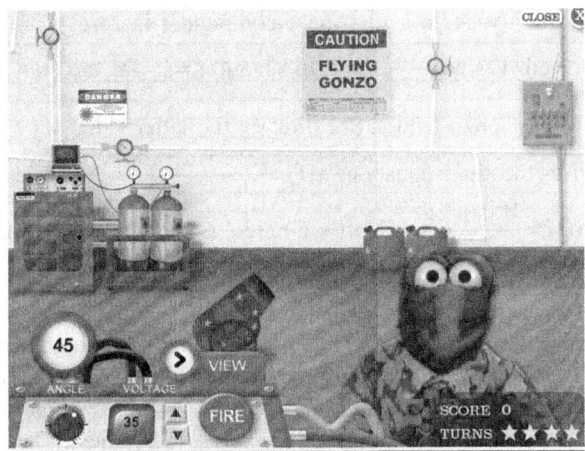

Fig. 3 Firing Gonzo to land in the bucket of water

to gravity. Noting that the *y*-component is zero when Gonzo lands, Geoff then solved $Vt\sin\vartheta - \frac{gt^2}{2} = 0$ to obtain $t = 0$ (at the start of flight) or $t = \frac{2V\sin\vartheta}{g}$ (when Gonzo lands). Substituting the latter value for *t* into the equation for $x(t)$ gives the range equation, $x(t) = \frac{2V^2\sin\vartheta\cos\vartheta}{g}$ or, using the double angle formula, $x(t) = \frac{V^2\sin 2\vartheta}{g}$. Geoff's Grade 12 students were to test a range of angles and velocities to predict the range, with the aim of landing Gonzo in the bucket of water. Because the real velocity and range were unknown, students instead recorded the cannon 'voltage' and estimated the range by counting the number of tiles on the wall in the screen background during Gonzo's flight. They entered all these values into a spreadsheet and compared the predicted range, calculated from the range equation, with the actual range expressed in 'tiles'. They then averaged the ratio of predicted to actual range to produce a constant factor (~140) that could be applied to subsequent tests to accurately predict Gonzo's range.

When she tried out the game at the professional development workshop that Geoff presented, Susie wondered whether she could adapt the mathematical content to suit her Grade 10 class. She devised an assessment task in which students used their graphics calculators or *TI-Interactive* software to tabulate and plot data that would allow them to find a mathematical model for the relationship between the range and the muzzle velocity. Algebraic methods were then to be used to determine the best cannon settings for Gonzo to hit a target at a given distance. Students were given a low voltage setting and high voltage setting. Keeping each constant in turn, they fired Gonzo at eight different angles and recorded the range for each trial. They then entered the data into their graphics calculators or *TI-Interactive* and found a quadratic model that gave the best fit. Note that when voltage (velocity) is kept constant the model is trigonometric rather than quadratic because the range varies with the angle of projection. Susie could perhaps have designed the task differently, to keep angle constant and vary the voltage, which would yield a true quadratic model. Nevertheless, a quadratic model fitted to the data as collected gives a good approximation and allowed students to practise finding critical points (intercepts and turning points) algebraically.

Interview and lesson observation data suggest that Susie was interested in having students use technology for mathematical exploration, and not just for checking calculations or making graphing quicker. In terms of the framework for teaching and learning roles introduced earlier, she was working with technology as a *partner* to develop students' understanding of mathematical concepts. Susie's and Geoff's use of a computer game to develop mathematical understanding of quadratic functions and projectile motion connects with Jorgensen and Lowrie's (2011) argument that immersion in digital game environments engages learners and reshapes their thinking. Although the game provided a dynamic image of Gonzo's motion, the effect was similar to 'panning' a camera so that the background seemed to move while Gonzo stayed in the centre of the computer screen. Students had to visualise his parabolic path and find an efficient method for measuring the horizontal distance travelled, which required many repetitions of the game. As Jorgensen and Lowrie noted, the purpose of this repetition was not to achieve fluency with taught skills as is often the case with practice on textbook exercises, but to gain a better understanding of the problem situation and solution strategies.

Susie's Developing Pedagogical Identity

During interviews, Susie referred to a range of people and environmental influences that shaped her development as a teacher of mathematics. Unlike Geoff, who found alignment between his mathematics learning experiences as a school student and the practices promoted by the pre-service teacher education programme, Susie's understanding that mathematics is learned and taught through memorisation and practice was challenged by her pre-service experience. It seemed that there was enough overlap between Susie's ZPD, representing her possibilities for development, and the zone of promoted action offered by the university teacher education programme to canalise her development towards more student-centred, exploratory approaches as she began her teaching career. But a teacher's identity trajectory is also influenced by the relationship between ZPA and ZFM and the meanings ascribed to different aspects of the school environment by the people who organise that environment. Development can be constrained when the environment seems not to permit teaching actions that are ostensibly promoted. However, this seemed not to be the case at Susie's school.

When Susie started working at the school she came under the influence of the Head of the Mathematics Department, who had developed a culture where mathematics was taught as much as possible in *context*, where students worked *collaboratively* and available *technologies* were used extensively. He had been the driving force behind the introduction of technology to the school during the 1990s, before the external curriculum had made the use of technology mandatory. When interviewed, he said he was able to achieve this cultural change because the school administration supported his teaching philosophy and provided funds for resources. Initially, however, even though he developed technology-based activities and provided teachers with professional development, there was not a great uptake of digital technologies by the mathematics teaching staff. To overcome this inertia he introduced technology into assessment tasks that had to be implemented by all teachers:

> You actually had to design activities that you ask all teachers to do or you build it into assessment and teachers will tend to engage a bit more because they always want their students to do the best they can. And it took a long time before it got to the point where it is now where people just pick it up and use it and there are still people that resist anything that's new, even in that culture.

Thus Susie started her teaching career in a school where there was a strong culture within the mathematics department that emphasised integrating digital technologies into everyday classroom practice, resulting in an expectation that she would teach in the same way. Susie described the approach at her school as "This is what we do here". She said it made sense to her, and she "learned so much in the first year about [her] personal understandings of maths, let alone to do with the teaching of it, but also the different approach to it". At this stage of her development, lesson observations indicated that Susie's main mode of working with digital technologies was as a *partner* in providing new ways for students to develop understanding of mathematical concepts.

The zone of free movement offered by the school supported technology innovation through an organisational culture that expected teachers of mathematics to make use of the substantial resources in which the school had invested. Students in Grades 9–12 had constant access to graphics calculators obtained through the school's hire scheme, there were additional class sets of CAS calculators for senior secondary classes, and data logging equipment compatible with the calculators was freely available. Computer software was also used for mathematics teaching; however, as is common in many Australian secondary schools, computer laboratories were difficult to access and had to be booked well in advance. Susie preferred to use graphics calculators so that students could access technology in class whenever they needed it. The data projector installed in her classroom also made it easy for her to display the calculator screen for viewing by the whole class.

The ZFM/ZPA complex that influenced Susie's development as a teacher featured an expansive zone of free movement with few constraints limiting her choice of actions and a zone of promoted action set up by the school administration and Head of Department that encouraged her to explore the resources that were available to her. As Susie explained, "Anything I think of that I would really like to do [in using technology] is really strongly supported". Susie's pedagogical identity was taking shape as she constructed meaning from her person-environment relationship. It seemed that the ZFM/ZPA set up by the school mapped exactly onto her ZPD, so much so that she evaluated the external ZPAs offered by formal professional development workshops in terms of how well they matched the teaching approaches permitted by her environment and promoted by the people who organised that environment. She had attended many conferences and workshops in the 3 years since beginning her teaching career, but found that most of them were not helpful "for where I am". She explained: "Because we use it [technology] so much already, to introduce something else we'd have to have a really strong basis for changing what's already here".

One of the risks in continually judging the fit of an external ZPA in terms of its match with existing people-environment relationships within a school is that it may limit possibilities for envisioning and adapting to change. A school's organisational culture and resources can change over time, as can the teaching approaches promoted if there is turnover of key staff. Susie was already aware that not all of the mathematics teaching staff were enthusiastic users of digital technologies. One experienced teacher who had been a longstanding staff member at the school expressed concerns that sometimes technology could be used "just because it's there" and cited as an example the use of dynamic geometry software in junior secondary classes at the expense of using concrete materials: "I think it's good to draw things and measure things". This teacher was willing to question the value of using technology in certain circumstances, and Susie acknowledged the teacher's influence in making her more discerning in her own use of technology with her classes. When the Head of the Mathematics Department left the school, Susie was promoted to the position of coordinator of the junior secondary mathematics programme. Now she noticed that some of the more recently appointed mathematics teachers were neutral and passive in their attitudes towards technology. Although they were willing to use technology in their teaching if pressed or shown how to, they rarely asked questions or engaged in discussions about improving existing tasks and technology-based teaching practices. Thus the ZPA implicitly set by the example of colleagues was contracting, and one might predict that Susie's identity trajectory of 'becoming' a creative user of digital technologies – perhaps as an *extension of self* – would be impeded unless she deliberately sought out external ZPAs consistent with her pedagogical beliefs and goals.

Conclusion

This chapter has focused on how mathematics teachers develop new practices in technological environments. Mathematics education researchers have been interested in the mathematical potential of technology and its effects on student learning for at least the last 30 years (Hoyles and Lagrange 2010), but only recently has there emerged a trend towards investigating how technology changes the professional work of mathematics teachers (Artigue 2010). The research reported in this chapter examined relations between factors known to influence ways in which teachers use digital technologies to enrich secondary school mathematics. Based on sociocultural theories that view learning as increasing participation in practices and constructing identities in relation to these practices, two frameworks were used to analyse the development of teachers' pedagogical identities as users of technology. The first framework classifies different ways of working with technology and provides evidence of 'what' changes in teachers' practice, while the second allows for investigation of teacher-environment relationships to explain the 'how' and 'why' of developing practice in terms of Valsiner's (1997) zone theory.

The analysis of two cases of beginning teachers illustrated several issues related to identity development. The first issue concerns the temporal dimension of identities, in that teachers are on a trajectory of 'becoming' a different practitioner. Zone theory is useful for conceptualising not only possibilities for development, but also the ongoing process of development as changing relationships between the zone of free movement, zone of promoted action, and zone of proximal development. Other issues are related to how trajectories of teacher development are constrained rather than fully determined. Teachers' knowledge and beliefs, on their own, do not determine how they will approach the classroom use of digital technologies. Neither does it make sense simply to 'add up' the positive or negative effects of institutional constraints or professional development opportunities to predict whether teachers will embrace or resist technology. Instead, an analysis is called for that gives attention to relationships amongst Valsiner's (1997) three zones, bearing in mind that the developing person is able to re-negotiate these relationships to some extent to achieve their emerging goals.

Susie and Geoff were regarded as innovative users of technology; however, they differed in the degree of fit between their respective ZPDs, ZFMs and ZPAs. The zone of free movement offered by their schools was important in allowing them some leeway to explore technology-enriched teaching approaches consistent with their pedagogical knowledge and beliefs. In both schools there was good access to most forms of technology, and an externally mandated mathematics curriculum that made it obligatory for teachers to use computers or graphics calculators. Yet, despite the availability of appropriate material resources, other institutional constraints worked against technology integration. In Geoff's case, constraints arose from the school's conservative academic culture and somewhat inflexible organisational structures that were not wholly conducive to experimentation with new technologies. For Susie, passive resistance from other mathematics teachers was beginning to undermine an organisational culture that had previously supported innovative technology integration. The zones of promoted action set up for and accessed by these two teachers also differed. Susie found that her school's ZPA enabled her to fully exploit the possibilities provided by the ZFM, although these circumstances were changing due to staff turnover at the time of the research study. In contrast, Geoff's school-based ZPA did not provide him with enough opportunities to develop and extend his teaching repertoire. Instead he sought an external ZPA through varying combinations of formal and informal professional development. This analysis shows that neither professional learning experiences, time, resources, curriculum mandates, nor supportive organisational structures and cultures are sufficient, on their own, to lead to a higher level of technology integration in mathematics classrooms. Instead, it is the dynamic relationships between these factors, and the teacher's active reshaping of their professional environment, that develop their professional identities as users of technology.

The extent of overlap between the ZFM/ZPA complex and the ZPD may be critical in supporting beginning teachers in further developing the innovative practices they typically encounter in pre-service programmes. Susie and Geoff experienced different combinations of factors known to influence technology integration, but both had a 'region' of overlap between their respective ZPD, ZFM and ZPA where they were able to find sources of assistance that supported their ongoing development as teachers of mathematics, and this in turn enabled them to integrate technology into their professional practice in a variety of ways. Some of these uses of technology went beyond the familiar applications of computer software and graphics calculators to incorporate elements of mathematical performance and digital gaming that may offer new ways of learning mathematics. Susie and Geoff developed these activities themselves without any intervention from the researcher, and the account in this chapter of how they implemented these activities provides an authentic picture of what is possible in a typical independent secondary school classroom. With new generations of students coming to school familiar with using digital technologies to organise their daily lives, provide entertainment, find information, and maintain social networks, mathematics education research needs to find better ways to understand the impact of such technologies on teachers' professional work and learning.

Police Crime Scene Coffee Analyser

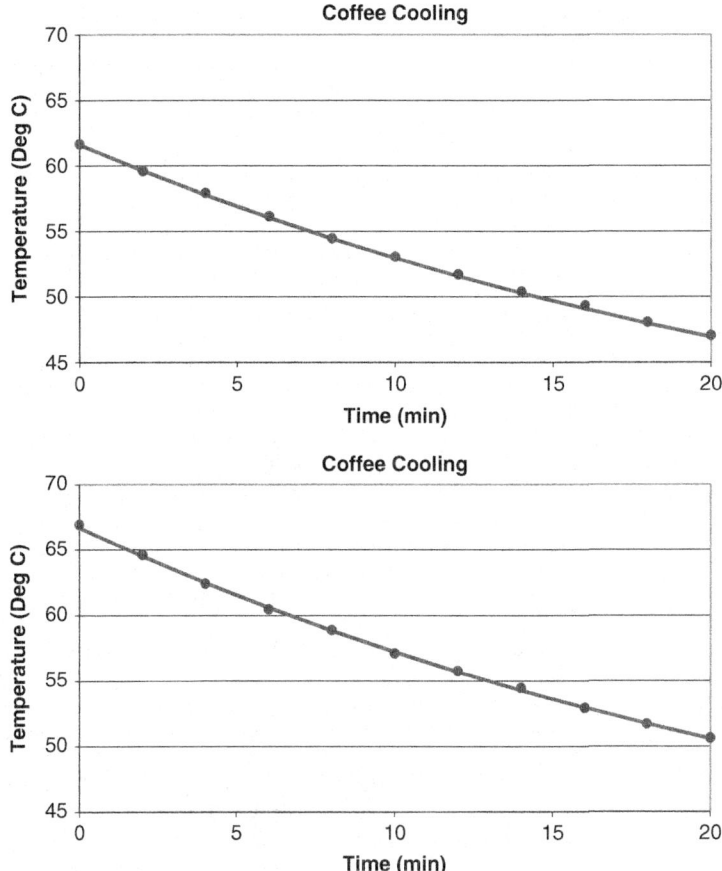

References

Artigue, M. (2010). The future of teaching and learning mathematics with digital technologies. In C. Hoyles & J.-B. Lagrange (Eds.), *Mathematics education and technology: Rethinking the terrain: The 17th ICMI study* (pp. 463–475). New York: Springer.

Burrill, G., Allison, J., Breaux, G., Kastberg, S., Leatham, K., & Sanchez, W. (2003). *Handheld graphing technology in secondary mathematics: Research findings and implications for classroom practice*. Michigan State University.

Cobb, P., McClain, K., Lamberg, T., & Dean, C. (2003). Situating teachers' instructional practices in the institutional setting of the school and district. *Educational Researcher, 32*(6), 13–24.

Confrey, J., Hoyles, C., Jones, D., Kahn, K., Maloney, A., Nguyen, K., Noss, R., & Pratt, D. (2010). Designing software for mathematical engagement through modeling. In C. Hoyles & J.-B. Lagrange (Eds.), *Mathematics education and technology: Rethinking the terrain: The 17th ICMI study* (pp. 19–45). New York: Springer.

Cuban, L., Kirkpatrick, H., & Peck, C. (2001). High access and low use of technologies in high school classrooms: Explaining an apparent paradox. *American Educational Research Journal, 38*, 813–834.

Ellington, A. (2003). A meta-analysis of the effects of calculators on students' achievement and attitude levels in precollege mathematics classes. *Journal for Research in Mathematics Education, 34*, 433–463.

Forgasz, H. (2006). Factors that encourage and inhibit computer use for secondary mathematics teaching. *Journal of Computers in Mathematics and Science Teaching, 25*(1), 77–93.

Gadanidis, G., & Borba, M. (2008). Our lives as performance mathematicians. *For the Learning of Mathematics, 28*(1), 44–51.

Goos, M. (2005). A sociocultural analysis of the development of pre-service and beginning teachers' pedagogical identities as users of technology. *Journal of Mathematics Teacher Education, 8*, 35–59.

Goos, M. (2009a). A sociocultural framework for understanding technology integration in secondary school mathematics. In M. Tzekaki, M. Kaldrimidou, & H. Sakonidis (Eds.), *Proceedings of the 33rd conference of the International Group for the Psychology of Mathematics Education* (Vol. 1, pp. 113–120). Thessaloniki: PME.

Goos, M. (2009b). Reforming mathematics teacher education: Theorising teachers' and students' use of technology. In C. Ng & P. Renshaw (Eds.), *Reforming learning: Concepts, issues and practice in the Asia-Pacific region* (pp. 43–65). New York: Springer.

Goos, M. (2011). Technology integration in secondary mathematics: Enhancing the professionalisation of prospective teachers. In O. Zaslavsky & P. Sullivan (Eds.), *Constructing knowledge for teaching secondary mathematics: Tasks to enhance prospective and practicing teacher learning* (pp. 209–225). New York: Springer.

Goos, M., & Bennison, A. (2002). *Building learning communities to support beginning teachers' use of technology.* Proceedings of the annual conference of the Australian Association for Research in Education, Brisbane. Retrieved 21 June 2012 from http://www.aare.edu.au/02pap/goo02058.htm

Goos, M., Galbraith, P., Renshaw, P., & Geiger, V. (2000). Reshaping teacher and student roles in technology-enriched classrooms. *Mathematics Education Research Journal, 12*, 303–320.

Hoyles, C., & Lagrange, J.-B. (2010). Introduction. In C. Hoyles & J.-B. Lagrange (Eds.), *Mathematics education and technology: Rethinking the terrain: The 17th ICMI study* (pp. 1–11). New York: Springer.

Jorgensen, R., & Lowrie, T. (2011). Digital games: Creating new opportunities for mathematics learning. In J. Clark, B. Kissane, J. Mousley, T. Spencer, & S. Thornton (Eds.), *Mathematics: New traditions and [new] practices. Proceedings of the 23rd biennial conference of the Australian Association of Mathematics Teachers and the 34th annual conference of the Mathematics Education Research Group of Australasia* (pp. 406–413). Adelaide: AAMT & MERGA.

Lerman, S. (2001). A review of research perspectives on mathematics teacher education. In F. Lin & T. J. Cooney (Eds.), *Making sense of mathematics teacher education* (pp. 33–52). Dordrecht: Kluwer.

Lowrie, T. (2005). Problem solving in technology rich contexts: Mathematics sense making in out-of-school environments. *The Journal of Mathematical Behavior, 24*(3–4), 275–286.

Olive, J., Makar, K., Hoyos, V., Kor, L. K., Kosheleva, O., & Straesser, R. (2010). Mathematical knowledge and practices resulting from access to digital technologies. In C. Hoyles & J. Lagrange (Eds.), *Mathematics education and technology: Rethinking the terrain: The 17th ICMI study* (pp. 133–177). New York: Springer.

Penglase, M., & Arnold, S. (1996). The graphics calculator in mathematics education: A critical review of recent research. *Mathematics Education Research Journal, 8*, 58–90.

Queensland Board of Senior Secondary School Studies. (2000). *Mathematics C senior syllabus.* Brisbane: Author.

Simonsen, L. M., & Dick, T. P. (1997). Teachers' perceptions of the impact of graphing calculators in the mathematics classroom. *Journal of Computers in Mathematics and Science Teaching, 16*, 239–268.

Tharp, M. L., Fitzsimons, J. A., & Ayers, R. L. B. (1997). Negotiating a technological shift: Teacher perception of the implementation of graphing calculators. *Journal of Computers in Mathematics and Science Teaching, 16*, 551–575.

Thomas, M. O. J. (2006). Teachers using computers in the mathematics classroom: A longitudinal study. In J. Novotná, H. Moraová, M. Krátká, & N. Stehliková (Eds.), *Proceedings of the 30th annual conference of the International Group for the Psychology of Mathematics Education* (Vol. 5, pp. 265–272). Prague: PME.

Valsiner, J. (1997). *Culture and the development of children's action: A theory of human development* (2nd ed.). New York: Wiley.

Wallace, R. (2004). A framework for understanding teaching with the Internet. *American Educational Research Journal, 41*, 447–488.

Wenger, E. (1998). *Communities of practice: Learning, meaning and identity*. Cambridge: Cambridge University Press.

Windschitl, M., & Sahl, K. (2002). Tracing teachers' use of technology in a laptop computer school: The interplay of teacher beliefs, social dynamics, and institutional culture. *American Educational Research Journal, 39*, 165–205.

Teaching Roles in a Technology Intensive Core Undergraduate Mathematics Course

Chantal Buteau and Eric Muller

Abstract We discuss the dual teaching roles of university mathematics tutors, as teachers and policy makers, in relation to the classroom implementation of technology while guided by departmental policies. The main contribution of this chapter is the exemplification of these roles in an undergraduate mathematics programme, called *Mathematics Integrated with Computers and Applications* (MICA), with systemic technology integration. The current classroom practices of tutors in one of the MICA core courses for mathematics majors and future teachers of mathematics are examined. The role of the tutors in this course is to carefully guide the students' instrumental genesis of programming technology for the investigation of both mathematics concepts and conjectures, and real-world applications. Acting as a mentor, the tutor encourages students' mathematical creativity as they design, program, and use their own interactive mathematics *Exploratory Objects*.

Keywords University mathematics education • Technology integration • Tutors' roles as teacher and policy maker • Mathematics department • Programming • Exploratory objects/microworlds • Instrumental integration • Creativity

---

C. Buteau (✉) • E. Muller
Brock University, St. Catharines, Canada
e-mail: cbuteau@brocku.ca; emuller@brocku.ca

Introduction

Since the advent of technologies many publications have focused on changes in tutors'[1] approaches to the teaching of undergraduate mathematics. For example, the National Science Foundation (1996) report entitled *'Shaping the Future: New Expectations for Undergraduate Education in Science, Mathematics, Engineering, and Technology'* has motivated the individual tutors and department as a whole to explore new ways of teaching mathematics. More recently, the International Commission on Mathematical Instruction (ICMI) Study 11 on *The Teaching and Learning of Mathematics at the University Level* (Holton 2001) reports on advances in practice, research, technology and other areas that impact undergraduate mathematics education. The chapter by Keynes and Olsen (2001), entitled 'Professional Development for Changing Undergraduate Instruction', discusses the challenges that a mathematics department faced when it implemented systemic changes in its undergraduate mathematics education. Their method was to use a team approach that involved "senior faculty, graduate and undergraduate assistants (TAs) and teaching specialists" (p. 113). This suggests that at the university level, the role of the tutors for practices in the mathematics classroom is twofold: as a teacher, and also as a policy maker within his/her department. The 2004 Mathematical Association of America (MAA) report entitled 'Undergraduate Programs and Courses in the Mathematical Sciences: CUPM Curriculum Guide' addresses all its recommendations to the department and only indirectly to the individual tutors. The department is responsible for developing, supporting and sustaining its programmes. Only the department can ensure the continuity required for systemic integration of changes in mathematics education at this level.

Moreover, two ICMI studies that focussed on technology in mathematics education have addressed some specific issues faced by tutors in their roles and responsibilities in undergraduate mathematics instruction. However, whereas the ICMI 1 study (Howson and Kahane 1986) entitled *The Influence of Computers and Informatics on Mathematics and its Teaching* was principally devoted to undergraduate mathematics education, the most recent ICMI 17 study *Digital Technologies and Mathematics Teaching and Learning: Rethinking the Terrain* (Hoyles and Lagrange 2010) included only a few contributions addressing undergraduate mathematics education. This turns out to reflect the shift over the years of mathematics educational research concerning the integration of technology to mostly one of focus on school level (Lagrange et al. 2003; Lavicza 2010). Nevertheless, the most recent ICMI 17 study addresses many issues about the use of technology in mathematics teacher education. This is of interest in university mathematics education as, for example, many Canadian mathematics departments have recently devoted significant effort and resources towards the education of future teachers of mathematics (Mgombelo et al. 2006).

---

[1] In this chapter that focuses on teaching roles, the term 'department' will be used to denote a university administrative unit that has the responsibility to set curriculum, develop the department's philosophy, etc., and the term 'tutor' will be used to denote a person who has a full-time position in the department, and is responsible for teaching university courses.

The important issue of what part technology should play in undergraduate mathematics education competes for attention with all the other challenges that tutors face in their teaching; for example, Mason (2002) in his book *Mathematics Teaching Practice – A Guide for University and College Lecturers* discusses many of these challenges and provides constructive ways to address them. There are many publications that can be classified as practitioner reports and that describe the implementation of technology in specific undergraduate courses (e.g. Calculus, Linear Algebra, etc.), for a given group of students, and within departmental constraints. An early example of a collection of these can be found in the MAA Notes 24 *Symbolic Computation in Undergraduate Mathematics Education* (Karian 1992). Few publications address the challenge of continuity, namely, the integration of technology over many years, the involvement of different tutors rotating through the same course, departmental decisions for on-going integration of technology in specific courses or programmes, etc. In other words, the tutor's role of teacher in technology integration in the classroom is most often solely considered, leaving out his/her role as policy maker. But among the few publications that address the issue of continuity is one by Schurrer and Mitchell (1994) who discuss how their mathematics department, acting on a whole departmental initiative, accomplished a systemic integration of technology in their programme.

In the first section of this chapter we discuss the use of technologies in university mathematics teaching, and explore substantial issues that impact upon both tutors and mathematics departments as they integrate these technologies in the teaching and learning of mathematics. The second section describes the evolving role of tutors in an innovative undergraduate first-year core mathematics course that integrates programming technology with interactive interfaces. Here we will explore how the use of this technology in undergraduate teaching changes the traditional 'lecturer' role of the tutor. Whereas the first section elaborates on the tutor's role as policy maker, this section describes in depth the tutor's role as teacher in the classroom while guided by departmental policy. The third section briefly touches on the intersection of students' mathematical creativity and their use of technology. The Chapter concludes with a few suggestions for future studies to inform both tutors and mathematics departments about alternative ways to integrate technology in university mathematics teaching and learning.

Technologies in University Mathematics Classrooms: Moving from Individual Innovations to Departmental Implementations

From his international comparative research study involving the survey of 1,103 mathematicians from Hungary, United Kingdom and United States of America, Lavicza (2010) found that "a large proportion of mathematicians use CAS [Computer Algebra Systems] and other technologies for both their research and teaching" (p. 112). The more recent Canadian extension of the survey corroborates Lavicza's findings (Buteau et al. forthcoming). For example Buteau et al. found that 85 % of the survey participants (N = 302) use a least one technology[2] in their teaching, in particular 69 % of them use CAS.

Overall it seems that the great majority of tutors who integrate technology into their mathematics teaching do so by their own volition; for example 81 % of the mathematician participants in the Canadian survey study agreed to "*I can freely choose whether or not I use CAS in my teaching*" (Buteau et al. forthcoming). Similarly many individuals have described their experiences in scientific papers, conference communications, etc. There is no shortage of practical ideas for integrating technologies into undergraduate mathematics education. However, despite the many examples provided by tutors, it seems that mathematics departments provide them with little support. For example, Buteau et al. (forthcoming) found that 54 % of mathematician participants in their Canadian survey indicated the lack of departmental support as a factor hindering the CAS integration in teaching. To promote consistent delivery of mathematics programmes there need to be some departmental policies on the integration of technology. The Mathematical Association of America CUPM curriculum Guide (2004) stresses, in its *Recommendations for Departments, Programs, and All Courses* section, that individual initiatives to substantial changes to a programme are not enough:

> No program will long survive if it represents the work of a single individual. For long-term sustainability, initiatives must be team efforts, with faculty in supporting roles who can be prepared to expand or take over the leadership of the program. (p. 25)

In what follows, we provide some insights into why such departmental policies on technology integration appear so difficult to achieve.

Within university settings the generation of a departmental policy that focuses on teaching is a challenging task. Informal discussions with colleagues at other universities confirm that university mathematicians contribute willingly to debates on the curriculum but many, often a majority, do not participate in discussions about the teaching of mathematics. Mathematicians tend to have established views about how mathematics should be taught at the university level and these are likely to be based mainly on their own undergraduate and graduate experiences. Introducing technology into the infrequent departmental discussions on teaching and learning mathematics adds new formidable challenges. For example, technologies that have potential to assist, enrich and improve the teaching and learning of mathematics evolve at a pace that few, if any departments, have been able to keep up with. The many published case studies mainly report on what has worked for the author in a specific course and a given set of students. For example, in their literature review (326 contributions) on CAS use in postsecondary education, Buteau et al. (2010) found that 67 % of the contributions was of this kind; 27 % addressed a grouping of courses (e.g., first-year courses or calculus courses) and only 6 % discussed a programme-wide implementation within a department.

---

[2] CAS, dynamic geometry software, programming, discrete mathematics software, simulation software, and/or statistical analysis software; i.e., excluding communication technologies, such as emails, text editors, LaTeX, online fora, etc.

It will be some time before one is able to summarise the specific technologies used in which particular ways are most likely to be successful for specific classes of students and for the tutors who are charged with teaching them. By the time we are able to do so, it is very likely that the technologies in question will have evolved or been superseded. For some tutors, requiring the use of technology into mathematics programmes conflicts with their view of academic freedom. Mason (2002) explores a number of issues and tensions that tutors face in their roles as teachers; for example, "There is an endemic tension between technique and concepts, between understanding and facility, and hence a pedagogy based on ideas and a pedagogy based on practice" (p. 170). Technology in mathematics education is an issue that introduces new teaching tensions and these are unlikely to be resolved to the satisfaction of all mathematics tutors. Building on Mason's views and moving the perspective from the tutors to the department, there is a profound tension between a pedagogy based on integrating technology in a systemic way and one that leaves the responsibility to the individual tutor.

There are examples where departments have overcome significant challenges and have acted to integrate technology into their undergraduate mathematics education. Noss (1999) reports and analyses how mathematics departments in two major UK research universities (he called NU and SU), that had received substantial funding, decided "to exploit new technologies in the teaching of mathematics" (p. 375). Both departments decided to use *Mathematica* (Wolfram Mathematica n.d.) but they did so in very different ways. Noss describes,

> At NU the team centered their design on the production of eighty or ninety screens of hypertextual information, written in an appropriate authoring language and typically including animated examples of techniques. Students were given exercises, in the form of multiple choice questions, which were marked by the computer… (p. 375)

He further comments,

> SU's approach differed substantially. The SU team sought to employ *Mathematica* as a means for students to express their mathematical knowledge via programming. Sequences of tasks were presented as *Mathematica Notebooks*, i.e. textual information on screen, carefully sequenced, and capable of being executed as *Mathematica* code at suitable points. (p. 377)

Here Noss has provided two examples of many opportunities that technology offers in undergraduate mathematics. Another example of integration of technology in undergraduate mathematics education as a result of decisions by a mathematics department is reported by Ralph (2001) and Muller et al. (2009). The Mathematics Department at Brock University introduced in 2001 a policy of systemic integration of technology in its undergraduate courses and the development and implementation of the innovative core undergraduate mathematics programme called MICA (*Mathematics Integrated with Computers and Applications*), which is still operational to this day.

The situations in which mathematics departments find themselves vary from one university to another, most importantly in their tutors who have the power to impact and change the way mathematics is taught. Our understanding of the role that technology can play in students learning mathematics will evolve from the choices of models that departments select to integrate technology in their mathematics programmes. When different approaches in the use of technology are implemented over significant time, their results will provide pointers as to (1) what technologies are most likely to enhance mathematics learning; (2) when and how they can be implemented; (3) for which students they are most beneficial; and (4) which tutors are most likely to integrate them in their teaching. From the experiences reported by tutors and from the experimental programmes and initiatives developed by mathematics departments, a pedagogy of systemic technology integration may emerge.

We reiterate that Noss (1999) noted that "The SU team sought to employ *Mathematica* as a means for students to express their mathematical knowledge via programming" (p. 377). When the mathematics department at Brock University was developing its new MICA undergraduate philosophy, the introduction of computer programming was one of the most discussed and contentious issues. Some tutors were opposed to introducing programming because it would take up valuable time that could be better devoted to other mathematical topics. Others, who used programming in their research, needed little convincing about the power of programming when exploring new mathematical ideas.

Some of the benefits that accrue from using programming for learning school-age mathematics have been identified by a few scholars. Elliot (1976) mentions, "Papert… points out that the process of writing a program forces one to consider possible misunderstandings and ambiguities in a discourse… This may force a student to clearly understand the problem themselves" (pp. 448–449). Abrahamson et al. (2006) stress the benefits and skill development that students may experience when engaged in a mathematics programming activity:

> [The Students] checked the algorithm and the code several times and then formulated preliminary theories to explain the graph. That is, once they were satisfied that they had debugged the code, they reluctantly turned to debug their own thinking – the computer model they had themselves created now constituted an epistemic authority that forced them to reconsider their prior assumptions. (pp. 42–43)

Dubinsky and McDonald (2002, p. 279) use the Actions-Processes-Objects-Schemas (APOS) theoretical framework to comment on the role of programming,

> It is important to note that in this pedagogical approach, almost all of the programs are written by the students. One hypothesis that the research investigates is that, whether completely successful or not, the task of writing appropriate code leads students to make the mental construction of actions, processes, objects, and schema proposed by the theory. (p. 279)

Dubinsky and Tall (1991), in their paper on *Advanced Mathematical Thinking and the Computer*, explore the benefits, the concerns and the challenges of using programming within mathematics courses. They summarise,

> These experiences, both positive and negative, tell us that the issue in using programming to help students learn mathematical concepts is not whether it should be done, nor is it the particular language that is used. The main consideration is how the instructional treatment uses the language through the design of the programming tasks for the students. (p. 242)

Mathematics technologies, including programming, have the potential to bring benefits to teaching and learning mathematics at the postsecondary level.

For example, King et al. (2001) in the publication emanating from the ICMI Study on 'the Teaching and Learning of Mathematics at the University Level' quote Hoyles: There is considerable evidence of the computer's potential to:

- Foster more active learning using experimental approaches along with the possibility of helping students to forge connections between different forms of expression, e.g. visual, symbolic;
- Provoke constructionist approaches to learning mathematics where students learn by building, debugging and reflection, with the result that the structure of mathematics and the ways the pieces fit together are open to inspection;
- Motivate explanations in the face of 'surprising' feedback: that is, start a process of argumentation which can (with due attention) be connected to formal proof;
- Foster cooperative work, encouraging discussion of different solutions and strategies; computer work is more visible and more easily "conveyed" between lecturer and students;
- Open the window on student thought processes: students hold different conceptions of mathematical ideas, which are hard to access, even in the case of articulate adults. How students interact with the computer and respond to feedback can give insight into their conceptions and their beliefs about mathematics and the role of computers. (p. 350)

As an example where university students reaped some of these benefits, consider the study by Chae and Tall (2001) where they report that, after a computer-based experiment involving the period doubling of the logistic function, two thirds of a class were able to link numerical approximation and visual representation, and then link it with theory. Even three students who did not have the desired pre-requisite knowledge of geometric convergence were successful. They stress the change in the role taken by the tutors. To improve effective experimentation, the tutors responded to students by "providing support and explaining the phenomenon of period doubling. Sometimes the supervisor offered advice by providing directed questions to keep the students going if they were stuck" (p. 204), or in other words, "the supervisor acts as a mentor, using various styles of questioning to provoke links between different ideas" (p. 199).

For the mathematics department at Brock University, the use of programming in its MICA programme, besides many other technologies, was to meet a very specific goal – an education of mathematics majors and prospective teachers of mathematics that would empower them to develop, implement and use their own interactive mathematical objects. This goal guides a rich technology use in the sense that it addresses each of the five potentials suggested by Hoyles (King et al. 2001). Furthermore, it is our experience that to meet this goal tutors are often required to move away from a traditional teaching lecturer role to a more 'mentor' role.

In the next section we discuss the evolving roles of tutors assigned to teach one of the first year courses, called MICA I, where students are introduced to programming for mathematics learning. Our decision to focus on this course is based on two factors. For the students joining the Mathematics Department at Brock University, this course plays a central role in their transition from school to a university mathematics programme that integrates technology in an innovative way. For the tutors, it is the course that most explicitly affords an embodiment of the Department's philosophy.

The Evolving Roles of Tutors in an Innovative First-Year Course (MICA I) Based on Programming Technology with Interactive Interfaces

In this section, we discuss examples of teaching practice from the experience of tutors teaching the first-year course, MICA I, in the Brock University undergraduate core mathematics programme MICA. This programme was established in 2001, and aspects of it have been described in a number of publications. The philosophy and objectives of the Department that spurred the development of this programme are described by Ralph (2001). A review of all (traditional) courses of the programme under the umbrella of the new philosophy and principles was undertaken. For example, it was decided that Maple (Maplesoft n.d.) would be the standard CAS used overall in the MICA programme, and the introductory Calculus would entail the use of Maple and an engaging, interactive software called *Journey Through Calculus* (Ralph 1999). In addition, three innovative core courses, called MICA I, II, III, were introduced, which most concretely embed two of the programme's principles: (i) to encourage mathematical creativity, and (ii) to develop mathematics concepts hand-in-hand with computers. A fuller description of the programme and review can be found in Ben-El-Mechaiekh et al. (2007). Muller et al. (2009) have described how students enrolled in MICA I-III courses learn mathematics as they engage in designing, programming and using interactive, computer-based environments, called *Exploratory Objects* that are "interactive and dynamic computer-based model[s] or tool[s] that capitalise on visualisation and [are] developed to explore a mathematical concept or conjecture, or a real-world situation" (p. 65).

Marshall (2012b) has explored the differences and similarities between Exploratory Objects and Microworlds at the university level (e.g., Wilensky 1995). Marshall notes similarities of the two as a learning activity rather than as an end product by noting that,

> [a]n important aspect of the evolution of microworld idea, is that "although initial focus was on microworld as a digital object it quickly became apparent that it made much more sense to discuss the term in association with the kinds of activities emerging from the use of microworlds and the scope of each microworld with respect to the conceptual field it was designed to embody" (Healy and Kynigos 2010, as given by Kynigos 2012, p. 4). (p. 51)

He further notes that,

> Exploratory Objects are distinguishable from microworlds by the fact that the latter are designed (by tutors, teacher) with a didactical purpose for other (student) users to learn mathematics, whereas *Exploratory Objects* are entirely designed (by students) for self-use to do mathematics – students are provided with no initial object to base their constructions. (p. 51)

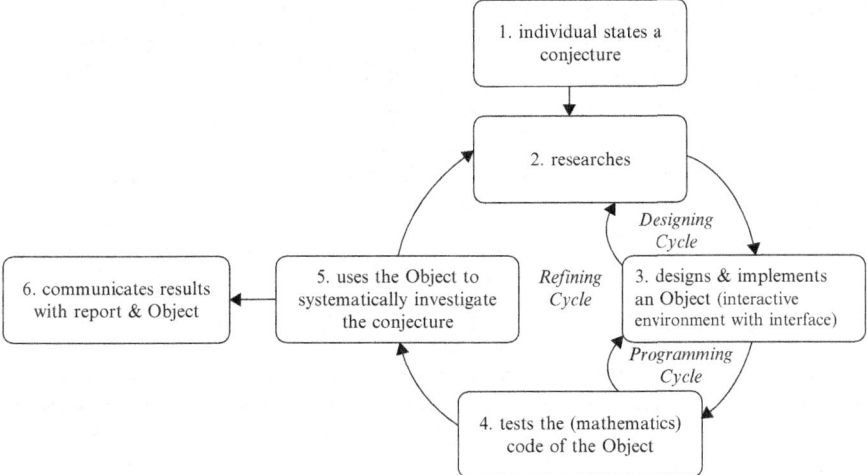

Fig. 1 Student development process of a mathematics Exploratory Object for the investigation of a conjecture as illustrated in Buteau and Muller (2010, p. 1113)

A preliminary task analysis led Buteau and Muller (2010) to provide insights of the learning activity through their student development process model of creating and using Exploratory Objects: see Fig. 1. Based on a literature review and using some empirical data for validation, Marshall (2012b) refined this diagram by (1) adding a step between Step 5 and Step 6 which he termed 'integrates results of the investigation with mathematical knowledge', and (2) adding an arrow from Step 5 to Step 1 to better describe the *Refining Cycle*.

To illustrate the student development process model, we select Profetto's (2005) Exploratory Object,[3] entitled *The Mandelbrot Set*. After having had a discussion with his tutors, second-year student, Adam Profetto, enrolled in MICA II course, was interested in investigating the Mandelbrot set. His initial aim was to visualise the set and its self-similarities, together with its related Julia Set. Since the Mandelbrot Set is defined by use of a recursive quadratic polynomial, Adam set his investigation to consider also higher degree polynomials (Step 1 in Fig. 1).

As mentioned in the *Special Thanks and Credit* section of his Exploratory Object, Adam researched his topic by use of a textbook (Step 2). He developed his Exploratory Object to visualise sets (overview) and to graphically explore the similarities through a manual zoom-in access: with the mouse or with manually entered complex plane coordinates. The related Julia Sets are displayed through a mouse-click on a point in the Mandelbrot set (Steps 3–4). Adam writes, "[This Exploratory Object project] opened up a world of interest for me and pushed my programming and patience to the max." Adam mentioned in an informal discussion that whilst exploring graphically the Mandelbrot Set with a systematic increase of the integer

---

[3] Both the Object and a summary of the written report (based on the original report submitted as an assignment for the course) are accessible via a web site (Brock Math n.d.).

exponent parameter (Step 5), he became intrigued as to how it affected the area of the bounded regions defined by the sets. He thereafter extended his Exploratory Object (Refining cycle in Fig. 1) to systematically calculate the area of the bounded regions, as the exponent increases, by repeating a Monte Carlo Integration approximation method.[4] Adam chose to have the results represented in a two-dimensional graph in *Observations and Findings* section of his Exploratory Object.[5] Adam proudly describes his findings in this section:

> Firstly, the area of the Mandelbrot Set was found to be Pi/2!! This is amazing.... This is where the magic starts to happen. As the exponent n is increased the areas of the generated regions will also start to increase. However, shortly after, they start to converge back to a fixed value. Can you guess what that fixed value is? Yes, that's right. Pi/2!!

His excitement was utmost after he searched on the Web about this result but said he could not find evidence of it as a known result. Adam (Profetto) therefore summarised this result as 'Profetto's Conjecture' not only in his written report, but also in his Exploratory Object (Step 6).

Students enrolled in MICA core courses I to III create 4–5 of such Exploratory Objects per term, including at least one about a topic of their own choice such as Adam's *Mandelbrot Set* Exploratory Object. Whereas MICA I concentrates on introducing students to learn mathematics by creating and using Exploratory Objects, MICA II and III[6] focus on furthering students' skills for more sophisticated mathematics Exploratory Objects and related mathematical investigations. In short, the MICA courses I to III could be recognised as courses embracing approaches of: experimental mathematics; inquiry-based learning; learning by using/modifying mathematics simulation; and learning mathematics by programming (Marshall 2012b). Programming their own exploration fosters experiences that initiate students into developing an intelligent partnership with technology (Martinovic et al. 2013).

Muller et al. (2009) describe a "different mathematics teaching paradigm that engages Brock University students from their first year into personalised original mathematics work" (p. 63). In this section we provide further analysis of the evolving role of tutors to assist students to reach a level of knowledge and confidence as demonstrated by Adam, to design, implement and use their own Exploratory Objects. Our analysis will focus specifically on the MICA I course in which first-year students are introduced to the technology (computer programming). It is organised according to a (12 week) course format: 2 weekly hours of lecture and 2 weekly hours of computer laboratory. The last 2 weeks are devoted to the final project, which is to produce an original Exploratory Object. In the analysis we will refer to the student's development process model (Fig. 1), and explain the tutors' roles as they have evolved over more than 10 years of implementation. The following classification of the role of tutors is based on insightful observation by the authors.

---

[4] This method had previously been covered in an Exploratory Object assignment.

[5] The web-version of Adam's original EO contains only a static graph summarising this experiment due to the intense computations involved and the time required to produce the results.

[6] MICA III has now evolved into two one-term optional courses recommended in the applied mathematics stream.

Muller was involved with the original design and implementation of the MICA programme. Buteau has taught the MICA I course for 8 years and, in this period has conducted and published studies with Muller on diverse issues about the course. For example, the discussion about the learning experiences of students when creating Exploratory Objects (Muller et al. 2009) and the description of the student process when creating an Exploratory Object as shown in Fig. 1 (Buteau and Muller 2010).

The Role of Tutors in MICA I Lectures

Tutors engage with students during MICA I course lectures mostly in direct relation to Step 1 (conjecturing), Step 2 (researching topic) and Step 6 (communicating results) of the development process model (Fig. 1). The mathematics content is introduced during lecture time as a means to support these three steps, but also as source and motivation for the learning of mathematics programming during the computer lab sessions. The pace is slow and highly interactive, which is not a traditional lecture format as it aims to maximise student engagement. Typically, topics in elementary number theory (e.g., prime numbers) and discrete dynamical systems are selected as they enable the purpose of the lab sessions to be served well, that is for students to learn a programming technology for investigating mathematics. Due to an increase in the student enrolment in the course over the years[7] and the requisite student engagement during lectures and lab sessions, the Department has decided that the only way to meet the objectives of the course is to limit the enrolment at 35 students per section. In the following we describe the lecture activities and identify teaching criteria related to all three steps (1, 2 and 6) of the student development process model.

Throughout the lectures, tutors prompt students, directly or indirectly, to conjecture, to reword conjecture(s), to reflect critically on what is of interest, and to evaluate the potential of finding evidence through the use of technology, whilst developing proper terminology in the realm of mathematical experimentation (conjecture, proof, counter-example, theorem, etc.). Conjecturing (Step 1 in Fig. 1) is at the heart of the MICA I course. In fact, it has long been promoted as an important activity in the mathematical problem solving process. For example, Mason et al. (1985) have argued that specialising, generalising, conjecturing and convincing are the components of thinking in problem solving. In another publication Mason (2002) points to the in-class tone that tutors need to provide,

> The essence of a scientific debate and of a conjecturing atmosphere is that people are eager to try out ideas and neither embarrassed nor ashamed to make a mistake: everything said is offered as a conjecture, with the intention of modifying it if necessary (in contrast to an ethos in which things are only said when the sayer is confident they are correct). (p. 78)

In the case of MICA I course, Martinovic et al. (2013) report on the tutor's role in developing a conjecturing atmosphere:

---

[7] In 2012: three sections of MICA I, for a total of 80 students. In 2002: there were nine students in total enrolled in the course.

> To help students over- come such a recipient role and to take a more active role "as mathematical thinkers" (Goos & Cretchley, 2004; Jonassen, 1996, 2006; Pea 1987; Willis & Kissane, 1989), class time is assigned to activities where students develop their competency in question-posing and conjecturing in mathematics. (p. 92)

Sometimes the conjecturing task is initiated in small groups or directly as a whole class discussion. In this dialogue, all ideas are welcome, treated as important, and are criticised in a constructive way, leading to a well-stated conjecture as the result of a whole class effort. At times the discussion evolves to consider potential experiments that could be designed to test the conjecture, and throughout which the mathematical experiment terms (conjecture, counter-example, proof, evidence, etc.) are carefully used (Step 6 in Fig. 1). Muller et al. (2009) have commented further on the tutor's role:

> Since students do not have any experience in raising open-ended mathematical questions, the professor's first goal is to set up fertile circumstances for this to happen. Faculty look for contexts that are experientially real for the students and can be used as starting points for progressive mathematisation (Gravemeijer 1999, p. 158) and introduce students to easily described mathematical problems that admit many possible points of view generating many different lines of inquiry. (p. 65)

For tutors with little prior teaching experience who are assigned to teach this course, this teaching approach can be a rather challenging teaching situation (Buteau and Muller 2006).

The goal in MICA I is to expand the student conjecture horizons by the potential use of technology. Other researchers have recently focused their attention on student conjecturing in undergraduate mathematics courses; for example Morselli (2006) reports on the 'Use of examples in conjecturing and proving' in the context of elementary number theory, while Burtch (2003) explores 'The evolution of conjecturing in a differential equations course'. In the MICA courses, the focus 'the individual states a conjecture' (Step 1 in Fig. 1) highlights the action that each student selects a conjecture that is of interest to him/her and is to be investigated with technology. The student's selected conjecture will be at the heart of his/her investigative process, and will be the focus of his/her productive and creative mathematics activity. As such, tutors must ensure that students develop their abilities to state conjectures, and to assess critically their relevance, which often involves some researching about the topic (Step 2 in Fig. 1). In MICA I, researching mostly involves using the internet, and the role of the tutor has evolved to address in lecture sessions the trust-worthiness, usefulness, and relevance of reference sources. Although there is some discussion about programming in the lectures the students develop their skills and knowledge of programming in their lab sessions, which is what we now move to consider.

The Role of Tutors in MICA I Lab Sessions

The overall purpose of the computer lab sessions within the MICA I course is for students to learn the basics of mathematics programming (Steps 3 and 4 of the student development process model; Fig. 1.), and for them to then use their programming skills to undertake mathematical investigations (Step 5). The Department takes care to stress that its students' primary interest is mathematics and not programming. Therefore as tutors teaching in MICA I: "We are conscious not to overload the students' cognitive effort required to learn the programming for their exploration of mathematics. This we do by motivating but limiting the amount of programming language for the mathematics task at hand" (Muller et al. 2009, p. 66). But how is this principle implemented in practice by tutors?

To guide our discussion in this section, we use the theoretical framework of *instrumental integration* (Assude 2007), based on the *instrumental approach* (Artigue 2002) that encompasses four growing stages of students' technology use in the classroom. "Instrumental integration is a means to describe how teachers organise the conditions for instrumental genesis of the technology proposed to the students and to what extent (s)he fosters mathematics learning through instrumental genesis" (Goos and Soury-Lavergne 2010, p. 313). Briefly, these four stages are: (i) *instrumental initiation* (stage 1) where students are engaged in strictly learning how to use the technology; (ii) *instrumental exploration* (stage 2) where mathematics problems serve as motivation for students to further learn to use the technology; (iii) *instrumental reinforcement* (stage 3) where students solve mathematics problems with the technology, but that requires students to extend their technology skills; and (iv) *instrumental symbiosis* (stage 4) takes place when the student's fluency with technology scaffolds the mathematical task resulting in an improvement of both the student's technology skills and his/her mathematical understanding. The identification of the stages throughout the lab activities will help characterise the course design decisions and the learning objectives that guide the tutors in these activities. We will structure the analysis around the three course assignments (Exploratory Object projects), that evolve in complexity, and the 2-h weekly labs that are pre-planned to prepare the students for each of these assignments.

The first four labs introduce Visual Basic.NET programming to students (stages 1–2), and to provide them with minimal, though sufficient, programming skills for their first experience at creating and using an Exploratory Object (stage 3) in their assignment 1. It is worth noting that in the first year that the MICA programme was offered the Department required a Computer Science Java course (stage 1). Students found it difficult to see the relevance of this course as it included few applications to mathematics. The following year the Java requirement was dropped and the Department revised the MICA I course to include programming in Visual Basic. NET. The development environment Microsoft Visual Studio (n.d.) supporting Visual Basic.NET language, was selected by the Department due to its user-friendliness. With this change students now progress through stages 1–3 of Instrumental Integration concurrently, and eventually to stage 4, which we describe next.

Most of our students come to the MICA I course without any programming background, which causes anxiety for many. In the first lab the tutor demystifies programming by having students first recreate a simple, fun interactive program. A detailed introductory Visual Basic.NET programming textbook (e.g., Halvorson 2010) is used by students throughout the course for the introduction to the basics of programming (stage 1). The first lab requires students to follow step-by-step, from the textbook, the code to re-create a user interface of the program. They use a user-friendly menu (both to select objects and change their properties) and re-type line by line the programming code (involving a decision structure, global and local variables, and a randomise command), which comes with explanation for each line (stage 1). The tutors and his/her two TAs, who help throughout the 12 labs, walk around the lab and talk to each individual. In this very first lab, they mostly acknowledge the good work, encourage the enjoyment of the task, and may need to help. Over the years, we have observed the surprise, pride and somewhat relief of students as they walk out of their first MICA I programming lab session.

The subsequent three lab sessions introduce students to the first concepts of mathematics programming: variables, decision structures, and loops. For example, students create a program to test whether a positive integer is prime and are asked to use and modify it through some (strictly mathematical) exercise questions. These labs all have a similar format (as do lab sessions 5–9) in that the tutor gives a brief introduction to the whole class about the main programming concepts in a mathematical context, thus establishing an initial connection between mathematics and the new programming concepts (stage 2). Each student then works individually on the details and exercises by reading through the textbook (stage 1). Also the tutor provides additional programming exercises that have been put in a mathematical context (stage 2). The use of the textbook encourages students to develop as independent learners, as does the use of the debug tool (embedded in Visual Studio), which the tutor introduces to the students early in the course. Tutors are aware that programming provides a means of self-assessment for the students. Most computer programming software is unforgiving in that the code is either correct or the system records an error, so a knowledge of how to use the debug tool correctly can prevent much (technology related) frustration.

The mathematical focus of the course is stressed through the tutor's whole class interventions at various times during the labs. These constantly emphasise the mathematical motivation, aim and context for programming (stages 2–3). In particular, tutors frequently remind students to check whether the results match expectation (stage 2). TAs provide help to students not only with their technical questions (stage 1), but they are also asked to frequently remind students of the course's mathematical purpose (stage 3).

The first four lab sessions lead to the first assignment, when it is expected that the students have learned enough programming concepts to engage in an original mathematical experiment with technology (stages 2–3). This assignment aims to strengthen Steps 1, 3–4 (Fig. 1), and provides an initiation of the whole process shown in Fig. 1. Students each select a question or conjecture that has been raised and discussed in class that is of particular interest to them. Using their Visual Basic.NET programming abilities, they explore the question or conjecture and report their findings. If needed, the tutor assists students on an individual basis to select a conjecture and also ensures that the level of programming required for the exploration is within the student's capabilities. Here the tutor's aim is to stimulate an exciting first Exploratory Object experience for students and avoid a negative and frustrating one. This individual guidance by tutors for the first assignment, although time demanding and considered as a non-traditional role for first-year courses, is crucial to keep and build further students' confidence and motivation in learning programming for the purpose of doing mathematics. When the tutor dialogues with individual students, s/he also stresses the importance of their written report to communicate their results, reiterating that the assignment is a mathematical one (stage 3).

The second series of lab sessions (labs 5–7) aims at strengthening and furthering students' basic programming skills within a mathematical context (stage 2) with an emphasis on arrays, functions and procedures, and modules. At this stage the tutor usually selects the topic of Modular Arithmetic and RSA encryption that, due to its high computational demand, requires the use of technology for any meaningful application. The tutor provides students with mathematics problems that are more conceptually difficult to implement (stage 2). The second assignment aims to strengthen Steps 3 and 4 of the development process model (Fig. 1) through an implementation of the RSA encryption method, with application (stage 3).

Labs 8–10 are designed around Steps 3 and 5 of the development process model. The foci in this part of the lab sessions are (i) the need for, implementation, and use of multiple representations of concepts, and (ii) the ability to investigate mathematics systematically. At this juncture tutors usually select the study of discrete dynamical systems. Students learn the basics of programming graphics (stage 1): to pay attention to integrating parameters in their computer environments; to create rich user interfaces appropriate to gather relevant data (stage 2); and to approach investigations systematically. Students reinforce their graphics programming skills by creating an Exploratory Object for the numerical and graphical investigation of the dynamical system based on the logistic function (stage 2). The last lab session (lab 10) is converted into a whole-class interactive session of mathematics investigation by use of the Exploratory Objects they created. The tutor guides students step-wise in the investigation through the use of questions such as: What do you expect?; Re-write in the form of a conjecture; How should you perform the experiment (how do you record results? What to do with the parameters? etc.); What could be interesting to note? How do you write a result (what is the proper wording)? The third assignment aims to strengthen Steps 3 and 5 of the development process model (Fig. 1) through the investigation of a discrete dynamical system involving three parameters, and requiring adapting the range of the functions (stage 3).

To summarise, the roles of the tutors in the MICA I labs, complemented by the assignments, are:

(a) To provide students with an inspiring, meaningful mathematical context that fosters the need and motivation to learn the programming basics (i.e. to move quickly from stage 1 towards stage 2);
(b) To keep emphasising the aim of course, which is to do mathematics with technology (stages 3 and 4).

The tutors work to provide an environment in which students interact constantly with one another, with TAs, and with tutors to develop their programming skills and to build their confidence to explore mathematics using technology.

Role of Tutors in Final Projects Phase

The last 2 weeks of the course are completely dedicated to the students' original final projects. Individually or with a partner the students select their own topic of investigation, and create and use an Exploratory Object about their topic (i.e., students experience the whole process described in Fig. 1), as did Adam with his exploration of Mandelbrot Set.[8] For the students, the final project is the highlight of the course. Lecture and lab sessions all take place in a computer laboratory. The role of tutors at this stage is mainly to provide individual guidance. The tutors motivate each student or pair of students to search for a topic of great interest to them and discuss with them the level of difficulty of programming requirements for their planned Exploratory Object. The tutors explicitly encourage creativity and uniqueness of their mathematical work in terms of mathematics, programming, and artistry. Tutors also inspire them to reach beyond their skill and knowledge boundaries (programming-wise, this points to stage 3, possibly stage 4). For the tutors this also involves prompting students to think about, rather than trying to avoid, multiple representations of the mathematical concepts.

At this stage the students would have covered all of the basics of mathematics programming. However, from our experience, a large majority of students feel the need to extend their programming skills in order to fulfill their own vision of their mathematical investigation (stages 3 and 4). Whereas the tutors and TAs may provide assistance to students about programming issues, often students deal with the latter either by using external resources (textbook, help features embedded in Visual Studio, Internet) or by asking other students. Overall, at the final project stage, the tutors continue to encourage the students' development of their intelligent partnership with technology in doing and learning mathematics (Martinovic et al. 2013).

The creation and use of Exploratory Objects by students, for their mathematics exploration, requires not only programming know-how, but also the design and use of interactive, dynamic graphical interfaces. For the students to succeed they need to be creative as they develop mathematics that is new to them and they must apply their creativity, particularly in their design of interfaces. This topic, creativity, informs the next short section.

Creativity, Technology and Educating Students for Their Future

Ervynck (1991) in his chapter on *Mathematical Creativity* advises that,

> Creativity plays a vital role in the full cycle of advanced mathematical thinking. It contributes in the first stages of development of a mathematical theory when possible conjectures are framed as a result of the individual's experience of the mathematical context. (p. 42)

---

[8] For their final project, future teachers may choose to create and test a so-called *Learning Object* for the learning of school mathematics concepts (Muller and Buteau 2006; Muller et al. 2009; Buteau and Muller 2010). See the web site (Brock Math n.d.) for other examples of student projects.

In addition, Holton (2005), when reflecting on the nature of mathematics and its teaching at the university level, proposes,

> [Mathematics] is a living and breathing entity that students can participate in. It is a subject that can involve their creative abilities and a subject where discussion is valuable [...] It seems to me axiomatic that university lecturers must take the lead in this movement to 'redefine' mathematics, to make the creative side more visible to our students. (pp. 306–7)

In line with Ervynck and Holton, we note that one of the guiding principles of the MICA programme is 'Encourage Creativity and Intellectual Independence' (Ralph 2001, p. 1). For the mathematics department at Brock University, creativity is seen as a human talent that needs to be developed when students are learning mathematics. In the MICA programme technology is used to motivate students to be creative by building on their existing mathematical knowledge as they create mathematics that is new to them. We briefly elaborate on the type of environment and student experiences provided by tutors in the MICA I course that aim to stimulate the student creative mathematical abilities. To guide our discussion, we use Ervynck's (1991) proposed three stages for development of mathematical creativity, namely:

1. A preliminary technical stage.
2. An algorithmic activity stage.
3. A creative (conceptual, constructive) activity stage.

It is our opinion that by the time students complete their final project in MICA I, they will have been mentored and have followed a number of phases that can be mapped onto these three stages. In a preliminary technical stage (1) students will have raised and explored, as a class and individually, problems in areas of mathematics that they can tackle with their school mathematical knowledge. In that stage they will also have started investigating the problems using simple programming. In an algorithmic stage (2) Ervynck envisages that the students will perform mathematical techniques. In MICA I the students will have implemented their own programs, thereby using the computer to perform many of the mathematical tasks. It is at the end of this stage that the 'what if' questions arise and students modify their programs to extend their mathematical understanding. Students enter a creative (conceptual, constructive) activity stage (3) as they start to question their mathematical conjecture, to query the results that their program has generated, to use their Exploratory Object to reinforce or modify their mathematical understanding, etc. This stresses that for students, this is more than the mere use of programming technology to solve a particular mathematics problem. The interactive, dynamic user-friendly interface that they design for themselves, enables them to enter a dialogue with technology, that they control from the programming code, to serve their evolving (through the refining cycle in Fig. 1) mathematical investigations. This involves, for example, addition/deletion of parameters, modification of initial conjecture, new representations of results, etc. (Buteau and Muller 2010). The Exploratory Object that the students have created becomes a trace of the development of their understanding of the mathematics involved.

Within MICA I, one way that tutors can be successful is to allow the students to develop their own creative mathematical abilities by mentoring and tutoring the students as they work through the various steps in the development model (Fig. 1). This course, and MICA II-III, can provide the students with an insight into how some mathematicians do research. A graduate of the MICA programme expressed it as follows: "Conjecturing, designing mathematical experiments, running simulations, gathering data, recognising patterns and then drawing conclusions are things many modern mathematicians do as part of their research" (Marshall 2012a, p. 72). Furthermore, the MICA programme goes further and engages students also to communicate their understanding in a creative environment, namely through their designing and implementing of interactive mathematical objects.

Conclusion

At the university level, tutors play a dual role in teaching as teachers and as policy makers within a departmental structure. When integrating technology into their courses tutors can do so on an individual basis but they should also be concerned about the continuity of the student experience within their mathematics programmes (MAA 2004). The latter and the support for systemic change or innovation in university mathematics education is the responsibility of the mathematics department, i.e., of tutors in their roles as policy makers (MAA 2004). As previously noted, in the Canadian survey study about the integration of CAS in university teaching (Buteau et al. forthcoming), 54 % of participants expressed the view that the lack of departmental support is a factor hindering CAS integration in teaching. This indicates the need for research into the roles that a department should play to ensure continuity of student mathematics learning experiences while using technology.

We have used the MICA programme as an example of a departmental initiative that led to an innovative integration of programming technology in a number of core courses. In their survey study, Buteau et al. (forthcoming) found that programming technology[9] was the second most used mathematics technology by tutors in their research work after CAS. However surprisingly, they also found that amongst all the surveyed mathematics technologies,[10] only programming was not integrated to the same extent in research and in teaching. It was reported being used about twice as often in research (43 %) as it is in teaching (18 %). Buteau et al. (forthcoming) continue,

> This naturally raises the question as to why programming is relatively absent in mathematics teaching [...] One could argue that the learning curve for students' use of programming is much steeper than for many other technologies that feature more user-friendly graphical interfaces. (p. 13)

---

[9] The survey question indicated, "Programming (Java, C++, Fortran, ...)".
[10] See footnote 3.

For programming, the MAA (2004) recommendation that, "Using the same software in several different courses also can shorten the total technology learning curve" (p. 23) may be even more important than for other technologies. As a consequence a departmental policy could be key for individual integration of programming in teaching. Since programming can play a significant role in students' learning in mathematics (Dubinsky and Tall 1991; Wilensky 1995; Dubinsky and McDonald 2002; Abrahamson et al. 2006), we contend that mathematics departments should explore ways in which programming could be integrated in their programmes.

The teaching roles of tutors, during lectures and lab sessions, in a particular first year core mathematics course (MICA I at Brock University) that integrates programming technology are non-traditional. The different teaching paradigm is dictated by the nature of the course that cannot be described by a list of mathematical concepts and techniques to be covered, but rather through a method that is made possible by technology. The approach is for students to do mathematics by designing, programming, and using mathematics Exploratory Objects to explore conjectures, concepts, or real-world situations. Unlike traditional university mathematics courses, MICA I lecture notes are unquestionably insufficient when passing on the baton to another tutor: firstly, the means whereby students learn mathematics 'partnering' with technology has to be well understood, and secondly, the teaching aims and strategies used, both in lecture and in lab sessions, need to be carefully described. In particular, the description should stress the shift of the role of the tutor from 'lecturer' to 'mentor'. With the experimental aspect of the learning activity, i.e., conjecturing, designing and conducting experiments to test conjectures, etc., mathematics and programming technology need to be carefully intertwined, thereby providing an opportunity for students to develop their creativity both in mathematics and in their communication of their understanding of mathematics. In his reflections on the many contributions in the book *Advanced Mathematical Thinking*, Tall (1991) writes,

> We therefore arrive at a possible new synthesis in teaching and learning advanced mathematics which offers a more complete cycle of advanced mathematical thinking to students, even those with modest abilities. The active participation in thinking is essential for the personal construction of meaningful concepts. Students need to be challenged to face the cognitive reconstruction explicitly, through conjecture and debate, through problem solving, and they may be assisted in the acquisition of insights at higher levels by selectively sharing the construction with the computer. (p. 258)

The MICA programme provides one example where students are challenged to face the cognitive reconstruction explicitly through conjecture and debate, through problem solving and simulation and where they do share the construction of their mathematical object with the computer. However, MICA provides only one example of a mathematics programme where a department is addressing the role of technology in mathematics teaching and learning. Other examples can be found but many more examples of departments that integrate technology in a systemic way are needed. This will make it possible for departments and tutors to evaluate, compare and decide what path to follow in their own university.

An interesting observation from the Canadian CAS survey study (Buteau et al. forthcoming) is that a large majority (93 %) of mathematician participants, who are active in research, indicated that they used at least one form of mathematics technology[11] in their research work. Furthermore, the use of CAS in research was found to be the strongest factor influencing the use of CAS in teaching (Lavicza 2010; Buteau et al. forthcoming). If the results of the Canadian CAS survey can be projected over time, then any growth in the number of mathematicians who use technology in their research should be reflected in an increase of mathematics courses where technology is integrated.

We have touched on a few, of many, issues that face mathematics tutors and their departments as they integrate technology into undergraduate mathematics education. One issue that we have not raised is how technology may force a redefinition of mathematics. Noss (1999), in the study we have previously referred to, concludes,

> But the issue is not simply one of pedagogic innovation, it is one of what counts as mathematics. As stated at the outset, the opportunity to examine two different design decisions and their outcomes actually revealed fundamental differences in the way mathematical knowledge was conceived. This, perhaps is the main contribution of new technology in mathematical teaching and learning: it provides us with an opportunity to reassess not simply only how we teach, or even how students learn, but what it is that we teach them and why. (p. 388)

In the same vein, Lavicza (2010) in his international survey study on CAS use in university mathematics instruction (on which the Canadian CAS survey was built), recommends:

> [M]athematicians accept that CAS is part of the literacy, but at the same time they are reluctant to accept that CAS shapes mathematical knowledge. This disparity is possibly derived from the mismatch between mathematicians' CAS-related and mathematical beliefs... a closer examination of th[e] relationship between these conceptions would be beneficial. (p. 111)

The questions raised by both Noss and Lavicza are in the minds of tutors who integrate CAS and other technologies in their classes. Tutors who use technology extensively in their research and who communicate their research methods and results in their courses are likely to be the ones who exemplify how technology is reshaping mathematical knowledge.

References

Abrahamson, D., Berland, M., Shapiro, B., Unterman, J., & Wilensky, U. (2006). Leveraging epistemological diversity through computer-based argumentation in the domain of probability. *For the Learning of Mathematics, 26*(3), 19–45.

Artigue, M. (2002). Learning mathematics in a CAS environment: The genesis of a reflection about instrumentation and the dialectics between technical and conceptual work. *International Journal of Computers for Mathematics Learning, 7*(3), 245–274.

---

[11] See footnote 2.

Assude, T. (2007). Teachers' practices and degree of ICT integration. In D. Pitta- Pantazi & G. N. Philippou (Eds.), *Proceedings of the fifth congress of the European Society for Research in Mathematics Education* (pp. 1339–1348). Larnaka: Department of Education, University of Cyprus.

Ben-El-Mechaiekh, H., Buteau, C., & Ralph, W. (2007). MICA: A novel direction in undergraduate mathematics teaching. *Canadian Mathematics Society Notes, 39*(6), 9–11.

Brock Math. (n.d.). http://www.brocku.ca/mathematics-science/departments-and-centres/mathematics/undergraduate-programs/mica/student-learning-objects

Burtch, M. (2003). The evolution of conjecturing in a differential equations course. Retrieved from http://mcli.maricopa.edu/book/export/html/1024

Buteau, C., & Muller, E. (2006, December 3-8). Evolving technologies integrated into undergraduate mathematics education. In L. H. Son, N. Sinclair, J. B. Lagrange, & C. Hoyles (Eds.), *Proceedings for the seventeenth ICMI study conference: Digital technologies and mathematics teaching and learning: Revisiting the terrain* (8 pp.). Hanoi: Hanoi University of Technology (c42)[CD-ROM].

Buteau, C., & Muller, E. (2010). Student development process of designing and implementing exploratory and learning objects. *Proceedings of the sixth conference of European Research in Mathematics Education* (pp. 1111–1120). Lyon. Retrieved from http://www.inrp.fr/editions/editions-electroniques/cerme6/working-group-7

Buteau, C., Marshall, N., Jarvis, D., & Lavicza, Z. (2010b). Integrating computer algebra systems in post-secondary mathematics education: Preliminary results of a literature review. *International Journal for Technology in Mathematics Education, 17*(2), 57–68.

Buteau, C., Jarvis, D., & Lavicza, Z. (forthcoming). On the integration of computer algebra systems (CAS) by Canadian mathematicians: Results of a national survey. Accepted for publication in *Canadian Journal of Science, Mathematics and Technology Education*.

Chae, S., & Tall, D. (2001). Construction of conceptual knowledge: The case of computer-aided exploration of period doubling. *Research in Mathematics Education, 3*(1), 199–209.

Dubinsky, E., & McDonald, M. (2002). APOS: A constructivist theory of learning. In D. Holton (Ed.), *The teaching and learning of mathematics at university level: An ICMI study* (pp. 275–282). Dordrecht: Kluwer.

Dubinsky, E., & Tall, D. (1991). Advanced mathematical thinking and the computer. In D. Tall (Ed.), *Advanced mathematical thinking* (pp. 231–248). Dordrecht: Kluwer.

Elliott, P. (1976). Programming – an integral part of an elementary mathematics methods course. *International Journal of Mathematical Education in Science and Technology, 7*(4), 447–454.

Ervynck, G. (1991). Mathematical creativity. In D. Tall (Ed.), *Advanced mathematical thinking* (pp. 42–53). Dordrecht: Kluwer.

Goos, M., & Cretchley, P. (2004). Teaching and learning mathematics with computers, the internet and multimedia. In B. Perry, G. Anthony, & C. Diezmann (Eds.), *Research in mathematics education in Australasia 2000–2003* (pp. 151–174). Flaxton, Queensland, Australia: Post Pressed.

Goos, M., & Soury-Lavergne, S. (2010). Teachers and teaching: Theoretical perspectives and classroom implementation. In C. Hoyles & J.-B. Lagrange (Eds.), *ICMI Study 17, technology revisited, ICMI study series* (pp. 311–328). New York: Springer.

Gravemeijer, K. (1999). How emergent models may foster the constitution of formal mathematics. *Mathematical Thinking and Learning, 1*(2), 155–177.

Halvorson, M. (2010). *Microsoft Visual Basic 2010 step by step* (p. 579). Washington, USA: Microsoft Press.

Healy, L., & Kynigos, C. (2010). Charting the microworld territory over time: Design and construction in mathematics education. *ZDM, 42*(1), 63–76. doi:10.1007/s11858-009-0193-5.

Holton, D. (Ed.). (2001). *The teaching and learning of mathematics at university level* (ICMI study series: New ICMI study series, Vol. 7, p. 560). Dordrecht/Boston/London: Kluwer Academic Publishers.

Holton, D. (2005). Tertiary mathematics education for 2024. *International Journal of Mathematical Education in Science and Technology, 36*(2–3), 305–316.

Howson, A. G., & Kahane, J. P. (Eds.). (1986). *The influence of computers and informatics on mathematics and its teaching* (ICMI study series, Vol. 1, p. 155). Cambridge, UK: Cambridge University Press.

Hoyles, C., & Lagrange, J.-B. (Eds.). (2010). *Mathematics education and technology – rethinking the terrain: The 17th ICMI study* (p. 494). Springer: New York.

Jonassen, D. H. (1996). *Computers in the classroom: Mindtools for critical thinking*. Englewood Cliffs, NJ: Prentice-Hall.

Jonassen, D. H. (2006). *Modeling with technology: Mindtools for conceptual change* (3rd ed.). Upper Saddle River, NJ: Merrill.

Karian, Z. A. (Ed.). (1992). *Symbolic computation in undergraduate mathematics education* (MAA Notes, Vol. 24, p. 200). Washington, DC: Mathematical Association of America.

Keynes, H., & Olson, A. (2001). Professional development for changing undergraduate mathematics instruction. In D. Holton (Ed.), *The teaching and learning of mathematics at the university level: An ICMI study* (pp. 113–126). Dordrecht: Kluwer.

King, K., Hillel, J., & Artigue, M. (2001). Technology – a working group report. In D. Holton (Ed.), *The teaching and learning of mathematics at university level: An ICMI study* (pp. 349–356). Dordrecht: Kluwer.

Kynigos, C. (2012). Constructionism: Theory learning or theory of design? *Proceedings of the 12th International Congress on Mathematical Education* (ICME 12), 8–15 July 2012, Seoul (Korea). 24 pp.

Lagrange, J. B., Artigue, M., Laborde, C., & Trouche, L. (2003). Technology and mathematics education: Multidimensional over- view of recent research and innovation. In F. K. S. Leung (Ed.), *Second international handbook of mathematics education* (Vol. 1, pp. 237–270). Dordrecht: Kluwer.

Lavicza, Z. (2010). Integrating technology into mathematics teaching at the university level. *ZDM, 42*, 105–119.

Maplesoft. (n.d.). http://www.maplesoft.com/

Marshall, N. (2012a). Simulation and Brock University's MICA Program – reflections of a graduate. In E. R. Muller, J-P. Villeneuve & P. Etchecopar (Eds.), Using simulation to develop students' mathematical competencies – post secondary and teacher education, *Proceedings of Canadian Mathematics Education Study Group/GCEDM 2011 meeting* (pp. 59–76).

Marshall, N. (2012b). *Contextualizing the learning activity of designing and experimenting with interactive, dynamic mathematics exploratory objects*. Unpublished M.Sc. thesis, Brock University, St.Catharines.

Martinovic, D., Muller, E., & Buteau, C. (2013). Intelligent partnership with technology: Moving from a mathematics school curriculum to an undergraduate program. *Computers in the Schools, 30*(1–2), 76–101.

Mason, J. H. (2002). *Mathematics teaching practice – a guide for university and college lecturers*. Chichester: Horwood Publishing.

Mason, J., Burton, L., & Stacey, K. (1985). *Thinking mathematically*. Wokingham: Addison-Wesley.

Mathematical Association of America. (2004). In W. Barker et al. (Eds.), *Undergraduate programs and courses in the mathematical sciences: CUPM curriculum guide*. Washington, DC: Mathematical Association of America.

Mgombelo, J. R., Orzech, M., Poole, D., & René de Cotret, S. (2006). Report of the CMESG working group: Secondary mathematics teacher development (La formation des enseignants de mathématiques du secondaire). In L. Peter (Ed.), *Proceedings of the annual meeting of Canadian Mathematics Education Study Group (CMESG)*, University of Calgary, Calgary

Microsoft Visual Studio. (n.d.). http://www.microsoft.com/visualstudio/fr-ca

Morselli, F. (2006). Use of examples in conjecturing and proving: An exploratory study. In Novotnà et al. (Eds.), *Proceedings of the 30th conference of the international group for the Psychology of Mathematics Education* (Vol. 4, pp. 185–192), Prague.

Muller, E., Buteau, C., Ralph, B., & Mgombelo, J. (2009). Learning mathematics through the design and implementation of Exploratory and Learning Objects. *International Journal for Technology in Mathematics Education, 16*(2), 63–73.

National Science Foundation. (1996). *Shaping the future: New expectations for undergraduate education in science, mathematics, engineering, and technology*. Arlington: National Science Foundation.

Noss, R. (1999). Learning by design: Undergraduate scientists learning mathematics. *International Journal of Mathematical Education in Science and Technology, 30*(3), 373–388.

Pea, R. D. (1987). Cognitive technologies for mathematics education. In A. H. Schoenfeld (Ed.), *Cognitive science and mathematics education* (pp. 89–122). Hillsdale: Lawrence Erlbaum.

Profetto, A. (2005). *Mandelbrot Set* Exploratory Object. http://www.brocku.ca/mathematics/resources/learningtools/learningobjects/mathobjects/mandelbrot/executable/mainpage.swf

Ralph, B. (1999). *Journey through Calculus*. Pacific Grove: Brooks/Cole/Thomson Learning. (CD)

Ralph, B. (2001). Mathematics takes an exciting new direction with MICA program. *Brock Teaching, 1*(1), 1. Retrieved October 30, 2011 from http://www.brocku.ca/webfm_send/18483

Schurrer, A., & Mitchell, D. (1994). Technology and the mature department. *Electronic Proceedings of the 7th international conference on Technology in Collegiate Mathematics*. Retrieved June 29, 2012 from http://archives.math.utk.edu/ICTCM/VOL07/C002/paper.txt

Tall, D. (1991). Reflections. In D. Tall (Ed.), *Advanced mathematical thinking* (pp. 251–259). Dordrecht: Kluwer.

Wilensky, U. (1995). Paradox, programming, and learning probability: A case study in a connected mathematics framework. *The Journal of Mathematical Behavior, 14*(2), 253–280.

Willis, S., & Kissane, B. (1989). Computer technology and teacher education in mathematics. In *Department of employment, education and training, discipline review of teacher education in mathematics and science* (Vol. 3, pp. 57–92). Canberra, Australia: Australian Government Publishing Service.

Wolfram Mathematica. (n.d.). http://www.wolfram.com/mathematica/

Part II
Instrumentation of Digital Resources in the Classroom

Digital Technology and Mid-Adopting Teachers' Professional Development: A Case Study

Paul Drijvers, Sietske Tacoma, Amy Besamusca, Cora van den Heuvel, Michiel Doorman, and Peter Boon

Abstract The integration of digital technology into secondary mathematics education is not yet a widespread success. As teachers are crucial players in this integration, an important challenge is not only to attract early adopters, but also to support mid-adopting teachers in their professional development on this point. The questions addressed in this Chapter are: which practices such mid-adopting teachers develop when starting to use technology in their mathematics classroom; and how these practices change over time while engaging in a project with colleagues and researchers. To answer these questions, theoretical notions of instrumental orchestration, TPACK and community of practice underpin the case study of two mathematics teachers from a group of twelve, who engaged in a project on technology-rich teaching. The data includes lesson observations, blogs and results from questionnaires. The results show the type of teaching practices the teachers develop and the changes in these practices. Even if these changes are modest and the impact of the community is limited, the teachers clearly became more confident in integrating technology in their teaching.

Keywords Algebra • Community of practice • Digital resources • Geometry • Instrumental orchestration • Professional development • TPACK

P. Drijvers (✉) • S. Tacoma • A. Besamusca
C. van den Heuvel • M. Doorman • P. Boon
Freudenthal Institute for Science and Mathematics Education,
Utrecht University, Utrecht, Netherlands
e-mail: p.drijvers@uu.nl

Introduction

Nowadays, digital technology plays an important role in both personal and professional life. For several decades its potential for mathematics education in particular has been widely recognised. For example, NCTM's position statement claims that "Technology is an essential tool for learning mathematics in the 21st century, and all schools must ensure that all their students have access to technology" (NCTM 2008).

In spite of this, the integration of digital technology into secondary mathematics education lags behind the high expectations that many researchers and educators may have had in the past. It seems that the integration of digital technology into mathematics education is not at all to be taken for granted and that its success depends on several, sometimes complex and subtle factors (Artigue et al. 2009). One of these factors is the teacher. Teachers are considered as crucial players in education, and their ability to exploit the opportunities that technology offers determines to a high extent the success of the integration of digital technology in mathematics education. While integrating technology, teachers are confronted with new, sometimes destabilising situations, which challenge their existing teaching practices and may invite the development of a new repertory of appropriate teaching practices for these technology-rich settings (Doerr and Zangor 2000; Lagrange and Ozdemir Erdogan 2009; Ruthven 2007).

Of course, there are skilled and enthusiastic teachers who easily assimilate new technological developments in their teaching, who are able to deal with technological obstacles, and who are the early adopters of new tools as well as designers of new pedagogies. These 'frontline teachers' form an important minority for the design of teaching materials and the development of good practices. Meanwhile, the main challenge for integrating technology in regular mathematics education is not to attract these early adopters but, rather, to disseminate their experiences and to convince and support mid-adopting teachers, who are less experienced and less convinced of the benefits of ICT. For a widespread integration, these mid-adopters are the critical group.

The issue at stake, therefore, is how mid-adopting teachers may engage in a process of professional development concerning the integration of digital technology and the development of appropriate teaching techniques.

Theoretical Framework

The study's theoretical framework consists of three main components: the notion of instrumental orchestration to describe teachers' practices, the TPACK model to describe the teachers' skills, and the theory on communities of practice to investigate the impact of participating in a collegial community on teachers' professional development.

Instrumental Orchestration

The notion of instrumental orchestration emerges from the so-called instrumental approach to tool use, in which artefacts are expected to mediate human activity in carrying out a task. To describe the teacher's role in guiding students' acquisition of tool mastery and their learning processes, Trouche (2004) introduced the metaphor of instrumental orchestration. An *instrumental orchestration* is the teacher's intentional and systematic organisation and use of the various artefacts available in a learning environment – in this case a computerised environment – in a given mathematical task situation, in order to guide students' instrumental genesis (Trouche 2004). Within an instrumental orchestration, we distinguish three elements: a didactic configuration, an exploitation mode and a didactical performance (Drijvers 2012; Drijvers et al. 2010).

A *didactical configuration* is an arrangement of artefacts in the environment or, in other words, a configuration of the teaching setting and the artefacts involved in it. In the musical metaphor of orchestration, setting up the didactical configuration can be compared with choosing the musical instruments to be included in the band, and arranging them in space so that the different sounds result in polyphonic music, which in the mathematics classroom might come down to a sound and converging mathematical discourse.

An *exploitation mode* is the way the teacher decides to exploit a didactical configuration for the benefit of his or her didactical intentions. This includes decisions on the way a task is introduced and worked through, on the possible roles to be played by the artefacts and on the schemes and techniques to be developed and established by the students. In terms of the metaphor of orchestration, setting up the exploitation mode can be compared with determining the partition for each of the musical instruments involved, bearing in mind the anticipated harmonies to emerge.

A *didactical performance* involves the ad hoc decisions taken by teaching on how to actually perform in the chosen didactic configuration and exploitation mode: what question to pose, how to do justice to (or to set aside) any particular student input, how to deal with an unexpected aspect of the mathematical task or the technological tool, or other emerging goals. In the metaphor of orchestration, the didactical performance can be compared to a musical performance, in which the actual interplay between conductor and musicians reveals the feasibility of the intentions and the success of their realisation.

In a study on the use of applets for the exploration of the function concept in grade 8, the instrumental orchestration lens was used to describe observed teaching practices (Drijvers 2012; Drijvers et al. 2010). Six orchestrations for whole class teaching were identified, and a seventh for the setting in which students work individually or in pairs with technology. As this categorisation, which does not claim completeness, is the point of departure for the study presented here, we now summarise the seven orchestrations.

- The *Technical-demo* orchestration concerns the demonstration of tool techniques by the teacher. It is recognised as an important aspect of technology-rich teaching (Monaghan 2004). A didactical configuration for this orchestration includes access to the technology, facilities for projecting the computer screen and a classroom arrangement that allows the students to follow the demonstration. As exploitation modes, teachers can demonstrate a technique in a new situation or task, or use student work to show new techniques in anticipation of what will follow.

- In the *Link-screen-board* orchestration, the teacher stresses the relationship between what happens in the technological environment and how this is represented in the conventional mathematics of paper, book and board. In addition to access to the technology and projection facilities, the didactical configuration includes a board and a classroom setting so that both screen and board are visible. The teachers' exploitation modes may take student work as a point of departure or start with a task or problem situation they set themselves.
- The *Discuss-the-screen* orchestration concerns a whole-class discussion about what happens on the computer screen. The goal is to enhance collective instrumental genesis. A didactical configuration once more includes access to the technology and projecting facilities, preferably access to student work and a classroom setting favourable for discussion. As exploitation modes, student work, a task, a problem or an approach set by the teacher can serve as the point of departure for student reactions.
- The *Explain-the-screen* orchestration concerns whole-class explanation by the teacher, guided by what happens on the computer screen. The explanation goes beyond techniques and involves mathematical content. Didactical configurations can be similar to the Technical-demo ones. As exploitation modes, teachers may take student work as a point of departure for the explanation, or start with their own solution for a task.
- In the *Spot-and-show* orchestration, student reasoning is brought to the fore through the identification of interesting student work during the preparation of the lesson and its deliberate use in a classroom discussion. Besides previously mentioned features, a didactical configuration includes access to the students' work in the technological environment during lesson preparation. As exploitation modes, teachers may have the students whose work is shown explain their reasoning, and ask other students for reactions, or may provide feedback on the student work.
- In the *Sherpa-at-work* orchestration, a so-called Sherpa student (Trouche 2004, 2005) uses the technology to present his or her work, or to carry out actions the teacher requests. A didactical configuration includes access to the technology and projecting facilities, preferably access to student work and a classroom setting favourable for interaction. The classroom setting should be such that the Sherpa student can be in control of using the technology, with all students able to follow the actions of both Sherpa student and teacher easily. As exploitation modes, teachers may have work presented or explained by the Sherpa student, or may pose questions to the Sherpa student and ask him/her to carry out specific actions in the technological environment.
- In the *Work-and-walk-by* orchestration, the didactical configuration and the corresponding resources basically consist of the students sitting at their technological devices, and the teacher walking by in the classroom. In some cases a data projector or whiteboard may be available. As exploitation mode, the students work individually or in pairs on the tasks. The teacher answers students' questions and monitors their progress. In answering questions, the teacher may use the board or the projector, but often there is just individual interaction between teacher and student.

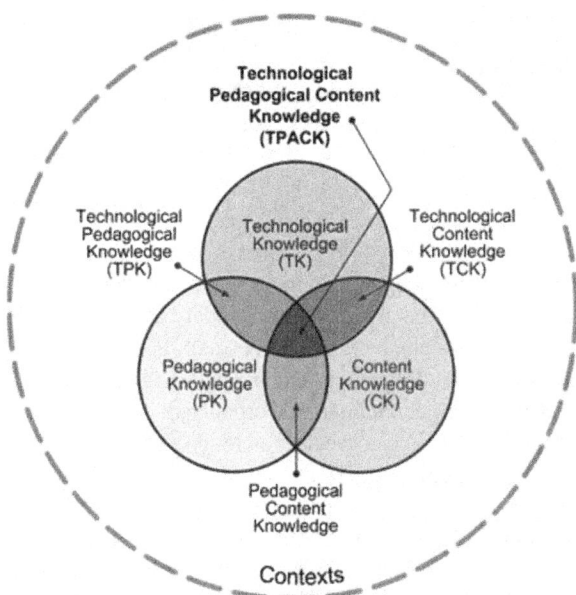

Fig. 1 The TPACK model (www.tpack.org)

In the study presented here, the instrumental orchestration perspective is used in two ways. First, we use it to describe and analyse the techniques that teachers use. Second, the instrumental orchestration model is presented to the participating teachers to help them reflect and report on their lessons. The model guided the design of a blog template described in the method section. As six out of the above seven orchestrations concern whole-class teaching, we expect that the study's outcomes will inform a further development of the seventh orchestration, Work-and-walk-by, which seems to be quite common in Dutch mathematics education.

The TPACK Perspective

The acknowledgement that teachers need to go through a process of professional development to find ways to successfully integrate digital technology in their teaching led to the development of the notion of technological pedagogical content knowledge, abbreviated as TPACK. The TPACK framework is an extension of the concept of pedagogical content knowledge (Shulman 1986). Shulman distinguishes content knowledge CK (in the case of mathematics teaching mathematical knowledge) and pedagogical knowledge PK. Pedagogical content knowledge (PCK) forms the intersection of the two and includes domain-specific pedagogical insights. The need to address technological knowledge led to the development of TPACK, which is the coherent body of knowledge and skills that is required for the implementation of ICT in teaching (Koehler et al. 2007). Figure 1 shows the different components of professional knowledge and skills in the TPACK model with their relations and intersections.

While definitions of the TPACK concepts vary in different publications (Cox and Graham 2009; Graham 2011; Voogt et al. 2012), we take the following descriptions provided by Mishra and Koehler (2006, p. 1021, 1026–1028) as points of departure. Pedagogical knowledge (PK) is knowledge about the processes and practices or methods of teaching and learning. Content knowledge is knowledge about the actual subject matter that is to be learned or taught. In the case of digital technologies, technological knowledge (TK) includes knowledge of operating systems and computer hardware, and the ability to use standard sets of software tools such as word processors, spreadsheets, browsers and e-mail. Pedagogical content knowledge (PCK) represents the blending of content and pedagogy into an understanding of how particular aspects of subject matter are organised, adapted and represented for instruction. Technological pedagogical knowledge (TPK) is knowledge of the existence, components and capabilities of various technologies as they are used in teaching and learning settings, and conversely, knowing how teaching might change as the result of using particular technologies. Technological content knowledge (TCK) is knowledge about the manner in which technology and content are reciprocally related. For example, it includes insight into the relationship between the viewing window of a graphing tool and the mathematical notions of domain and range of a function. Technological pedagogical content knowledge (TPACK), finally, includes an understanding of the representations of concepts using technologies; pedagogical techniques that use technologies in constructive ways to teach content; knowledge of what makes concepts difficult or easy to learn and how technology can help redress some of the problems that students face; knowledge of students' prior knowledge and theories of epistemology; and knowledge of how technologies can be used to build on existing knowledge and to develop new epistemologies or strengthen old ones.

The TPACK model has the virtue of simplicity and accessibility; at the same time, it is criticised for its ambiguities and the limited clarity of its construct definitions, including the ways in which these constructs are related to each other (Cox and Graham 2009; Graham 2011; Voogt et al. 2012). This particularly seems to hold for the 'intersections' in the TPACK diagram, the PCK, TCK, TPK and TPACK categories (Ruthven 2013). In spite of these limitations, we do believe the TPACK perspective can contribute to this study and we have thus used it as a model to analyse the skills and knowledge involved in the teachers' practices.

Teachers in Communities of Practice

Wenger (1998) advocates an emphasis on collective learning. This collective learning results in "practices that reflect both the pursuit of our enterprises and the attendant social relations" (Wenger 1998, p. 45). A community in which these practices are central can be defined as a community of practice. Communities of practice can be described using three dimensions: Mutual engagement, a joint enterprise and a shared repertoire. Together these three dimensions encompass a process in which negotiation of meaning is central. Wenger uses the term negotiation of meaning to characterise the process through which we experience the world and our engagement in it as meaningful.

Communities of practice provide a context for the notion of Community Documentational Genesis (Gueudet and Trouche 2012), which is an extension of the notion of documentational genesis (Gueudet and Trouche 2009). Documentational genesis is the process through which an individual uses a certain resource within his or her scheme of utilisation and, in so doing, turns it into a document. This process is dynamic and ongoing. A document comprises resources which can be associated with others and involved in the development of other documents. Within this model the terms instrumentalisation and instrumentation are used to denote, respectively, the constitution of the schemes of utilisation of the resources, and the way in which a subject (in our case a teacher) shapes the resources. When we consider documentational genesis within a community of practice we speak of Community Documentational Genesis (CDG). Gueudet and Trouche coin the expression CDG "for describing the process of gathering, creating and sharing resources to achieve the teaching goals of the community" (Gueudet and Trouche 2012, p. 309). The result of this process is community documentation: a repertoire of shared resources, associated knowledge and practices. Sabra (2011) elaborates on this idea in his study on the development of two communities of practice and shows how individual professional genesis is closely related to documentational processes within the community.

In this study, the notion of community of practice is used to monitor the teachers' professional development in relation to their participation in a collegial community.

Research Questions

The theoretical framework allows us to better phrase the issue informally presented in the introduction. The following three research questions are addressed in this paper:

1. In which ways do mid-adopting teachers with limited experience in the field of technology in mathematics education orchestrate technology-rich activities?
2. How does this repertoire of orchestrations and the corresponding TPACK skills change during a professional development process?
3. Can the teachers' individual professional development be explained by the participation in a collegial community?

Method

To address the above research questions, we carried out a case study focussing on two out of twelve mathematics teachers who participated in a collegial community project on the use of digital technology in grade 8. We now describe the digital technology involved, the design of classroom interventions, the participants, the instruments, the data and the data analysis.

Fig. 2 Snapshot of the project's digital environment in Moodle

Digital Technology

In this study, two types of digital technologies are used: digital technology for teaching mathematics and technology for supporting the collaborative work within the community of teachers and researchers. The technology for teaching mathematics is the Freudenthal Institute's Digital Mathematics Environment (DME), which integrates a content management system, a learning management system and an authoring environment.[1] The content consists of online modules in the form of Java applets or Geogebra applets. The learning management system offers means to distribute content among students and to monitor the students' progress. In the authoring environment one can adapt existing online modules or create new ones, based on existing materials and basic tools such as graphing and equation editing facilities.

The second type of technology involved is an online environment to support collaboration within the participating teachers and researchers. Available services include options for blogging, discussion and file exchange. For reasons of user friendliness, costs and accessibility, we decided to set up a project environment in Moodle (see Fig. 2).

Classroom Intervention Design

To facilitate and support the teachers' integration of digital technology in their lessons, the research team, consisting of four researchers/designers, designed three interventions for mid- to high-achieving grade 8 classes (14 year old students). The interventions consist of online modules for students accompanied by tests and teacher guides delivered through the Moodle environment. The topic of the first intervention was geometry, with a focus on perpendicular bisectors, altitudes and

---

[1] See www.fi.uu.nl/dwo/en/.

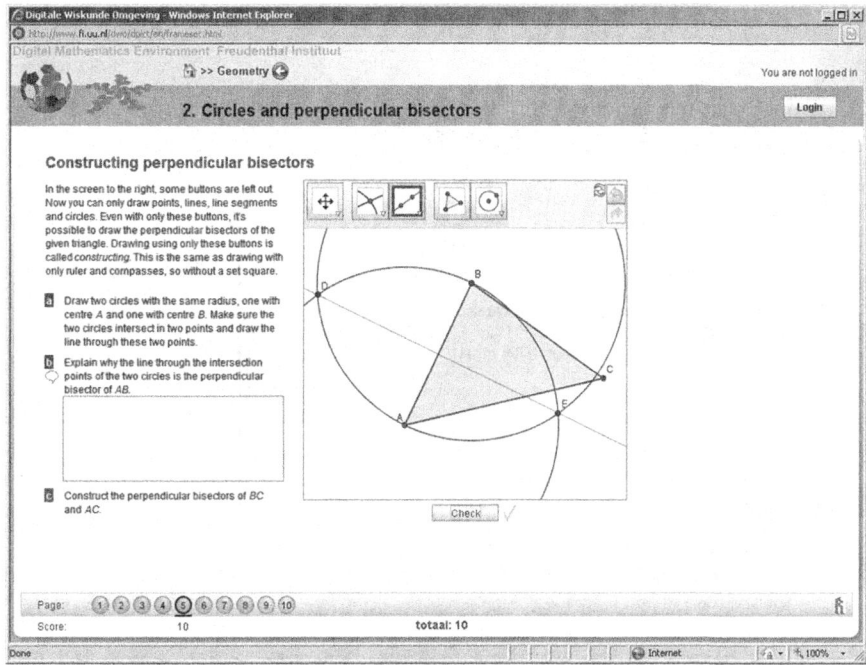

Fig. 3 An exemplary online task from the geometry module

medians of triangles. The second intervention was on linear equations, with a focus on the balance strategy to solve them. The third intervention was on quadratic equations. Figures 3 and 4 provide exemplary tasks in the online modules; the full modules can be accessed through the internet.[2]

The design of the interventions was guided by different design principles, such as the emergent modelling perspective, the option to practice skills using randomisation and feedback, and progressive formalisation. For more details on the design principles, we refer to Boon (2009) and Doorman et al. (2012). The online modules were intended to replace the regular text book chapters, even if teachers could decide to include paper-and-pencil work in their lessons.

Participants

The study's participants are six pairs of mid-adopting mathematics teachers and four designer-researchers. The 12 teachers volunteered to participate. As a criterion for being considered as mid-adopter, the teachers were only admitted if they had taught less than 20 h in a mathematics class with technology during the previous school

---

[2] See www.fi.uu.nl/dwo/en/.

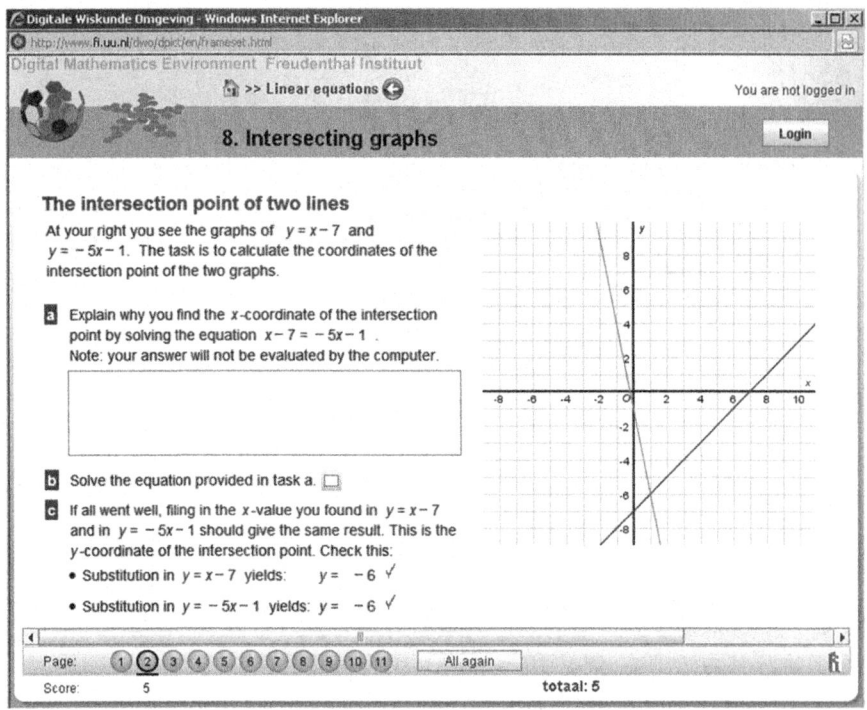

Fig. 4 An exemplary online task from the linear equations module

year. During the school year 2011–2012 these teachers implemented the digital interventions in their grade 8 classes, while being supported by five face-to-face community workshops and the online Moodle platform for virtual collaboration.

In this article we focus on the teaching practices and professional development of two of the twelve mathematics teachers. The two are colleagues from a Christian school in a small town in the centre of the Netherlands. We chose this pair because of their difference in background. Teacher A is a female teacher who has a teaching license for students up to 18 years old and has 18 years of experience in teaching students 12–18 years old. Before participating in this project, she used computers according to the suggestions made in the closing sections of regular textbook chapters. Teacher B is a male teacher with a teaching license for students from 12 to 15 years old. He has been teaching this age group for over 25 years and had never entered the computer room with his classes before the project.

Instruments

In this paper, the following research instruments play a role:

- A blog template that provides teachers with a format for the self reports on their lessons. The headings of this template are Prepare the lesson, Carry out the

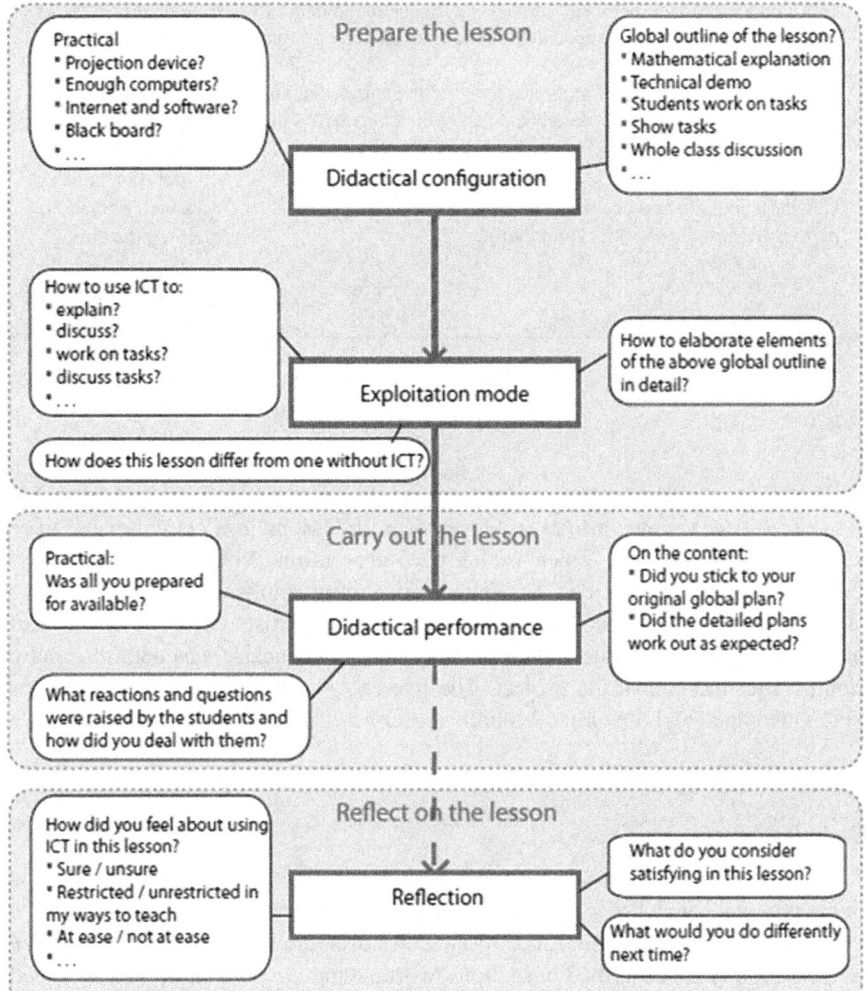

Fig. 5 The 'Orchestration chart' linking blog template and orchestration model

lesson, and Reflect on the lesson. The rationale for this template and the relation with the orchestration model is provided through the so-called orchestration chart shown in Fig. 5.

- An ICT questionnaire for teachers on their views and opinions on the role of technology in mathematics education. This questionnaire was based on the one developed by Reed et al. (2010). It consists of 37 questions on a five point Likert scale.
- A post-project questionnaire on the teachers' retrospective reflection on the benefits of their participation.

Table 1 Research questions and corresponding data

Research question	Data
1. In which ways do teachers with limited experience in the field of technology in mathematics education orchestrate technology-rich activities?	Lesson observations
2. How does this repertoire of orchestrations and the corresponding TPACK change during a professional development process?	Lesson observations ICT questionnaires Post-project questionnaire
3. Can these individual processes of change be explained by the participation in the collegial community?	Lesson observations Self reports through blogs Community workshops Moodle activities

Data

Table 1 shows the data on the two teachers in relation to the research questions. For the lesson observations, a total of eleven 50-min lessons in a computer lab were observed and videotaped, 2 per teacher per intervention (with one lesson less for teacher B's third intervention). The self reports through blogs were submitted to the Moodle environment. The ICT questionnaire was administered twice; at the start and at the end of the project. The post-project questionnaire was administered 6 months after the end of the project. The five face-to-face community workshops were videotaped and the online Moodle platform activities were collected.

Data Analysis

Qualitative data analyses were carried out using appropriate software[3] and with the lenses provided by the theory. For the lesson observations, the typology of seven orchestration types described in section 'Instrumental Orchestration' was extended with new types, particularly for individual settings. For the latter, we initially identified seven categories. However, as the inter-rater reliability was problematic for a cluster of three of them, we merged them into one category, which will be called Guide-and-explain in the results section.

In addition, the TPACK model was used to identify the teachers' skills and knowledge involved and a video clip was coded with one of the TPACK model components if that type of knowledge and skill was involved. Also, a researcher's judgement on the effect was attached: a '+' if the attributed TPACK skills led the student to understand the issue or to be able to continue the work, a '0' if this is not

---

[3] We used Atlas ti, see www.atlasti.com.

clear from the data, and a '−' if the TPACK application by the teacher led to misunderstanding or miscommunication. In line with the criticism on TPACK that we discussed in section 'The TPACK Perspective', we acknowledge that this coding was not straightforward, but we were able to assign these codes in a satisfactory way after some discussions and improvements of the codes. The analyses of the ICT questionnaire were guided by the TPACK model as well, and the community workshops and Moodle activities were analysed on the topics addressed. The different types of coding were partially repeated by a second coder and cases of disagreement were discussed until consensus was reached.

Concerning the third research question, the face-to-face community meetings were analysed with respect to the main topics addressed. The teacher's blogs and questionnaire results were analysed as well. Next, we tried to establish links between these community topics and the individual teacher data.

Results

This result section is organised along the lines of the three different research questions (see section 'Research Questions'), each with its own theoretical background described in sections 'Instrumental Orchestration', 'The TPACK Perspective', and 'Teachers in Communities of Practice', respectively.

Teachers' Orchestrations

The first research question addresses the ways in which mid-adopting teachers with limited experience in the field of technology in mathematics education orchestrate technology-rich activities. The lesson observations took the seven orchestrations described in section 'Instrumental Orchestration' as points of departure. For the six whole-class orchestrations, this categorisation suited most of the observed practices. Two new whole-class orchestrations were defined: the Guide-and-explain orchestration and the Board-instruction.

The *Guide-and-explain orchestration* shares with Explain-the-screen and Discuss-the-screen a didactical configuration of access to the technology and projecting facilities, preferably access to student work, and a classroom setting favourable for students to follow the explanation. The exploitation mode, however, straddles Explain-the-screen and Discuss-the-screen. On the one hand, the teacher provides a somewhat closed explanation based on what is on the screen. On the other, there are some, often closed questions for students, but this interaction is so limited and guided that it cannot be considered as an open discussion.

The *Board-instruction* orchestration is the traditional one of a teacher in whole-class teaching in front of the board. The board can be a chalk board, a whiteboard or an interactive whiteboard, but in any case it is just used for writing. No connections are made to the use of digital technology. The didactical configuration is the classical one of the teacher in front of the classroom working with the board. Different exploitation modes are possible, with different degrees of student involvement and interaction; however, no use of or reference to digital technology is made. We added this orchestration as we felt the need to also include the regular teaching in our analysis.

For the individual Work-and-walk-by orchestration, which was quite frequent in the case of the two teachers in this case study, it was clear that a closer look was needed and that similarities with whole-class orchestrations could be noticed. This led to a refinement of the Work-and-walk-by orchestration into five sub-orchestrations. These all share the didactical configuration, that is, the students sitting individually or in pairs in front of their technological devices that provide access to their online work and the teacher walking by in the classroom, but they differ in exploitation modes. Within this setting, the following individual orchestrations are identified and, when appropriate, named according to corresponding whole-class orchestrations:

- Individual Technical-support
  In this orchestration, in which technical issues play a central role, the teacher supports the student in technical problems that go beyond the DME technology, such as login difficulties, software bugs or hardware issues.
- Individual Technical-demo
  The didactical configuration is exploited for the individual demonstration of techniques for using the digital content by the teacher. The goal is to avoid obstacles that emerge from the students' technical inexperience in using the digital environment.
- Individual Guide-and-explain
  The exploitation of this orchestration involves an individual exchange between teacher and (a pair of) student(s) in which the teacher takes the position of the instructor through providing guidance and instruction to the student, explains mathematical concepts or methods based on what happens on the screen, or raises questions to make the student reflect on his actions and results.
- Individual Link-screen-book
  In the student-teacher interaction that characterises this orchestration, the didactical configuration is exploited by the teacher for connecting the representations and techniques encountered in the digital environment and their conventional paper-and-pencil and textbook counterparts. The goal is to link the mathematics on the screen and the mathematics of the regular paper-and-pencil. As an extra requirement for the didactical configuration, the setting should allow switching between screen, notebook and textbook. This is not self-evident in computer labs that are often (too) full.
- Individual Discuss-the-screen
  In this orchestration, the phenomena on the screen lead to a discussion between teacher and student(s). This discussion may start with a question from the student or with a remark made by the teacher. The goal of the discussion may not be clear beforehand and the student has considerable impact on the direction and the content of the talk by, for example, expressing his/her difficulties.

In Table 2 the frequencies of the whole-class and individual orchestrations for the observed lessons taught by the two teachers for the three modules are shown. The low whole-class orchestration frequencies can be explained by the fact that the observed lessons took place in a computer lab, which neither teacher considered very suitable for whole-class teaching. In spite of this, teacher A did exploit some whole-class orchestrations in the computer lab, but she also sometimes split up the lesson in two parts: one part in the regular classroom for whole-class teaching, and the other part in the computer lab for individual work. As for teacher B, he tried to prepare for and benefit from the students' computer experiences in the lessons before and after the computer lessons, to avoid whole-class teaching in the computer lab.

As the two teachers privileged individual work in the computer lab, the individual exploitations of the setting were more frequent. The data in Table 2 shows that the Guide-and-explain orchestration accounts for the majority of the observations (144 out of 222 cases, which is 65 %), followed by Technical-support and Technical-demo. Therefore, the global image that emerges from the data is that the two teachers, once technological issues are solved, walk by the students to engage in more or less interactive, teacher-driven forms of instruction on the mathematics provoked by the digital technology.

In Table 3, the results of the application of the TPACK categories and the researchers' judgement of the success of this are shown. Most frequent categories are PACK+ and TPACK+ (108 and 53 cases, respectively, out of a total of 235), with TK in third position. We interpret these findings as follows. As Table 3 refers to the same set of video clips as Table 2, most codes apply to individual orchestration settings. In many of these clips, the teachers use their pedagogical content knowledge, often in combination with technological skills. This implies that the researchers identify the combination and integration of the different TPACK components as being used in many cases. In the majority of these cases, the judgement is positive, suggesting that the teachers are able to integrate these components in a satisfying and effective way. The relatively high scores for TK, in combination with the '0' and '–' occurring relatively frequently, suggests that teachers' technological knowledge and skills are important, and may be an issue.

Changes During the Project Period

The second research question refers to the changes of the teachers' repertoire of orchestrations and the corresponding TPACK change during the project period. A first way to answer this question is to look at Tables 2 and 3, and compare the three different interventions that took place subsequently throughout the project's school year; as such, they may reveal change over time. In Table 2's individual orchestrations, we notice a decrease of Technical-demo and Technical-support from the first intervention on geometry to the third on quadratic equations. Meanwhile, Guide-and-explain frequencies are increasing. Apparently, the technology itself

Table 2 Orchestration frequencies for the two teachers over the three modules

	Teacher A			Teacher B			Total
	Geometry	Linear equations	Quadratic equations	Geometry	Linear equations	Quadratic equations	
Whole-class orchestrations							
Board-instruction	0 (0 %)	1 (10 %)	6 (100 %)	0 (0 %)	1 (100 %)	0 (0 %)	8 (30 %)
Technical-demo	2 (25 %)	1 (10 %)	0 (0 %)	0 (0 %)	0 (0 %)	0 (0 %)	3 (11 %)
Guide-and-explain	0 (0 %)	1 (10 %)	0 (0 %)	0 (0 %)	0 (0 %)	0 (0 %)	1 (4 %)
Link-screen-board	2 (25 %)	4 (40 %)	0 (0 %)	1 (50 %)	0 (0 %)	0 (0 %)	7 (26 %)
Discuss-the-screen	1 (13 %)	0 (0 %)	0 (0 %)	0 (0 %)	0 (0 %)	0 (0 %)	1 (4 %)
Explain-the-screen	1 (13 %)	1 (10 %)	0 (0 %)	1 (50 %)	0 (0 %)	0 (0 %)	3 (11 %)
Spot-and-show	0 (0 %)	1 (10 %)	0 (0 %)	0 (0 %)	0 (0 %)	0 (0 %)	1 (4 %)
Sherpa-at-work	2 (25 %)	1 (10 %)	0 (0 %)	0 (0 %)	0 (0 %)	0 (0 %)	3 (11 %)
Total	8 (100 %)	10 (100 %)	6 (100 %)	2 (100 %)	1 (100 %)	0 (0 %)	27 (101 %)
Individual orchestrations							
Technical-support	12 (30 %)	3 (7 %)	5 (19 %)	4 (17 %)	6 (10 %)	2 (7 %)	32 (14 %)
Technical-demo	12 (30 %)	6 (14 %)	1 (4 %)	5 (22 %)	2 (3 %)	1 (4 %)	27 (12 %)
Guide-and-explain	14 (35 %)	29 (66 %)	17 (63 %)	13 (57 %)	47 (80 %)	25 (89 %)	145 (66 %)
Link-screen-board	0 (0 %)	4 (9 %)	2 (7 %)	1 (4 %)	1 (2 %)	0 (0 %)	8 (4 %)
Discuss-the-screen	2 (5 %)	2 (5 %)	2 (7 %)	0 (0 %)	3 (5 %)	0 (0 %)	9 (4 %)
Total	40 (100 %)	44 (100 %)	27 (100 %)	23 (100 %)	59 (100 %)	28 (100 %)	221 (100 %)

Table 3 TPACK knowledge and skills shown by teachers and judgements by the researchers

		Teacher A			Teacher B			Total
		Geometry	Linear equations	Quadratic equations	Geometry	Linear equations	Quadratic equations	
TK	+	12 (26 %)	4 (8 %)	2 (8 %)	0 (0 %)	5 (8 %)	1 (4 %)	24 (10 %)
	0	2 (4 %)	2 (4 %)	3 (12 %)	2 (9 %)	0 (0 %)	0 (0 %)	9 (4 %)
	−	0 (0 %)	2 (4 %)	0 (0 %)	1 (4 %)	0 (0 %)	1 (4 %)	4 (2 %)
PA	+	0 (0 %)	0 (0 %)	0 (0 %)	1 (4 %)	2 (3 %)	0 (0 %)	3 (1 %)
	0	0 (0 %)	0 (0 %)	0 (0 %)	0 (0 %)	1 (2 %)	0 (0 %)	1 (0 %)
	−	0 (0 %)	0 (0 %)	0 (0 %)	0 (0 %)	0 (0 %)	0 (0 %)	0 (0 %)
CK	+	1 (2 %)	0 (0 %)	0 (0 %)	0 (0 %)	0 (0 %)	0 (0 %)	1 (0 %)
	0	0 (0 %)	0 (0 %)	0 (0 %)	0 (0 %)	0 (0 %)	0 (0 %)	0 (0 %)
	−	0 (0 %)	0 (0 %)	0 (0 %)	0 (0 %)	0 (0 %)	0 (0 %)	0 (0 %)
TPA	+	4 (9 %)	2 (4 %)	0 (0 %)	0 (0 %)	1 (2 %)	0 (0 %)	7 (3 %)
	0	0 (0 %)	0 (0 %)	0 (0 %)	0 (0 %)	0 (0 %)	0 (0 %)	0 (0 %)
	−	0 (0 %)	0 (0 %)	0 (0 %)	0 (0 %)	0 (0 %)	0 (0 %)	0 (0 %)
TCK	+	3 (7 %)	0 (0 %)	1 (4 %)	1 (4 %)	1 (2 %)	0 (0 %)	6 (3 %)
	0	2 (4 %)	0 (0 %)	0 (0 %)	0 (0 %)	0 (0 %)	0 (0 %)	2 (1 %)
	−	1 (2 %)	0 (0 %)	1 (4 %)	0 (0 %)	0 (0 %)	0 (0 %)	2 (1 %)
PACK	+	9 (20 %)	23 (46 %)	13 (50 %)	6 (26 %)	40 (63 %)	17 (63 %)	108 (46 %)
	0	1 (2 %)	2 (4 %)	1 (4 %)	0 (0 %)	0 (0 %)	3 (11 %)	7 (3 %)
	−	1 (2 %)	0 (0 %)	1 (4 %)	0 (0 %)	0 (0 %)	0 (0 %)	2 (1 %)
TPACK	+	8 (17 %)	15 (30 %)	4 (15 %)	10 (43 %)	12 (19 %)	4 (15 %)	53 (23 %)
	0	2 (4 %)	0 (0 %)	0 (0 %)	1 (4 %)	1 (2 %)	1 (4 %)	5 (2 %)
	−	0 (0 %)	0 (0 %)	0 (0 %)	1 (4 %)	0 (0 %)	0 (0 %)	1 (0 %)
Total	+	37 (80 %)	44 (88 %)	20 (77 %)	18 (78 %)	61 (97 %)	22 (81 %)	202 (86 %)
	0	7 (15 %)	4 (8 %)	4 (15 %)	3 (13 %)	2 (3 %)	4 (15 %)	24 (10 %)
	−	2 (4 %)	2 (4 %)	2 (8 %)	2 (9 %)	0 (0 %)	1 (4 %)	9 (4 %)
Total		46 (100 %)	50 (100 %)	26 (100 %)	23 (100 %)	63 (100 %)	27 (100 %)	**235**

needed more attention in the first teaching sequence than in the others. This is because the students had to get used to the Digital Mathematics Environment and because the first module also involved the additional use of Geogebra. In the second and third module, the Guide-and-explain orchestration could be more frequent, as technical issues no longer played such an important role. Also, the mathematical topic may be a factor as, for example, solving linear and quadratic equations, the topics of the second and third module, are more algorithmic than the geometry tasks in the first module. The data in Table 3 confirms these findings. The teacher work needed isolated technological knowledge slightly less in the second and third intervention, whereas pedagogical content knowledge, eventually in combination with technological skills and knowledge, is more central in Guide-and-explain formats.

A second way to consider teacher development over the year is to analyse the results from the ICT questionnaire, which was administered twice, once at the start of the project and once at the end. We focused on the questions in which the teachers changed their opinion by at least two points on the five-point scale. For teacher A this led to a number of findings. Firstly, she became more convinced that the results of the students' work using ICT would improve in the short term. Apparently, she noticed learning effects from the ICT activities. Secondly, she changed her initial opinion that there was a big difference between what students learn while using ICT and while using paper-and-pencil. This can be explained on the one hand by the second module, which is aimed at transfer between online work and paper-and-pencil work, and on the other by this teacher's increasing skills to link and relate online and paper-and-pencil activities. Thirdly, she lost some belief in ICT being efficient for learning, compared to the traditional setting. We conjecture that her teaching skills were so much in a process of development that she was not yet able to make the ICT lessons efficient. Fourthly, she became more positive about the means ICT offers for student exploration. Even if the tasks in the online modules were fairly closed, apparently she experienced the opportunities to enable student exploration. Fifthly, she changed her opinion toward claiming that teachers do have enough time to integrate technology in their teaching, probably because she felt more experienced in preparing ICT lessons and she noticed that teaching time spent on using the technology also affected paper-and-pencil skills. Finally, she appreciated more than before that student work could be followed by the teacher. This might be due to the student monitor system that the teacher had access to in the DME. All together, teacher A's opinions of ICT use in her mathematics lessons became more positive during her project participation, even if she was not sure about the effectiveness of her ICT lessons.

Teacher B, however, hardly changed his opinions. The only question where a change of two points could be identified concerned the visibility of student work for the teacher. After the project, he was more positive about this than before. As was the case for teacher A, this may be due to the student monitoring facilities that the DME offers. In addition, teacher B's Work-and-walk-by orchestrations enabled him to regularly interact with the students and to oversee their work while walking around and watching the students' screens. All together, teacher B's opinions did not change much during the project.

Finally, teacher changes were also seen in their answers to the post-project questionnaire. Both teacher A and teacher B reported a more positive attitude toward, and an increased confidence in, using technology in the mathematics classroom as a main project outcome. Indeed, they both started new technology-rich teaching sequences in the new school year, without the project's support.

The Influence of the Community

The third question is whether the teachers' individual processes of change can be explained by the participation in the collegial community. Both teacher A and teacher B were very much involved in the project and the community. For example, they wrote 48 lesson blogs (26 by teacher A and 22 by teacher B), which is far more than the 15 blogs that the average participant posted. Also, they were active users of the community's Moodle, which they accessed 475 and 509 times, respectively, compared to an average number of 396 accesses. During the face-to-face community meetings, teacher A spoke a lot, whereas teacher B was less expressive, but clearly involved.

In three cases, we identified traces of relationships between the main topics addressed in the community meetings and the blogs the teachers wrote afterwards. In other cases, we were not able to trace such relationships, suggesting that the effect of the meetings was not manifest in the teachers' reflections on their lessons.

The topic *Computer-paper-classroom* concerns the balance a teacher chooses to make between computer work, paper work and classroom sessions. This topic was discussed frequently during the meetings. For both teachers, a thorough discussion of this topic during the first meeting was followed by a high emphasis on it in the blogs. For the following two meetings and periods of blogs, however, this relationship does not appear so clearly. Still, it is interesting to look at some quotations from the blogs. In the two passages below we see a clear relationship between a teacher's choice for computer use and classroom sessions and their view on student insight.

> The lesson went smoothly; my better students do appear to like this module most. The weaker students prefer a standard lesson. That is why I try to alternate, to get everyone up to the necessary end level. (Blog teacher A, 5 oct 2011, lesson 7 and 8 module 1)
>
> This week I will only go to the computer classroom twice. During the third lesson I want to work using paper to see who do and who don't understand the theory. (Blog teacher B, 28 sept 2011, lesson 4 module 1)

The topic *Degree of difficulty* concerns the difficulty of the modules. Contrary to the previous topic, this one shows an overall recurrence in the blogs related to the meetings. The teachers often mentioned the different degrees of difficulty of the subsequent modules, as the following quotations show:

> The tasks demanded a lot of insight. They were better suited for the high achieving students than for my mid achieving students. There were few repeating tasks. Fortunately this is different for the next module. (Blog teacher A, 12 jan 2012, after module 1)
>
> The students enjoyed it more as well, because they noticed that is was a lot less complicated (than the geometry module). (Blog teacher B, 11 jan 2012, lesson 1 module 2)

Finally, the topic *Planning the module* concerned the actual planning of the teaching sequences, which was the teachers' responsibility. This topic shows an almost overall recurrence in the blogs related to the meetings. Sometimes, the topic is related to student insight or behaviour, as is the case for teacher A's quotation, but this co-occurrence appeared sparsely. Most quotations coded in relation to this topic are short and matter-of-fact, like the following quotation of teacher B.

> The students need a lot of time for the tasks in paragraph 3. At the end of the lesson they had not finished it. This means they have to finish it as part of their homework as well as paragraph 4. (Blog teacher A, 13 jan 2012, lesson 2 module 2)
>
> During the first lesson the students are going to work on the first paragraph and maybe start on the second paragraph. (Blog teacher B, 11 jan 2012, lesson 1 module 2)

In the post-project questionnaire, the two teachers both rated the importance of the community aspects of the project (their colleagues' blogs, the background literature on the Moodle, and the Moodle forum) as neutral to reasonable, which were relatively low scores compared to other aspects of the project. This confirms the overall impression that we were able to trace some links between the community participation and the teachers' professional development, but only to a limited extent. These results suggest that the project was not really successful in establishing a community of practice.

Conclusion and Discussion

In this paper we set out to answer three questions, the first being: In which ways do mid-adopting teachers with limited experience in the field of technology in mathematics education orchestrate technology-rich activities? While answering this question, two new whole-class orchestrations were defined: the Guide-and-explain orchestration and the Board-instruction. A closer look at individual orchestrations led to a refinement of the Work-and-walk-by orchestration into five sub-orchestrations.

The data in Table 2 shows that the individual Guide-and-explain orchestration accounts for the majority of the observations. Therefore, the global image that emerges from the data is that the two teachers, once technological issues are solved, walk by the students to engage in more or less interactive but teacher-driven forms of instruction. In terms of TPACK skills, the teachers make use of their pedagogical content knowledge, often in combination with technological skills. In most cases, the teachers are able to integrate these components in a satisfying and effective way. Teachers' technological knowledge and skills are important, and may be an issue to them.

As a further conclusion on the first question, we note that the Drijvers et al. (2010) orchestrations served as a good point of departure, but led to the identification of additional orchestrations. The added descriptions of individual orchestrations not only offer an elaboration of the global Work-and-walk-by orchestration, but they also allow for a more detailed view on the relationships between whole-class and individual orchestrations, in that some of the exploitation modes and goals of

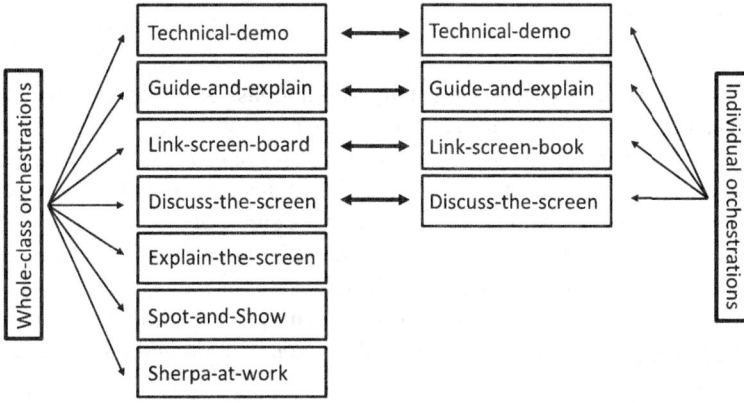

Fig. 6 Whole-class and individual orchestrations (Based on Van den Heuvel 2012)

whole-class orchestrations have similar counterparts in individual orchestrations. For example, the whole-class Link-screen-board and the individual Link-screen-book orchestrations clearly share similar teaching goals. Other orchestrations, such as Spot-and-show, are constrained to whole-class or individual settings. The resulting 'landscape' of whole-class and individual orchestrations, as well as the relationships between the two, is depicted in Fig. 6. As Board-instruction and Technical-support are not at the heart of this study's interest, we did not include them in the figure.

The second research question was: How does this repertoire of orchestrations and the corresponding TPACK skills change during a professional development process? We noticed that the teachers' orchestration preferences are changing, showing a decrease in Technical-demo and Technical-support, and Guide-and-explain becoming more frequent. This is explained both by the different nature of the three modules and by increasing professional development. This development also involves more complex teacher skills, with PACK and TPACK being the most frequently observed skills needed. More information on professional development is provided by the teachers' self reports in lesson blogs and ICT questionnaires, which in the case of Teacher A show a development in reflection on the skills and knowledge needed, and in the acquisition of these skills. The post-project questionnaire results suggest that the teachers' self-confidence increased through their participation in the project. In all, both teachers developed a more thoughtful and confident attitude to their use of technology in teaching.

The third research question was: Can the teachers' individual professional development be explained by the participation in the collegial community? The results suggest that the project was not successful in establishing a community of practice. The overall impression is that some traces between the community participation and the teachers' professional development were identified, but only to a limited extent. In addition to this, the post-project questionnaire reveals that the two teachers both rated the importance of the community aspects of the project as relatively low.

Discussion

In this discussion we first address the study's limitations. Of course, the observations of two teachers in eleven lessons cannot provide exhaustive and conclusive data on the complex issue of how mid-adopting teachers engage in a process of professional development concerning the integration of digital technology and the development of appropriate teaching techniques. Neither can we be sure that the two case studies are representative of mid-adopting teachers in the Netherlands or elsewhere. However, we do see the results from these case studies as useful in exploring the issue and in generating hypotheses as to how crucial steps can be made in the dissemination of technology in education, and in professional development for mid-adopting teachers in particular.

If we look back at the study's theoretical framework, we see that the instrumental orchestration model was useful in two ways. First, it helped us as researchers to set up the blog template for the teachers' lesson reports. Second, it provided us with a framework to identify and describe the observed orchestrations and teaching practices in the videotaped lessons. We recognise, however, that we were not very successful in discussing the orchestration framework with teachers in a way that was useful to them.

As for the TPACK model, it provided us with a framework to analyse teachers' blogs, as indicated in Table 3. While doing so, we acknowledge that coding teacher statements in terms of the TPACK model was not always straightforward, which is in line with the criticisms on TPACK constructs described in Graham (2011), Ruthven (2013), and Voogt et al. (2012). In addition to this, the model seemed to be less effective in supporting teachers' reflections and self-reports.

Concerning the idea of establishing a community of practice, we think that this is a powerful idea, but one that we were unable to fully exploit, probably due to a lack of ownership over the project by the participants. Also, the relationships between face-to-face meetings and virtual communications might have been too weak. We might conjecture, for example, that having regular virtual meetings might bridge the gap between face-to-face and online communication.

As a closing remark, we do believe the three theoretical lenses proved valuable, in spite of their limitations. We recommend their further elaboration, refinement and fine-tuning, probably in collaboration and comparison, as was done by Tabach (2011).

Acknowledgement We thank the participating teachers and their students for their collaboration, and Kennisnet for supporting the research study (project no. C/OZK/2131).

References

Artigue, M., Drijvers, P., Lagrange, J-b, Mariotti, M. A., & Ruthven, K. (2009). Technologies numériques dans l'enseignement des mathématiques, où en est-on dans les recherches et dans leur intégration? [Technology in mathematics education: How about research and its integration?] In C. Ouvrier-Buffet & M.-J. Perrin-Glorian (Eds.), *Approches plurielles en didactique des mathématiques; Apprendre à faire des mathématiques du primaire au supérieur: quoi de neuf?* [Multiple approaches to the didactics of mathematics; learning mathematics from primary to tertiary level: What's new?] (pp. 185–207). Paris: Université Paris Diderot Paris 7.

Boon, P. (2009). A designer speaks: Designing educational software for 3D geometry. *Educational Designer, 1*(2). Retrieved June 19, 2012, from http://www.educationaldesigner.org/ed/volume1/issue2/article7/

Cox, S., & Graham, C. R. (2009). Diagramming TPACK in practice: Using an elaborated model of the TPACK framework to analyze and depict teacher knowledge. *TechTrends, 53*(5), 60–69.

Doerr, H. M., & Zangor, R. (2000). Creating meaning for and with the graphing calculator. *Educational Studies in Mathematics, 41*, 143–163.

Doorman, M., Drijvers, P., Gravemeijer, K., Boon, P., & Reed, H. (2012). Tool use and the development of the function concept: From repeated calculations to functional thinking. *International Journal of Science and Mathematics Education, 10*(6), 1243–1267.

Drijvers, P. (2012). Teachers transforming resources into orchestrations. In G. Gueudet, B. Pepin, & L. Trouche (Eds.), *From text to 'lived' resources: Mathematics curriculum materials and teacher development* (pp. 265–281). New York/Berlin: Springer.

Drijvers, P., Doorman, M., Boon, P., Reed, H., & Gravemeijer, K. (2010). The teacher and the tool: instrumental orchestrations in the technology-rich mathematics classroom. *Educational Studies in Mathematics, 75*(2), 213–234.

Graham, C. R. (2011). Theoretical considerations for understanding technological pedagogical content knowledge (TPACK). *Computers in Education, 57*, 1953–1960.

Gueudet, G., & Trouche, L. (2009). Towards new documentation systems for mathematics teachers? *Educational Studies in Mathematics, 71*, 199–218.

Gueudet, G., & Trouche, L. (2012). Communities, documents and professional genesis: Interrelated stories. In G. Gueudet, B. Pepin, & L. Trouche (Eds.), *From text to 'lived' resources: Mathematics curriculum materials and teacher development* (pp. 305–322). New York: Springer.

Heuvel, C. van den (2012). *Méér dan het krijtje; Docentpraktijken bij gebruik van ICT in het wiskundeonderwijs.* [More than chalk; Teachers' practices while using ICT in mathematics education.] Unpublished Master thesis. Utrecht: Utrecht University.

Koehler, M. J., Mishra, P., & Yahya, K. (2007). Tracing the development of teacher knowledge in a design seminar: Integrating content, pedagogy and technology. *Computers in Education, 49*, 740–762.

Lagrange, J.-B., & Ozdemir Erdogan, E. (2009). Teachers' emergent goals in spreadsheet-based lessons: Analyzing the complexity of technology integration. *Educational Studies in Mathematics, 71*(1), 65–84.

Mishra, P., & Koehler, M. J. (2006). Technological pedagogical content knowledge: A framework for teacher knowledge. *Teachers College Record, 108*(6), 1017–1054.

Monaghan, J. (2004). Teachers' activities in technology-based mathematics lessons. *International Journal of Computers for Mathematical Learning, 9*, 327–357.

National Council of Teachers of Mathematics. (2008). *The role of technology in the teaching and learning of mathematics.* Retrieved May 31, 2011, from http://www.nctm.org/about/content.aspx?id=14233

Reed, H., Drijvers, P., & Kirschner, P. (2010). Effects of attitudes and behaviours on learning mathematics with computer tools. *Computers in Education, 55*(1), 1–15.

Ruthven, K. (2007). Teachers, technologies and the structures of schooling. In D. Pitta-Pantazi & G. Philippou (Eds.), *Proceedings of the V Congress of the European Society for Research in Mathematics Education CERME5* (pp. 52–67). Larnaca: University of Cyprus.

Ruthven, K. (2013). Frameworks for analysing the expertise that underpins successful integration of digital technologies into everyday teaching practice. In A. Clark-Wilson, O. Robutti, & N. Sinclair (Eds.), *The mathematics teacher in the digital era, mathematics education in the digital era 2.* Dordrecht: Springer Science+Business Media.

Sabra, H. (2011). *Contribution à l'étude du travail documentaire des enseignants de mathématiques: les incidents comme révélateurs des rapports entre documentations individuelle et communautaire.* [Contribution to the study of documentary work of mathematics teachers: Incidents as indicators of relations between individual and collective documentation.] Dissertation. Lyon: Université Claude Bernard Lyon 1.

Shulman, L. S. (1986). Those who understand: Knowledge growth in teaching. *Educational Researcher, 15*(2), 4–14.

Tabach, M. (2011). A mathematics teacher's practice in a technological environment: A case study analysis using two complementary theories. *Technology, Knowledge and Learning, 16*, 247–265.

Trouche, L. (2004). Managing complexity of human/machine interactions in computerized learning environments: Guiding students' command process through instrumental orchestrations. *International Journal of Computers for Mathematical Learning, 9*, 281–307.

Trouche, L. (2005). Instrumental genesis, individual and social aspects. In D. Guin, K. Ruthven, & L. Trouche (Eds.), *The didactical challenge of symbolic calculators: Turning a computational device into a mathematical instrument* (pp. 197–230). New York: Springer.

Voogt, J., Fisser, P., Pareja Roblin, N., Tondeur, J., & Van Braak, J. (2012). Technological pedagogical content knowledge – a review of the literature. *Journal of Computer Assisted Learning, 29*(2), 109–121.

Wenger, E. (1998). *Communities of practice: Learning, meaning, and identity*. New York: Cambridge University Press.

Teaching Mathematics with Technology at the Kindergarten Level: Resources and Orchestrations

Ghislaine Gueudet, Laetitia Bueno-Ravel, and Caroline Poisard

Abstract In this chapter we study the use of software in mathematics by French kindergarten teachers who are working with 5 and 6-year-old children. We retain the theoretical perspective of the documentational approach, considering that teachers interact with a variety of resources, including technology. These interactions lead to the development by the teachers of documents, associating resources and professional knowledge. We focus here on the way teachers organise the available resources, for a given mathematical objective through the orchestrations they choose. By focusing on three teachers in particular, we identify different types of orchestrations, evidencing teacher agency and a specific attention to individual children's differences. Teacher knowledge of different kinds (pedagogical knowledge, knowledge about curriculum material, knowledge about the teaching of numbers at kindergarten) influences the choice of orchestration.

Keywords Abacus • Documentational approach • Genesis • Kindergarten • Orchestration • Resources

---

G. Gueudet (✉) • L. Bueno-Ravel
CREAD, University of Brest, IUFM Bretagne, site de Rennes,
153 rue Saint Malo, 35000 RENNES, France
e-mail: Ghislaine.Gueudet@bretagne.iufm.fr

C. Poisard
CREAD, University of Brest, IUFM Bretagne, site de Quimper,
8 rue Rosmadec BP 301, 29191 QUIMPER cedex, France

Teacher Resources and Orchestrations

Instrumental Orchestrations, Orchestration Types

The concept of instrumental orchestration (introduced in Trouche 2004) develops the instrumental approach for the study of teaching and learning of mathematics to include a focus on technology integration. The instrumental approach (Verillon and Rabardel 1995) distinguishes between a given artefact (here, a mathematical software) and an instrument developed by the subject (here, a student) using this artefact. Along with one's use of the artefact, in different contexts for a similar aim, one develops knowledge about the artefact itself, about its use, and also other kinds of knowledge (in particular, here, mathematical knowledge). The instrument is composed of the artefact and the knowledge developed. The process of development of this instrument is called an instrumental genesis. Instrumental orchestration describes how a teacher guides the instrumental geneses of the children, using a given piece of software. It comprises two aspects: a didactical configuration and an exploitation mode. A didactical configuration is an arrangement of artefacts in the environment, while an exploitation mode refers to the way the teacher decides to exploit this didactical configuration. The configuration and the exploitation mode are not only material organisations, but they also encompass precise didactical objectives, in terms of the mathematical knowledge at stake in the situation.

Drijvers (2012) has refined and clarified the concept of orchestration. He introduces in particular a third element of orchestration, the didactical performance, in order to distinguish between what has been planned and what actually happens in class. The didactical performance "involves the ad hoc decisions taken while teaching on how to actually perform in the chosen didactic configuration and exploitation mode: what question to pose now, how to do justice to (or to set aside) any particular student input, how to deal with an unexpected aspect of the mathematical task or the technological tool, or other emerging goals" (Drijvers 2012, p. 266). In addition, having followed several teachers, Drijvers characterises seven different orchestration types: Technical-demo (demonstration of tools techniques by the teacher); Explain-the-screen (whole-class explanation of what happens on the screen); Link-screen-board (explanation by the teacher of the link between the screen and mathematics written on the board); Discuss-the-screen (whole-class discussion about what happens on the computer screen); Spot-and-show (showing interesting student's work); Sherpa-at-work (a student carries out actions requested by the teacher); and Work-and-walk-by (the children work individually on computers; the teacher observes their work and intervenes if necessary).

In our study, we adopt this definition of orchestration, and the idea of characterising orchestration types. We assume from the beginning that orchestration types observed at kindergarten will be different from the ones observed in previous studies, which all take place at the secondary school level. Some differences may arise due to the fact that the orchestrations are at the kindergarten level where, in France, there is much less whole-class teaching and less mathematics written on the board.

Orchestration Choices, Resources and Geneses

Orchestrations can be considered as the choices made by teachers about the use of technology in their classrooms. Which factors are likely to influence these choices? Ruthven (2012) proposes the following five dimensions related to classroom practices of a teacher integrating technology: the working environment; the resource system; the activity format; the time economy; and the curriculum script (described as a set of goals and actions). These dimensions can also enlighten the orchestration choices. The working environment, that is the material conditions of teacher work, is certainly important. In France, only a few kindergartens have a computer lab; in most cases, there are one or two computers in each classroom, and not always a digital projector. It certainly contributes to the rarity of technology use. But, as Ruthven demonstrates, this material aspect does not explain everything.

In our work, we focus mostly on the resource system and on the curriculum script, as well as, more generally, teachers' professional knowledge. We use the perspective introduced by the documentational approach (Gueudet and Trouche 2009). In their professional activity teachers interact with a great variety of resources, including curriculum material, children's work and software. Clark-Wilson (2010), when considering teachers' use of technology, demonstrates that teacher knowledge shapes their use of technology and that, simultaneously, the use of technology contributes to teacher learning. Instrumental geneses (Verillon and Rabardel 1995) also occur for teachers engaged in their professional activity, and using technology. Similar processes happen when teachers interact with textbooks (Remillard 2012). The development of the documentational approach is based on accounting for this multiple resource use and learning. This approach considers that, for a given professional aim, the teacher interacts with sets of resources such as textbooks, official texts, websites and software. If a teacher has already taught this topic, he/she certainly also uses previous notes and children' worksheets in preparing future lessons. When he/she creates the lesson in class, children' productions and reactions also constitute resources. All this belongs to what we call the *teacher's documentation work*. Teacher knowledge intervenes in this work. On the one hand, the knowledge influences the use of resources (this part of the process is called *instrumentalisation*, referring to the instrumental approach); on the other hand, the use of resources leads to evolution of the knowledge *(instrumentation)*. We call this process *a documentational genesis*. Within such geneses, for different teaching objectives, in different classes, the teachers constitute a resource system, which is an organised set of resources, transformed in the course of their use in class (Gueudet and Trouche 2012). We consider that orchestration choices are influenced by teacher professional knowledge and by their resource system. We especially focus on the knowledge linked with the mathematical content (but we do not refer to precise categories, like those proposed by Ball et al. (2008)).

Within this perspective, the research questions addressed in this paper are:

- Which orchestrations do kindergarten teachers choose when using technology in their teaching of mathematics? Is it possible to identify 'orchestration types'? If so, do these resonate with those that Drijvers (2012) identified at the secondary school level?

– Which of the following factors influence these choices of orchestrations: the resources and their features (the software in particular) within an instrumentation process? or the teacher knowledge within an instrumentalisation process (and if so, which kind of knowledge)?

In order to answer these questions, we draw on data gathered during a design and research project that we now present.

The 'Mathematical Package' Project

Our work takes place in the frame of a contract with the French education ministry and the French Institute for Education (IFÉ), 'the mathematical package' project, which aimed to design mathematical tools and situations for the teaching of mathematics at the kindergarten level and in grades 1 and 2. In France, kindergartens reside within primary schools and most children attend from the age of three. They comprise three classes: young section, middle section and older section. Our work for kindergarten concerns only the older section (children aged 5–6).

The study presented here draws on the work of a group of teachers and researchers where the teachers created lessons in their classes that were observed and videotaped by researchers and subsequently discussed during working group meetings. The teachers wrote descriptions of their lessons, with the aim of sharing them with other colleagues. As researchers, we participated in the group. We presented several pieces of software to teachers, including those considered in this paper, but we did not intervene in the teachers' choices or on their lessons with this software. Our intervention in the design concerned more the format of the lesson descriptions in particular, the relevant categories. At the same time we studied the material produced by the group with the aim of analysing orchestrations and documentational geneses. This process was supported by the completion of a questionnaire by the teachers that asked about their use of resources and, in particular, the technological ones.

The data we gathered for each teacher were:

- Notes from the group meetings (always taken by a researcher);
- Videos of the lessons, with accompanying field notes of the observing researcher;
- Resources used by the teacher in her preparation, produced by the teacher for the children, and produced by the children during the lesson;
- Lesson plans elaborated by the teacher for colleagues (some of these descriptions are written individually and others by several teachers working together);
- Questionnaires completed individually by the teachers;
- Children's work.

The data was analysed with two specific aims. Firstly we wanted to describe the orchestrations developed by the teachers. The observations, videos and lesson descriptions provided us with information about the teachers' configurations, exploitation modes and their didactical performances. An orchestration can correspond to a short amount of time spent in class (some of the orchestrations described by Drijvers (2012) lasted only 10 min). In our work we generally considered longer time periods because we were interested in the descriptions produced by the teachers. Naturally these descriptions cannot exist within a timescale of a few minutes.

Secondly, we wanted to understand the factors shaping these orchestrations, especially those linked with teacher professional knowledge and with teacher resource systems. Hence we analysed the teachers' questionnaire responses alongside their lesson descriptions and the other resources. The lesson descriptions were essential data because the teachers themselves wrote them, so they did not contain the researchers' interpretations. They also provided evidence of what the teacher thought important to emphasise to their colleagues. We consider that such descriptions are linked with a teacher's knowledge, and in particular the knowledge developed during the use of the software. Naturally, descriptions proposed by teachers carry specific biases as they correspond more to the view of the teacher on his/her teaching than to the actual practice. However, a comparison of the descriptions alongside the videos shed light on the teachers' orchestration choices.

We worked with seven teachers during the whole academic year 2011–2012. In this chapter we select the cases of three of these seven teachers; all of whom have 12-15 years experience at kindergarten level. None of them had used software in their mathematics teaching before the beginning of the research. One teacher used the abacus, both material and virtual and the other teachers both used a specially designed software program.

Instrumental Orchestrations: Two Case Studies at the Kindergarten Level

Learning Numbers with the Virtual Abacus

In this section we focus on Deborah, a kindergarten teacher and a member of the research group. The classroom work with the abacus (which we will refer to as the abacus-lessons) lasted 12 sessions and involved number sense. After presenting an outline of these lessons, we focus on specific aspects of it: the introduction of the virtual abacus and the interaction between the teacher and the children.

The Chinese abacus, both virtual and material, was the central resource in the lessons we followed. The virtual version used by Deborah was developed by Sésamath – IREM of Lille and is available online.[1] On the virtual abacus, the children can move one or several beads by clicking on a bead with the mouse. They have some feedback from the software as they can verify their work by using the icon 'see number', which is written in numeral form. One important feature of the software is that there is no possibility for the teacher to save student work. The teachers have three opportunities to find out what the children have done: they can observe them manipulating the abacus, they can ask a child to show a manipulation on the board using the digital projector or interactive white board (IWB, see below) or they can offer a paper and pencil task as the only way to keep a record of the work done.

---

[1] http://cii.sesamath.net/lille/exos_boulier/boulier.swf

Fig. 1 Deborah's class organisation before the IWB: teacher using a digital projector (*left*) and the children manipulating the physical abacus (*right*)

Fig. 2 Deborah's classroom organisation with the IWB and laptops (*left*) and the children manipulating the virtual abacus (*right*)

The Abacus-Lessons in Deborah's Class

During the sequence of lessons, the general classroom organisation was the same: the 24 children sat in three groups of eight and Deborah spent 20 min with each group, thereby repeating the session three times. The school had a computer room with an IWB but Deborah thought this solution would be too complicated for young children. The [http://cii.sesamath.net/lille/exos_boulier/boulier.swf] abacus-lessons took place in Deborah's regular classroom. However, the classroom equipment changed in April 2012, about half way through the lessons. From November to March, the classroom was equipped with a digital projector (session 1–7, Fig. 1), and from April to June with an IWB and eight laptops (session 8–12, Fig. 2).

This change in the material environment resulted in changes in Deborah's exploitation mode, which we identify as a process of instrumentation. During sessions 1–7, the children learned to show and read numbers on the Chinese abacus, with one physical abacus per student. They manipulated the physical abacus to show a number and the digital projector allowed the teacher to show a correction with the virtual abacus. Deborah also asked the children to read numbers shown on the virtual abacus. So, for the first seven sessions, the children did not manipulate the virtual abacus. For the last five sessions (8–12), the resources available were the virtual abacus, the IWB and also student worksheets (Appendix 2). The sessions were organised in two phases. The first phase was an introduction during which Deborah asked the children to show numbers on the virtual abacus displayed on the IWB, and a discussion about the different possibilities for showing numbers was organised. This phase is what we call an 'investigation approach' (Poisard and Gueudet 2010), when the children are asked to show their work and to argue about the validity of suggestions. During the second phase, the children were asked to show and read numbers on a hand-out given by the teacher. Each student has a computer and can use the virtual abacus as a help to answer questions (Appendix 2). During the last two sessions, Deborah introduced a paper 'abacus-book' in the class representing numbers on the abacus and the equivalent numbers in numeral. When a number was shown on the virtual abacus, the page concerning the number was put on the board as a record of different ways to show numbers when needed.

Introduction of the Software

We observed two important moments during the initial introduction of the software in the classroom. The first moment occurred during session 1, when the children worked with the physical abaci. They did not directly manipulate the software but they were asked to read numbers projected using the software. Some children had difficulties connecting the horizontal physical abacus and the vertical virtual one. Moreover, the gestures needed are different between the two abaci. For example, to show seven on a physical abacus, only one gesture is needed (beads are carried on the central bar, pinching them between the forefinger and the thumb) while two gestures are needed on the virtual abacus (two moves with the mouse, above and below the central bar). The children overcame these difficulties quickly.

During sessions 1–7, Deborah planned a didactic configuration and an associated exploitation mode, where the children used the physical abacus. After organising a discussion and argumentation session between the children, Deborah used the virtual abacus as a means of collective explanation and correction (because it permits a projection, visible for all, and also because of the 'display the number' facility). There was in this case, in session 1, a short Technical-demo orchestration, where the teacher presented the features of the virtual abacus. Then most of the abacus use during sessions 1–7 corresponded to an 'Explain-the-screen' orchestration (Drijvers 2012), since explanations given by Deborah exceeded the technical aspects and also comprised mathematical knowledge. The following extract corresponds to such an Explain-the-screen orchestration in Deborah's class (the French version of the extracts is provided in Appendix 3; this is our translation). This discourse took place at the beginning of session 3 and it was stimulated by the children discussed how to show 5 on the physical abacus. Deborah was sitting with the children and she moved to the screen to show the children' suggestions.

Deborah:	You suggested to me… Laurie, you suggested to me to move the five beads on the red rod, ok. Why? [*The unit rod is red on the virtual abacus and others are green. Deborah moves 5 as five 1-unit counters on units*].
Laurie:	Because, the ones above, they are useless.
Deborah:	Because you think the above beads, they are useless. Do you agree with Laurie's choice?

Some children:	No!
Deborah:	No?
A student:	Because the above beads, they mark 5.
Deborah:	We are going to check, What did I ask to show?
A student:	Five!
Deborah:	Very well. [*Deborah activates the icon "see number" and 5 appears on screen*]
Some children:	Five.
Deborah:	So, is Laurie's choice right?
Few children:	Yes!
Deborah:	Now, there is another possibility for five. Some children moved one 5-unit counter that means one above bead. [*Deborah shows on screen number 5 as one 5-unit counter, the icon "see number" is inactivated.*] So, how many ways can we show five? Maëlle?
Maëlle:	Two.
Deborah:	Yes, two ways. Either we activate the five lower beads, or I activate one of the higher beads. [*Deborah shows on screen the two possibilities.*]

In this extract, Deborah's didactical performance corresponds to what she had planned before the lesson. Her aim was to elicit the two ways to show five: with five 1-unit beads or with one 5-unit bead. The children participated in the discussion and Deborah used their suggestions; but their responsibility remains limited, which means that this orchestration is more teacher-centred. We observed an evolution towards the children having more responsibility during the second half of the abacus lessons.

Using the Software on Laptops and on the IWB

In session 8, the children started to manipulate the software using the laptops and the IWB and they discovered specific features of the virtual abacus software. We identify in session 8 a Technical-demo orchestration (Drijvers 2012), which involved the demonstration of tool techniques by the teacher (not for the whole class, but for the group of eight children involved). This was very quick as children were able to manipulate the virtual abacus with no technical obstacles. The only technical point concerned the use of the IWB pencil, which they overcame with ease.

During sessions 8–12, Deborah circulated amongst children and watched their individual work, on the laptops or on paper, providing help if needed. This corresponded to Drijver's 'Work-and-walk-by' orchestration, when the teacher follows the individual work of the children. We can also identify a 'Discuss-the-screen' orchestration as children were asked to come to the IWB to display and argue in support of their suggestions as to how to show eight, as in the following extract:

Deborah:	I would like you to show... Eight! We think about... How do we do eight? [*Some children want to immediately give answer*] Eight is? -
Some children:	Five and three!
Deborah:	Show me with your hands. Five and three! Kevin. [*Kevin goes to the board and he activates one 5-unit counters and three 1-unit counters (the third bead, one gesture).*] Five and three. Yes, you activated five and three. [*Deborah goes closer to the board to show the activated beads.*] Do you agree with his choice? Is there another solution? Another way to show number eight? Number eight?
Some children:	Yes.

Deborah:	Yes Anaïs. [*Anaïs goes to the board and takes the pencil*]. Go on, you did not push hard enough, I think. [*Anaïs activates three 1-unit counters, within three gestures: one, two, three, and one 5-unit counter.*] So, it is because you, you activated the beads one after the others. It is very well. Maëlle. [*Maëlle goes to the board and activates three 1-unit counters in the tens and one 5-unit counter in the units, it marks 35, the icon "see number" is activated.*] No. [*Maëlle tries with three 1-unit counters in the tens, and five 1-unit counters in the units, it marks 35 as well.*] You activated indeed eight beads, but did you show number eight?
Maëlle:	No. [*Looking at number 35 written in numeral on board*]
Deborah:	Did you understand your mistake?
Maëlle:	Yes, three and five, it makes 30 and 5!
Deborah:	Three and five, it makes 35. [*Showing the two different rods*]. And above all, they are not located on the same rod.

In this extract, Deborah's didactical performance partly corresponds to what she planned, which was to find different ways to display 8 on the abacus. Nevertheless, the children participated more than in the previous extract as they display the numbers for themselves on the board. We also observe that she reacts in the moment to a mistake (anticipated, on a general level, because it is a classical difficulty with the abacus) arising from a confusion between the number of beads and the value they represent. We will focus below on the interactions between Deborah and the children in terms of how children's reactions constitute resources for Deborah.

Interactions Between Children and the Teacher

Clark-Wilson (2010) demonstrates how hiccups, in mathematics lessons using technology, lead to evolutions in the teacher practice and thus form part of a teacher's professional development. She proposes a classification of such hiccups, several categories of which correspond to unplanned teacher-children' interactions. We do not use the concept of hiccup here. Nevertheless, we obtain similar results, which we interpret as documentational geneses, with the children's productions and reactions constituting central resources for the teacher. We presented above the example of a local adaptation to a student's answer. Over a longer timescale, Deborah also changed her plans for progression within the abacus-lessons as a result of her observations of the children's work.

Deborah first centred the tasks proposed to the children on 'show a number on the abacus'. Her observations of the children over several sessions led her to also propose work on the task 'read a number shown on the abacus'. From sessions 8–12, children were asked to achieve two tasks on paper (Appendix 2): to read numbers (from abaci images on printed hand-outs) and write them in numerals; and to draw beads on empty abaci, corresponding to a number written as a numeral. Deborah first thought to ask children to complete both paper-based tasks at each session. But it appeared to be too difficult for the children when the numbers were above 5, so Deborah chose to alternate the paper tasks in the following sessions (session 9–12). The computer was then used as a possible means of support to the children alongside this paper and pencil task. Hence, Deborah modified her plans as a result of the knowledge she gained from her interactions with children.

During sessions 8 and 9, she observed that some children were still encountering difficulties. For this reason, she elaborated an 'abacus book', which was first used within session 11. This book (on paper) presented all of the possibilities to show numbers up to 10 on the abacus. This kind of book, generally used as a record of class work, is commonly used at this level in French classrooms. This can be considered as part of the process of instrumentalisation. Deborah has professional knowledge about the possibility to use such books in order to keep a collective record of class work, which she applies to the abacus, as a result of the difficulties encountered by some children. After the introduction of this book, it appeared that most children were able to read numbers proposed by the teacher for paper work (without using the virtual abacus). A few children manipulated the virtual abacus to verify their results. More precisely, in session 11 all of the children were able to recognise a 15 shown in the three different ways on the abacus. Deborah considered that the abacus lessons were useful for the children in learning about numbers. She planned to use the abacus next year and even to dedicate more than 12 sessions to it. In her opinion, it should be an everyday tool for the children.

We now consider a second case study, which involved a different software program, used over a much shorter time period.

Learning Numbers with the 'Passenger Train'

This section considers another software program, Passenger Train [http://python.bretagne.iufm.fr/docenligne/marene/Train_des_Lapins_Online_2012-10-05.html], which was chosen by two other teachers, Chloe and Mia, who were also members of the project group. The related classroom observations began in January 2012. Mia had a 'double level' class, with eight children aged 3 (young section) and twenty children of 5–6 years (older section). Chloe had a class of older children (5–6 years). The Passenger Train program was designed as a game and it corresponds nevertheless to a precise mathematical learning situation. We claim that the children will develop mathematical meanings within this playful context, which is set within a perspective that is relevant for young children as evidenced by previous research (Van Oers 2010).

Main features of the Passenger Train program

This freeware program was designed to focus on the specific function of numbers as indicators of a position on a number line. The children's task was to seat one to three passengers (rabbits) in the same passenger car of an empty train and to match those in a reference train. The freeware program enables two modes of use: 'discovering' mode (the reference train remains visible, Fig. 3) and 'learning' mode

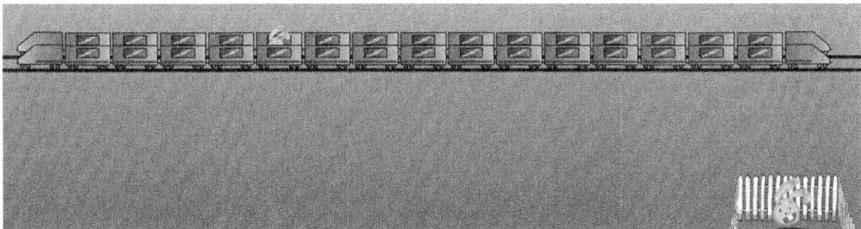

Fig. 3 The 'passenger train', 'discovering' mode

Fig. 4 Student 'score' sheet

(the reference train disappears when the empty train arrives). The discovering mode corresponds to the appropriation of the task, whereas the learning mode is designed for the learning of the mathematical knowledge.

In a previous research study (Bueno-Ravel and Gueudet 2009), we have shown that software features that provide teachers with *instrumented teaching techniques* for managing the students' heterogeneity promoted its integration into teachers' practices. Two main technical aspects supported this integration: the possibility to personalise through the choice of different settings for different children; and having access to the outcomes of the children's work on the computer. When working with the passenger train program the settings of the software can be customised by choosing: the number of passenger cars (from 10 to 30); the position of the rabbit in the train (near a locomotive, in the middle of the train, random); and the number of rabbits to place (from 1 to 3). Students' choices are not stored, so teachers do not yet have access to the outcomes of the children's activity unless they are observing them. Nevertheless, to progress from one attempt to another, the children have to fill in a score sheet (Fig. 4). The results of the last ten attempts appear in this score sheet, which provides some information for the teacher.

The Passenger Train Lessons

In Chloe and Mia's classrooms, the working environment and activity format (Ruthven 2012) were very similar. In both cases, the computer equipment was limited; one computer in Chloe's classroom, one to three computers in Mia's one and neither classroom had a digital projector. The activity format followed a typical approach in French kindergarten classrooms, involving whole-class discussion and group work. The groups, comprising of 4–6 children, practise the same activities in succession. There are two adults in the class, the teacher and an assistant so it is usual for two of the groups to have adult supervision, whilst the two remaining groups work by themselves. Chloe's and Mia's Passenger Train lessons are described in detail within Appendix 4. These lessons lasted five or six sessions. Each session was repeated four times during a week (in France, children go to school 4 days a week so teachers usually organise a group rotation each day for a same session, and a session lasts 1 week).

Introduction of the Software

Chloe's Class

Chloe introduced the mathematical situation, and the associated tools (paper materials and the software) simultaneously. She began by manipulating the software while the children watched. Then each student, in turn, tried to use the software, with the teacher nearby to help in case of difficulties. Within the software, each rabbit must be moved precisely using the mouse, which can be difficult for some children. When Chloe was not with a group of children in which some children had difficulty when moving a rabbit, she 'left' part of the responsibility of the technical knowledge to the children who had mastered the movement. They were in charge of helping others if needed.

Mia's Class

Mia introduced the mathematical situation with the paper material alone and she chose to start using the software in session 2. Even though she had organised group work for this session, she decided to introduce the software in a whole classroom setting. She sat near the computer and manipulated the mouse while all the children watched the computer screen, sitting on two rows of benches and on tables, in front of the screen, with the lights off (Fig. 5). Mia decided to introduce the software in a whole classroom setting because she thought that if she introduced it in a work group, the children working autonomously on other tasks would be more interested by the software than by the work they had to do.

For the introduction of the software (S1 for Chloe and S2 for Mia), the didactical configurations used were different, but we identify nevertheless the same two types

Fig. 5 Mia introduced the software in whole class (S2)

of orchestration *Technical-demo* and *Explain-the-screen* (Drijvers 2012): Our analysis points out that, contrary to the conclusions of Drijvers et al. (2010), the didactical configuration of the Technical-demo orchestration does not necessarily include "facilities for projecting the computer screen and a classroom arrangement that allows the children to follow the demonstration" (p. 219). What is clearly at stake, for this orchestration, is to design a didactical configuration allowing each student to follow the demonstration. Mia and Chloe both overcame the lack of a digital projector through their choices of classroom arrangement. In both cases, the main didactical objective that influenced the didactical configuration was that in order for the children to work independently of the teacher with the software, the teachers have to ensure that the children are 'technically' autonomous and that they understand how to realise the tasks on their own. So, as exploitation modes, teachers pay particular attention to the explanation of the tasks, the features of the interface and the related actions. The emphasis on the task explanation is necessary as the children are not able to read yet and must not see the software merely as a game.

We have shown that Mia and Chloe created different didactical configurations even though their working environment was similar. These choices were dependent on each teacher's professional knowledge about student behaviours and the pedagogical organisation of the classroom. Nevertheless, we notice that they both had anticipated the importance of guiding children's geneses by providing good explanations, even though the use of the software did not seem too complex.

Orchestrating Student Heterogeneity: Fostering Autonomy

We have identified two new orchestrations types in Chloe and Mia's classes: *Autonomous-use* and *Supported-use*. These orchestrations appeared more noticeably during sessions 3 and 4 in Chloe's class and during sessions 2 and 3 in Mia' class

(Appendix 4). Chloe and Mia designed these orchestrations in order to solve professional problems, which were how to manage the children's heterogeneity and, more precisely, how to provide support to the children most in need? In both cases, they responded by making themselves available for the children most in need. This implies that their main didactical objective in relation to these two new orchestrations was to design an orchestration that allows the teacher to be near the children most in need.

Autonomous-Use

As the children who do not have any difficulties are able to work autonomously with the software, the teacher is 'discharged' and can devote herself to the children who were most in need. The didactical configuration comprises at least one computer, a prior identification of the children who do not have difficulty through a diagnostic evaluation (instrumented or not) and individual or paired-work with the software. As an exploitation mode, the teacher mainly needs to anticipate and organise rotation (if needed) of the children working autonomously with the software. Of course, teachers do not leave the older children in complete autonomy. They occasionally check their work and intervene, when necessary, in relation to the mathematical or technical content. When teachers check the work of these children, the orchestration is similar to Drijver's (2012) 'Work-and-walk-by' orchestration type.

Supported-Use

For the children most in need the teacher stays nearby and provides help to the children as they work on the software. The didactical configuration comprises at least one computer, a prior identification of the children who do have difficulties through a diagnostic evaluation (instrumented or not) and teacher's presence (Fig. 6). In an exploitation mode, the teacher has to anticipate the main difficulties children can encounter and the remediation needed. These difficulties can concern the manipulation of the software, the mathematical content being addressed or the prerequisite mathematical content. In this orchestration, the teacher usually intervenes individually with children. However, we do not include this orchestration in a *Work-and-walk-by* type. Indeed, the didactical objectives of these two orchestrations differ significantly. For the *Supported-use* orchestration, the didactical objective is an explicit choice to support children most in need. For the *Work-and-walk-by* orchestration, Drijvers (2012) shows that the fact that "many teachers seem to prefer individual interactions to whole-class teaching" (p. 271) explains the high frequency of this orchestration. He does not take into account the possibility that handling the children's heterogeneity might explain such an orchestration choice.

The professional problem we have identified (how to deal with children's heterogeneity and precisely how to support children most in need) is very general. The pedagogical organisation of the class divided into four groups working in

Fig. 6 Mia is helping a student who has been identified as 'most in need', while a student on her left is working autonomously

parallel on different tasks can be interpreted as part of the teachers' answers to this problem. So *Autonomous-use* and *Supported-use* orchestrations are not specific of the use of software. However, we have identified two features of the Passenger Train program that facilitates such orchestrations:

- Passenger Train (paper or software) is a self-validating situation, supporting autonomous use;
- The Passenger Train provides the validation for the children's answers (Bueno-Ravel and Gueudet 2009) and allows many attempts, which also supports the autonomous work of children.

Autonomous-use orchestration leads Chloe and Mia to think about the design of resources that will enable them to keep a trace of the children's autonomous work as the Passenger Train software does not store their results. Within the teachers' joint description of their lessons that was written for colleagues, they proposed three resources that might overcome this problem:

- A worksheet on which children note each attempt, and whether they succeed or not;
- A road map on which children note each attempt, if they succeed or not, and indicate the parameters of the software (number of rabbits, number of passenger cars);
- An observation grid for teachers that highlights the main procedures and mistakes children might encounter.

We interpret this process as a genesis, encompassing intertwined instrumentation and instrumentalisation processes. Chloe and Mia know how important it is to have access to the children's work, particularly for children who cannot yet write.

Influence of a Digital Projector or IWB: Planned Orchestrations

In our analysis of the abacus-lessons, we have mentioned that the installation of an IWB in Deborah's classroom led her to modify her orchestration choices. (Chloe's and Mia's schools were not equipped with IWBs or digital projectors). We assume that had Chloe and Mia had access to projection facilities they would have modified their orchestrations. Within their lesson descriptions they suggested 'Sherpa-at-work' orchestrations for colleagues who had projection facilities. Indeed at interview, when asked how they would adapt the Passenger Train lessons, they said that having a digital projector or an IWB facilitated the introduction of the software, "At least one and a half hours and three work groups are saved during the software presentation phase if you have a digital projector [...] The large screen caught the attention of the children [...]" (our translation). However, most of their writings are centred on joint work concerning procedures or synthesis phases, instrumented by the Passenger Train software. "(digital projector and IWB) are interesting for synthesis and institutionalisation phases; the different strategies can be illustrated by the children, and, the attractiveness of the Passenger Train is a real context to facilitate talk [...]" (our translation). They developed three variations of *Sherpa-at-work* orchestration: (1) one Sherpa-student, 'guided' by the teacher, (2) one Sherpa-teacher, following on the computer actions that a student shows on the wide screen and (3) a Sherpa-pair, one student following on the computer actions that the other student shows on the wide screen (or the student near the wide screen follows the actions of the student on the computer).

The didactical objective of this orchestration is to foster verbal and non-verbal interactions (e.g. showing how to count the passenger cars with a finger) in the whole classroom setting in order to institutionalise expert procedures. Chloe and Mia have been led to develop such exploitation modes to make sure that children will learn something using this software and not only remember their successes or failures at the game: 'the teacher must be vigilant about the joint verbalisation (between the computer screen and the wide screen) and the reproduction of the counting mode on the wide screen [...]. Without such precautions, only the final result (the place of the rabbit and the validation) will be visible to the eyes of the whole class' (our translation).

Drijvers et al. (2010) have pointed out that, in a working context offering projection facilities, even though teacher guidelines describing *Sherpa-at-work* orchestration are given to teachers before their lessons, this type of orchestration is ignored 'to a certain extent'. They report that teachers' orchestration choices are consistent with 'their regular habits and their view on mathematics teaching'. In order to follow the lesson descriptions elaborated by Mia and Chloe it is necessary to be aware of research specific to the kindergarten school context. We will return to this perspective in our conclusion. In the next section, we discuss central issues emerging from both case studies.

Discussion

In this section we draw on the case studies in order to enlighten general issues about teachers' use of software in mathematics at the kindergarten level and, in particular, with 5–6-year-old children. Naturally, the scope of our study remains limited as we observed only a few teachers who are experienced and involved in a design and research group. Nevertheless, some of the facts that we observed are not restricted to their specific case. We first present issues which are directly linked with specific aspects of this level of schooling and then others concerning orchestration more generally.

Articulation of Resources

Kindergarten teachers in France usually use textbooks for their teaching of mathematics, which include a teacher's guide and specific worksheets for the children. These textbooks suggest exercises and some of them also include mathematical situations, developed in the course of collaborations between teachers and researchers. Moreover, diverse physical materials such as games, cards, tokens, cubes and figurines are also included. In previous studies we noticed that for the secondary school teachers' resource system, the articulation between a textbook and a software program is an important factor for technology integration (Gueudet and Trouche 2012). We observe the same here at the kindergarten level. For Chloe, the 'passenger train' is associated with a game on paper that she had already used before and for Deborah, two exercises from textbooks are associated with the work on the abacus (sessions S3 to S6). Moreover, many other kinds of material are included and the virtual abacus is naturally articulated with the physical abacus, whilst the Passenger Train software is used alongside the corresponding situation on paper. In fact the computer becomes one of many artefacts living in the classroom, which probably contributes to the richness and complexity of the didactical configurations, and thus of the orchestrations.

Types of Orchestration at Kindergarten Level

From the outset, we expected the orchestrations at this very early level to be significantly different from the orchestrations observed by Drijvers (2012) at the secondary school level. Some differences are linked with the available material. On the one hand, there are often only a few computers available in the classrooms; on the other hand, as mentioned above there is a wide variety of material used by the teachers at this level. So the computer is one resource, amongst many others. Nevertheless, the projection facilities seem to lead to a central role played by an image on the computer screen, projected for the whole class, or at least for a half-class. Differences are also linked with the importance of verbalisation in these class levels; it seems to foster orchestrations encompassing this verbalisation such as *Discuss-the-screen*.

The blackboard remains an important resource in the classroom, but for children aged 5–6 who cannot read, it is mostly filled with images, only some of these being connected to mathematics, which can explain the absence of the *Link-screen-board* orchestration. The material is not the only cause for the differences we observed. As mentioned in the previous section, there is much less whole classroom teaching and much more group work at this level. For the introduction of a new software program, the *Technical-demo*, *Explain-the-screen* and *Discuss-the-screen* orchestration types introduced by Drijvers (2012) are still present, if we assume that we can adapt them to groups of children (while they were introduced with a whole classroom presentation). But we also observed new orchestration types linked with the usual group work organisation, where children work on different tasks: the *Autonomous-use* and the *Supported-use*. We do not claim that these types of orchestrations are absent at the secondary school level. Nevertheless, they are probably much more frequent at the kindergarten level because of the usual *activity format* (Ruthven 2012), which includes group work on a regular basis and also because of the importance of managing heterogeneity at this school level.

The Software, Shaping the Orchestration Choices?

One of the questions raised by Drijvers in his work is the influence of the software's features on the orchestration choices. Our data clearly demonstrate such an influence, in several directions. Concerning the presence of the teacher with children working on the software, we noticed that it clearly depends on the feedback offered, or not, by this software. Naturally, if the software provides feedback, the teacher's presence is needed less. With the virtual abacus, the number inscribed is displayed, and children can compare it to the number they want to reach. Nevertheless, they will never receive a message such as 'wrong number displayed', so the teacher may still need to intervene. Another important aspect of the orchestration that depends on the software features is the presence of recording worksheets to complement the work on the computer. If the software provides access to the children's answers, these written notes are needed less, especially at this very early level. As the children cannot write, there is no objective linked with the writing of a mathematical procedure, but only the need for the teacher to have access to potential mistakes (if the rabbit has been misplaced, in which carriage was it? If a number has not been correctly inscribed, was there confusion between the rods etc.). We assume that a change in the software, permitting the recording of children's productions, would certainly change the orchestrations, requiring less written records.

Orchestration and Teacher Knowledge

Teacher Knowledge Shaping Orchestration Choices

The issues discussed above provide evidence that orchestration choices depend on the available material, the number of computers and the features of a software program. Orchestration choices also depend on the teacher's resource system and activity format. We argue that these are also linked with teacher knowledge and beliefs and give some important examples below. In these examples, we infer teacher knowledge by comparing the notes taken during the group's meetings, the lesson plans designed by the teacher, for herself and for colleagues, and the classroom videos.

Deborah has professional knowledge about number sense and its difficulty for children. In her orchestrations, she plans exploitation modes where she emphasises the different values of the beads depending on the particular rods, using the virtual abacus projected on a screen. This orchestration choice is clearly a consequence of teacher knowledge about number sense, about children's difficulties with it, and about the abacus.

For Chloe and Mia, their knowledge about the children's need to learn to use numbers as indicators of both quantity and position, determined their choice of the Passenger Train software, especially as only a few situations exist that enable work on the second aspect. Chloe and Mia placed great importance on the management of heterogeneity within the classroom. Thus they used the software as a resource to support this management. They developed a document, with a 'resource' part including the software, the associated children's worksheets, and all the material elements of the didactic configuration. This document also included professional knowledge about the management of heterogeneity as they considered that the skilled children could work by themselves whilst the teacher has to stay with the others, a choice that can also be considered as pedagogical knowledge.

Teacher Learning

During the lessons we observed unplanned elements within the teachers' didactical performances. We also observed evolutions in their creation of the orchestrations. The teacher-children's interactions were a major source for these evolutions alongside the major resource provided by the children's outputs. We consider this as evidence of teacher learning, involving different kinds of knowledge. The teachers certainly learned about children's reasoning. For example, in the case of the abacus, Deborah noticed that there were three possible explanations for how to display five on the abacus:

- Counting reasoning: Five one-unit counters are activated and displayed by five gestures by counting: one, two, three, four, five.

- Grouping reasoning: Five one-unit counters are activated and displayed in one gesture: the fifth bead is activated.
- Calculating reasoning: One five-unit counter is activated.

Deborah planned to revise her lesson plan for the abacus-lessons to include an additional objective to hold a discussion with the children to encourage them to reason in these three ways and make the connection with the calculations. In the case of the Passenger Train Chloe discovered that the children used two correct procedures, which were to use the number of the passenger car hosting the rabbit or to count the empty passenger cars on the left. In fact the use of number as memory of position is always linked with its use as memory of quantity. She also plans to discuss with the children these two possibilities during her next Passenger Train lessons.

In Deborah's class, the evolutions also concerned the responsibility given to the children, with a tendency towards giving them more responsibility. This evolution, from teacher-centred to children-centred orchestrations, can be a consequence of the participation of the teachers in the group (Drijvers observes that the teachers involved in his experiment proposed more student-centred orchestrations than in their usual practice). We consider anyway that it provides evidence for teacher learning and about the possibility to leave some responsibilities to the children when working in mathematics with technology.

Our focus here was not on teacher knowledge evolution and documentational geneses. Additional research would be needed to study these geneses and the links with the orchestration evolutions. Geneses are indeed long-term processes and a follow-up of teachers during several school years is necessary for their analysis. This will be the subject of another study; the same holds for the appropriation, by other teachers, of the resources designed during this project.

Conclusion: Orchestrations at the Kindergarten Level

In her review of research papers about the use of technology in the teaching of mathematics, Joubert (2013) identifies 'orchestrating learning' as the central theme, present in 74 % of the papers she considers. She also observes that less than 5 % of the papers concern primary school or kindergarten. Levy and Mioduser (2010) demonstrate that, in the context of kindergarten children learning mathematics with digital artefacts, the learning environment is of vital importance. Thus answering the question proposed here (§ 1.2) (which orchestrations are chosen by kindergarten teachers when using technology in their teaching of mathematics and which factors shape these choices of orchestrations?) is of central importance within mathematics education research.

Most of the orchestration types introduced by Drijvers from the secondary school context can also be observed at the kindergarten level, with adaptations resulting from the available material or from a usual activity format. In our work, the *Link-screen-board* orchestration does not intervene, but this might be a specific feature of the kindergarten context. We identified new types of orchestrations: *Autonomous-use* and *Supported-use*. These two types of orchestrations are linked with one of the teacher's objectives when orchestrating the teaching of mathematics, which is to take account of individual children's differences. The study of kindergarten, or primary school classrooms has shown that the teachers seem to focus their orchestrations more towards these differences. This issue resonates with a recommendation formulated by Hoyles et al. (2004), in their comments about the use of orchestration, to study the integration of digital technologies: "the individual difference is not something to be minimized or avoided, it is an inevitable part of orchestration itself" (p. 320). Our work certainly supports this claim.

The teachers also proposed variations on the *Sherpa-at-work* orchestration. We hypothesise that, at least in the French teaching context, the kindergarten (or primary school) level permits the development of more orchestration types than at the secondary school level. Kindergarten teachers follow the children for the whole day (a school day lasts 6 h), and are free to organise this time. They also teach several subjects which, in turn, certainly leads to different activity formats. This can foster the development of a specific teacher agency concerning orchestration; this hypothesis naturally needs to be investigated further.

Concerning factors shaping the orchestrations, we recorded influences from the material environment, the available resources (in particular from the software features), the usual activity formats, and from teacher knowledge, with all of these being tightly intertwined. Different kinds of teacher knowledge come into consideration: pedagogical knowledge about the management of heterogeneity; knowledge about curriculum material (mathematical exercises available in textbooks, for example); and knowledge about children's possible mistakes and difficulties. Knowledge about the teaching and learning of numbers at this school level (importance of number sense, distinction between number as memory of quantity and as memory of position) was especially important for the teachers' choice of a given software program. The kindergarten teachers involved in our study were not specialists in mathematics. In spite of this, the mathematical content and its didactical aspects were central in their choices. In this chapter we did not focus on professional development. Nevertheless, we point out that our observations are coherent with those of Erfjord et al. (2012), who comment that when involved in a research group concerning the orchestration of mathematical activities, kindergarten teachers adopt an inquiry stance. It seems to contribute, in particular, to the development of their awareness of the mathematical ideas involved. However, the identification of the specific interventions of different kinds of teacher professional knowledge, and how they articulate with instrumental knowledge about the possible use of a given software program, requires an additional study.

Much research about the use of technology for the teaching of mathematics to young children, in particular at the kindergarten level, is still needed. We consider that the perspective of the documentational approach is fruitful for these studies. Especially at the kindergarten level and in primary school, technology is only one teaching resource amongst many others, and investigating teachers' work and professional development requires taking into account their interactions with these resources.

Acknowledgements We warmly thank for their contribution all the members of the 'mathematical package' project. We especially thank Typhaine Le Méhauté and Michel Guillemeau for their support in the design of the Passenger Train software.

Appendix 1: The Abacus-Course in Deborah's Class (5–6 Years Old)

	Session title	Resources used by the teacher and children / Exploitation mode
S1	Discovering the abacus: setting numbers up to 6	Resources:
S2	Discovering the abacus: setting and reading numbers up to 6	Virtual abacus and digital projector (teacher)
S3	Setting and reading numbers (up to 12). Adding (1+2, 3+1, etc.) with the game greli-grelo (from teacher textbook): teacher has beads in hands and children are asked to say the total number of beads. Verification is made on physical abacus by children	Physical abacus (children) Teacher textbooks for preparation Exploitation: One physical abacus per student Children have to set numbers on the physical abacus and correction is made on the virtual abacus by teacher. Children do not use the virtual abacus before S8 Children have to read numbers set on the virtual abacus by teacher
S4	Setting and reading numbers (up to 25). Adding with the game greli-grelo	
S5	Setting and reading numbers. Adding with the treasure game (from teacher textbook): children win golden coins with a dice and they are asked to say the total number of coins. Verification is made on the physical abacus by children	
S6	Setting and reading numbers. Adding with the treasure game	
S7	Setting and reading numbers. Adding with the treasure game	
S8	Setting and reading numbers (0–5)	Resources: Virtual abacus and IWB for both teacher and children
S9	Reading numbers (5–10)	Pencil and paper work prepared for children Sessions 11 and 12: the 'abacus book' Exploitation: One computer per student with the virtual abacus (to help children to fill the paper work)
S10	Setting numbers (5–10)	
S11	Reading numbers (10–15)	
S12	Setting numbers (10–15)	Sessions' introduction on the virtual abacus by teacher and children (set numbers in different ways with the IWB) Individual work for children (pencil-paper and virtual abacus on computer)

Appendix 2: Examples of children' Work on the Chinese Abacus, Deborah's Class (5–6 Years Old)

Session 8

1. Read a number: the examples of 5 and 0

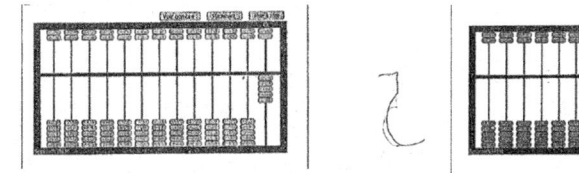 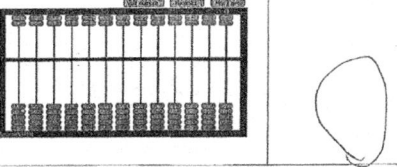

This student recognises number 5 set on the abacus but wrote it in 'a mirror form' which is usual at this level

Children learn to recognise a particular number that is 0

2. Set a number: the example of 5

This student drew the beads between two rods and after on the unit rods, activating five one-unit-counter

This student first wrote the numeral 5 on the unit rod and the teacher asked to draw the beads

This student drew the activated beads and the non activated as well. Most children drew only activated beads spontaneously

This student drew one five-unit-counter on the units rods to set 5

Session 9

1. Read a number: the example of 6

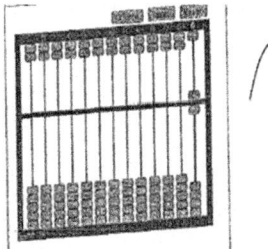 

This student recognised number 6 set on the abacus but wrote it in the wrong way (see session 8)

2. Set a number: the examples of 5 and 10 (for the same student)

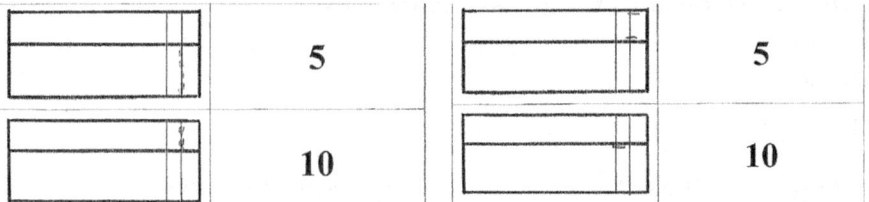

Here the number 5 is set as 5 one-unit counters and 10 as 2 five-unit counters. Only activated beads are drawn

Here the number 5 is set as 1 five-unit counter and a non activated bead is drawn. The number 10 is set in the economical inscription i-e 1 one-unit counter in the tens. To represent beads, this student draws a short line

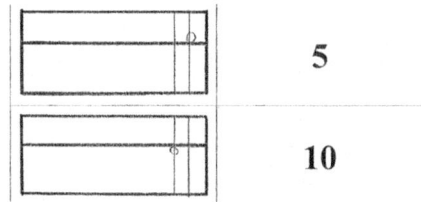
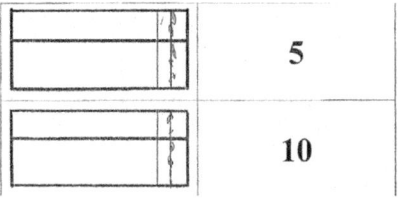

Here beads are circles (not filled) and the two numbers are set in the economical way

To set number 5, 5 beads are activated: there is a misunderstanding between quantity of activated beads and the quantity represented by the 5-unit counters. To set 10, it seems that the seven beads per rod are activated

Appendix 3: Deborah's Class Transcriptions in French

Session3: Beginning of the session. A discussion about how to set 5 on the physical abacus raised between children. Deborah was sitting with the children and goes to the screen to show children' propositions.

Deborah:	Vous m'avez proposé… Laurie, tu m'as proposé d'activer les cinq boules de la tige rouge, d'accord… Pourquoi ? *[La tige des unités est rouge sur le boulier virtuel, les autres sont vertes, Deborah active 5 comme cinq unaires dans les unités]*.
Laurie :	Parce que celles du haut, elles servent à rien.
Deborah:	Parce que tu penses que celles du haut, elles ne servent à rien. Est-ce que vous êtes d'accord avec le choix de Laurie?
Quelques élèves:	Non!

Deborah:	Non ?
Une élève :	Parce que les boules du haut, elles valent 5.
Deborah :	On va vérifier, j'avais demandé de faire combien ?
Une élève :	Cinq !
Deborah :	Très bien. *[Deborah active l'icône « voir nombre » et le chiffre 5 apparaît à l'écran]*
Plusieurs élèves :	Cinq.
Deborah:	Alors, est-ce que le choix de Laurie est juste?
Quelques élèves :	Oui !
Deborah :	Maintenant, il y a une autre possibilité par cinq. Certains élèves ont activé une quinaire, c'est-à-dire une boule du haut. *[Deborah montre à l'écran le chiffre 5 comme une quinaire, l'icône « voir nombre » est désactivé].* Alors, combien on a de possibilités pour inscrire cinq ? Maëlle ?
Maëlle :	Deux.
Deborah :	Oui, deux façons. Soit on active les cinq boules du bas, soit j'active une boule du haut. *[Deborah montre à l'écran les deux possibilités]*

Session 9. Children come to the IWB to set 8. « Discuss-the-screen » orchestration raises.

Deborah:	Je voudrais que vous activiez… Huit ! On réfléchit, comment est-ce qu'on fait huit ? *[Certains élèves veulent donner immédiatement la réponse].* Huit c'est ?
Quelques élèves :	Cinq et trois ! Montrez-moi avec vos mains. Cinq et trois ! Kevin. *[Kevin va au tableau, il active une quinaire et trois unaires (3ème boule, un geste)]*
Deborah :	Cinq et trois. Tu as bien activé cinq et trois *[Deborah s'approche du tableau pour montrer les boules activées].* Vous êtes d'accord avec son choix ? Est-ce qu'il y aurait une autre solution ? Une autre façon d'inscrire le chiffre trois ? Le chiffre huit ?
Quelques élèves :	Oui
Deborah :	Oui, Anaïs. [Anaïs va au tableau et prend la crayon]. Vas-y, tu n'as pas appuyé assez fort, je pense. [Anaïs active trois unaires, en trois gestes : un deux, trois, puis une quinaire]
Deborah :	Alors, c'est parce que toi, tu as activé les boules les une à la suite des autres. C'est très bien. Maëlle. *[Maëlle va au tableau et active trois unaires dans les dizaines et une quinaire dans les unités, ce qui fait 35, l'icône « voir nombre » est activée.]*
Deborah :	No *[Maëlle essaie avec trois unaires dans les dizaines et cinq unaires dans les unités, ce qui fait également 35.]*
Deborah : ?	Ah, tu as activé en effet huit boules, mais est-ce que tu as inscrit le nombre huit
Maëlle :	Non. *[Regardant le nombre 35 écrit en chiffres au tableau]*
Deborah :	Est-ce que tu as compris ton erreur ?
Maëlle :	Ah oui, trois et cinq, ça fait 30 et 5 !
Deborah :	Trois et cinq, ça fait 35. *[Montrant les deux différentes tiges.]* Et surtout elles ne sont pas situées sur la même tige.

Appendix 4: The 'Passenger Train' in Chloe and Mia's Classes (5–6 Years Old)

The 'passenger train' in **Chloe's class** (5–6 years old)

	Configuration/Exploitation mode	Resources used by the teacher
S1	**Presentation** of 'passenger train' on paper (by the teacher, for a group of six children)	Trains and rabbits on paper
	Presentation of the 'passenger train' on the computer, by the teacher, then successively by each student, working in a pair (the four other children are spectators)	Computer with the software
S2	**Learning phase** (30 min): The children are grouped by six and organised in pairs inside the group	Trains and rabbits on paper
	Synthesis: whole class, when all the groups have done the 'passenger train' situation	
S3	**Learning phase** (30 min): The children are grouped by six and organised in pairs inside the group	Trains and rabbits on paper
	Training on the software: individual work for advanced children	One computer with the software
	Synthesis: whole class, when all the groups have done the 'passenger train' situation	
S4	**Training on the software** (10 min): pair work for most of the children, in autonomy, rotation each 10 min. Morning	One computer with the software
	Scaffolding (30 min): group work for children who have difficulties, with the teacher. Afternoon	Trains and rabbits on paper + Computer with the software
	Synthesis: whole class, when all the groups have done the 'passenger train' situation	
S5	**Training on the software** (10 min): pair work (homogeneous), in autonomy, rotation each 10 min. Teacher chooses the difficulty level (numbers of passenger cars, numbers of rabbits) according to the level of the 'pair'	One computer with the software
S6	**Training on the software** (10 min): pair work (homogeneous), in autonomy, rotation each 10 min. Teacher chooses the difficulty level (numbers of passenger cars, numbers of rabbits) according to the level of the 'pair'	One computer with the software

The 'passenger train' in **Mia's class** (5–6 years old)

	Configuration/Exploitation mode	Resources used by the teacher
S1	**Diagnostic assessment:** children in groups of five or six and work individually	Trains and rabbits on paper One computer with the software (Diagnostic assessment was made on computer for one group of five children)

(continued)

(continued)

	Configuration/Exploitation mode	Resources used by the teacher
S2	**Presentation** of the 'passenger train' on the computer, by the teacher, for the whole class	One computer with the software
	Learning phase (30 min): children in groups of five or six. Children with no difficulties work in pairs taking turns on the computer (in autonomy). The others (one or two children) work individually, teacher is nearby	Three computers with the software
	Synthesis: whole class, when all the groups have done the 'passenger train' situation	
S3	**Training on the software** (30 min): pair work for most of the children, in autonomy, rotation each 10 min	Three computers with the software
	Scaffolding (30 min): group work for children who have difficulties, with the teacher	
S4	**Training on the software**: pair work (homogeneous), in autonomy, rotation. Teacher chooses the difficulty level (numbers of passenger cars, numbers of rabbits) according to the level of the 'pair'	Three computers with the software
		Training sheet created by Mia
	Training on the paper: individual work (homogeneous)	
S5	**Training on the software** (10 min): pair work (homogeneous), in autonomy, rotation each 10 min. Teacher chooses the difficulty level (numbers of passenger cars, numbers of rabbits) according to the level of the 'pair'	One computer with the software

References

Ball, D. L., Thames, M. H., & Phelps, G. (2008). Content knowledge for teaching: What makes it special? *Journal of Teacher Education, 59*(5), 389–407.

Bueno-Ravel, L., & Gueudet, G. (2009). Online resources in mathematics: Teachers' geneses and didactical techniques. *International Journal of Computers for Mathematical Learning, 14*(1), 1–20.

Clark-Wilson, A. (2010). *How does a multi-representational mathematical ICT tool mediate teachers' mathematical and pedagogical knowledge concerning variance and invariance?* Unpublished doctoral thesis, University of London, London.

Drijvers, P. (2012). Teachers transforming resources into orchestrations. In G. Gueudet, B. Pepin, & L. Trouche (Eds.), *From textbooks to 'Lived' resources: Mathematics curriculum materials and teacher documentation* (pp. 265–281). New York: Springer.

Drijvers, P., Doorman, M., Boon, P., Reed, H., & Gravemeijer, K. (2010). The teacher and the tool: Instrumental orchestrations in the technology-rich mathematics classroom. *Educational Studies in Mathematics, 75*(2), 213–234.

Erfjord, I., Hundeland, P. S., & Carlsen, M. (2012). Kindergarten teachers' accounts of their developing mathematical practice. *ZDM, the International Journal on Mathematics Education, 44*, 653–664.

Gueudet, G., & Trouche, L. (2009). Towards new documentation systems for mathematics teachers? *Educational Studies in Mathematics, 71*(3), 199–218.

Gueudet, G., & Trouche, L. (2012). Teachers' work with resources: Documentational geneses and professional geneses. In G. Gueudet, B. Pepin, & L. Trouche (Eds.), *From textbooks to 'Lived' resources: Mathematics curriculum materials and teacher documentation* (pp. 23–41). New York: Springer.

Hoyles, C., Noss, R., & Kent, P. (2004). On the integration of digital technologies in the mathematics classrooms. *International Journal of Computers for Mathematical Learning, 9*, 309–326.

Joubert, M. (2013). Using digital technologies in mathematics teaching: Developing an understanding of the landscape using three "grand challenge" themes. *Educational Studies in Mathematics, 82*(3), 341–359.

Levy, S. T., & Mioduser, D. (2010). Approaching complexity through planful play: Kindergarten children's strategies in constructing an autonomous Robot's behavior. *International Journal of Computers for Mathematical Learning, 15*, 21–43.

Poisard, C., & Gueudet, G. (2010). Démarches d'investigation : exemples avec le boulier virtuel, la calculatrice et le TBI. Journées mathématiques de l'INRP, Lyon. http://www.inrp.fr/editions/editions-electroniques/apprendre-enseigner-se-former-en-mathematiques

Remillard, J. (2012). Modes of engagement: Understanding teachers' transactions with mathematics curriculum. In G. Gueudet, B. Pepin, & L. Trouche (Eds.), *From textbooks to 'Lived' resources: Mathematics curriculum materials and teacher documentation* (pp. 105–122). New York: Springer.

Ruthven, K. (2012). Constituting digital tools and materials as classroom resources. In G. Gueudet, B. Pepin, & L. Trouche (Eds.), *From textbooks to 'Lived' resources: Mathematics curriculum materials and teacher documentation* (pp. 83–103). New York: Springer.

Trouche, L. (2004). Managing complexity of human/machine interactions in computerized learning environments: Guiding students' command process through instrumental orchestrations. *International Journal of Computers for Mathematical Learning, 9*, 281–307.

Van Oers, B. (2010). Emergent mathematical thinking in the context of play. *Educational Studies in Mathematics, 74*, 23–37.

Verillon, P., & Rabardel, P. (1995). Cognition and Artefacts: A contribution to the study of thought in relation to instrumented activity. *European Journal of Psychology of Education, 10*, 77–102.

Teachers' Instrumental Geneses When Integrating Spreadsheet Software

Mariam Haspekian

Abstract The spreadsheet is not *a priori* a didactical tool for mathematics education. It may progressively become such an instrument through the process of professional geneses on the part of teachers. This chapter describes the beginning of such a genesis, and presents some results concerning teachers' professional development with ICT by examining the outcomes of two different sets of data. Theoretical notions, such as instrumental distance and double instrumental genesis supported the analysis of data leading to a comparison of a teacher integrating spreadsheets, for the first time in her practices, with the practices of teachers who are more expert with spreadsheets. The similarities found in the ways they use the tool leads to some hypotheses on the importance of these common elements as key issues in teachers' ICT practices.

Keywords Mathematics teaching and learning • Teaching practices • ICT integration • Professional learning of mathematics teachers • Technology-mediated classroom practices • Spreadsheet • Professional/personal instrument • Double instrumental geneses (professional/personal) • Instrumental distance • Novice/expert teacher

Introduction

Around the 1980s, the idea that ICT could serve school learning, in particular mathematical learning, began to develop. Nowadays the use of ICT in classrooms is prescribed in the curricula of many countries and it includes detailed recommendations for teachers (Eurydice 2004, p. 24). However, many reports comment upon the

---

M. Haspekian (✉)
EDA, University Paris Descartes, Paris, France
e-mail: mariam.haspekian@parisdescartes.fr

poor integration of ICT in mathematics teaching. After an enthusiastic period in which the benefits of the use of ICT for learning mathematics have been claimed, researchers now describe a phenomenon of disappointment. It is a fact that the potential for ICT use in mathematics is rather poorly exploited and that ultimately, technology integration is very limited. For example, using data from PISA 2003, Eurydice (2005) reported that fewer than half of the students (from a more than 90 000 students survey) were familiar with activities with spreadsheets such as plotting a graph. One reason for this, which has been suggested by many studies, is the 'teacher barrier' (see for instance Ruthven 2007 or Balanskat et al. 2006). Hence it seems crucial to advance our knowledge of teachers' 'usual practices' alongside their technology-mediated ones: How do ICT practices develop and evolve in time? What do we know about the instrumental geneses with ICT and about teachers' resistances? My own doctoral research (Haspekian 2005a) led me to look for reasons beyond those that are often cited: lack of time, lack of training, lack of material, conservatism etc. Without denying these factors, my research claimed that there are deeper reasons for teachers' resistance, related to the *impact* that technology has on the mathematics to be taught, and the difficulty, for teachers in managing this impact. Therefore, it remains important to advance our understanding of this impact and the ways that teachers account for it.

With this purpose in mind, this chapter aims to provide an insight into teachers' practices with technology by comparing the results of different studies concerning a common technology, the spreadsheet (Haspekian 2005a, 2006, 2011). The first study formed different elements of my doctoral study. These were: an observation of a teacher, called Ann,[1] who was integrating spreadsheet for the first time in her practices; and an inquiry interviewing and comparing pre-service teachers with teachers who were considered 'experts' with spreadsheets.[2] The second and third studies resulted from a different research project observing ICT sessions in ordinary classrooms, during which I happened to return to Ann's classroom. Thus, I had the opportunity to observe her practice a year later. Consequently, these three studies provided an opportunity to make an interesting comparison concerning teachers' practices with spreadsheets at different stages of integration within mathematics teaching:

- Pre-service teachers that were novice in teaching and in using spreadsheets in mathematics teaching,
- Teachers who are expert with spreadsheets and teaching mathematics using spreadsheet;
- A teacher who is neither a novice, nor an expert with ICT in general.

This comparison involved two theoretical frameworks. The instrumental approach (Artigue 2002; Guin et al. 2004), which was developed around the concept

---

[1] The name taken in the initial French research is 'Dan'; in this chapter, it is translated to 'Ann' as the teacher is a woman.

[2] This term is explained in section 3.

of instrumental genesis, supported an analysis of the impact of the spreadsheet on mathematics. This led me to determine both the didactical potential of spreadsheets and the difficulties that might occur as the spreadsheet changed the mathematics to be taught. The second framework was the didactic and ergonomic approach (Robert and Rogalski 2002), which helped to describe teachers' activity. In the second section of this chapter, this is used alongside the instrumental approach to understand Ann's evolution over two years. The third section probes Ann's practices more deeply by comparing her evolution with the practices of the 'expert' teachers. This will highlight some results about the development of ICT use in teachers' practices concerning the way that their practices evolve and the difficulties they encounter when integrating spreadsheet technology.

ICT and Mathematics Education: The Case of the Spreadsheet

An increasing number of technologies can be found in today's mathematical school landscape, from pocket calculators adapted for the elementary school through to universities' virtual learning environments that include interactive exercises and complete courses for various domains of mathematics. In France, the spreadsheet is officially prescribed for use in junior high and high schools, especially for the teaching and learning of algebra. However, this tool was neither created for, nor has it been adapted to, mathematics learning. The origins of the spreadsheet are, quite remote from the educational world, in accountancy (see Bruillard and Blondel 2007 for a historical and economical approach of the creation of the spreadsheet). Yet, to know how to calculate with a spreadsheet, in particular by using a formula, is a competency required in the curricula of an increasing number of countries worldwide (Pelgrum and Anderson 2001). Prior to the existence of spreadsheets, the use of computer tools required competencies in programming and thus, the learning of a programming language. The spreadsheet provided, for the first time, a way to avoid the need to program, leading Baker and Sugden (2003, p. 18) to say, "Nowhere is its application becoming more marked than in the field of education". However, in spite of some isolated experiments to adapt them for education, the spreadsheet remains a tool for the business world, with an increasingly sophisticated set of functionalities that have been designed in response to business rather than educational demands.

The poor integration of spreadsheets within mathematics teaching contrasts with other educational software such as dynamic geometry software.[3] This seems to offer a contradiction in that, even if some researchers question the relevance of

---

[3] There is no research at world scale comparing integration of geometry software and spreadsheets, but all local studies that can be found indicate a better penetration of geometry software than spreadsheets (see the examples cited in Haspekian 2005a).

spreadsheets in mathematics education, the majority of the research highlights the potential benefits of the spreadsheet for students. A brief synthesis on this theme turns the attention to the teaching and learning of algebra. The next section examines these tendencies in the light of the instrumental approach in order to analyse further the characteristics and complex relations of the spreadsheet with mathematics.

Potential Uses of the Spreadsheet for Mathematics Learning: An Overview of Research Literature

I begin by asking "What mathematical topics can be engaged through the use of spreadsheets at school?" The field that comes to mind most naturally is that of statistics. However, a closer examination of the operations of the spreadsheet reveals the algebraic nature of such activity. Without going into technical details,[4] one can note that from a historical point of view, the relation with algebraic concepts had been long identified. According to Bruillard and Blondel (2007):

> le premier tableur connu serait le 'calcolatore tabulare meccanico automatico' ou calculateur tabulaire mécanique automatique de Giovanni Rossi (1870), qui a permis une avancée décisive dans la relation entre l'algèbre matricielle et les matrices comptables. (Cilloni and Marinoni 2006; Cilloni 2007)
>
> The first known spreadsheet would be the 'calcolatore tabulare meccanico automatico' or automatic mechanical tabular calculator from Giovanni Rossi (1870), who permitted a key advance in the relationship between matrix algebra and financial matrices. (Cilloni and Marinoni 2006; Cilloni 2007)

The ability to link cells by formulas is an effective feature of the spreadsheet that many research studies have affirmed to offer potential to support the learning of algebra (algebraic objects, modes of treatment, problem solving) by analysing the new opportunities that spreadsheets offer alongside the operational constraints of their use. The new possibilities concern:

- The interactivity, allowing feedback richer than paper and pencil (for example, the numeric feedback of a formula helps students to conjecture or detect errors);
- The capacity for calculation (automatic recopying of formulas, and instantaneous display of results);
- The articulation of multiple registers of representation (natural language, formulas, numbers and graphics).

The benefits of spreadsheets, which can derive from the constraints of use, relate to both the symbolic language and the methods for solving mathematical problems with them. The symbolic requirement is due to the tool itself as opposed to didactic contract that is usually entered into when students begin to encounter algebra involving the unmotivated use of letters that competes with non-algebraic

---

[4] The reader can find a brief explanation of the basic functionalities, in a didactic approach, in Haspekian 2005a, pp.18–23.

strategies.[5] Spreadsheets also compel students to plan their work, organise their worksheet and, in doing so, anticipate the possible feedback from the technology.

For most researchers (Ainley et al. 2003; Arzarello et al. 2001; Capponi 2000; Dettori et al. 2001; Rojano and Sutherland 1997), these potential benefits place the spreadsheet between arithmetic and algebra. This intermediate position is seen to be ideal for the learning of algebra. For instance, Rojano and Sutherland (1997) conclude that the spreadsheet supports a smooth transition for pupils' initial numeric methods towards algebraic ones. In a previous study I showed that by comparing arithmetic, algebraic and spreadsheet solution methods for the same problem,[6] the spreadsheet adds some algebraic characteristics to an arithmetic procedure (Haspekian 2005b). For others, spreadsheets could help to overcome the semantic/syntactic difficulties of algebra. In Arzarello et al. (2001), the complexity of algebra is interpreted as a difficulty for pupils to enter the 'game of interpretation' between the algorithmic and symbolic functions of algebra. The various registers of representation of the spreadsheet are then seen as a tool helping the pupils to enter this 'game' through the construction and interpretation of formulae.

These potential benefits of spreadsheets contrast with the previous discussion of their weak integration. In the reality of the classroom, after having been introduced to them within the study of algebra, students use them rarely during their time at secondary school. The results of the DidaTab project (Bruillard et al. 2008) showed that the high school students from regions where the spreadsheet is most used do not have higher competences than average, except for the competencies of selecting and formatting cells. More generally, the research concludes that all of the 288 students involved in the study:

> seem to manage the 'surface' components, such as formatting the cells and the tables, but the mastery of the essential functioning of the spreadsheet, the writing of formulas, and the knowledge of its constituent elements (operators, operands, references, functions...) is not demonstrated by the large majority of students.

Capponi (2000) adopts a more moderate position about the potentiality for spreadsheets. His view is that the intermediate position of the spreadsheet between arithmetic and algebra may allow the pupil remain entirely on the arithmetic side without ever noticing the algebraic aspects.[7] Capponi quotes, for example, the display or editing of a formula which centres the user on the numeric aspects (computation results, designation of numbers) to the detriment of the underlying algebraic aspects (formulas, and cell references that play the role of variables).

So the question becomes, how can we support pupils to build algebraic knowledge with this tool? All of the above-mentioned researchers underline the

---

[5] One can see in Coulange (1998) at which point the algebraic methods rest on rules of didactic contract and remain fragile for pupils ages 15–16 who, facing atypical problems, provide correct answers in rupture with the algebraic rules of the didactic contract.

[6] Analyse/synthesis, trial/refinement and equations.

[7] because the algebraic character of the formulas is restricted to their utility in carrying out and automating calculations, the focus is not on providing an operational language to analyse and handle relations (Capponi and Balacheff 1989).

importance of the didactical design of the situations but say little about these situations, such as how to create them, and on which variables to focus the teaching. In many spreadsheet resources that have been published on professional websites one can identify the mathematical variables used, while the 'instrumental' variables (the tool features) remain mostly implicit. Yet, if these elements are not examined, they may generate misunderstandings, resulting in the pupils using spreadsheets in ways 'other than' what is expected. The organisation of the teaching (didactical and mathematical), the way the tool is introduced, its links to mathematics, the techniques taught, their links with the mathematical techniques already learned (or to be learned) in paper and pencil environment, the role of the teacher and her didactic managements are all elements that must be created by the teacher. For instance, how and when does the teacher introduce into the lesson the important technical specificities of spreadsheets, such as the functionality of dragging? How does the teacher structure the teaching so that the ideal didactic potential of the spreadsheet becomes actual? Again, the question of linking the tool features with mathematical concepts arises, revealing that the work will be different from work in the paper and pencil environment. What exactly are these differences and what impact could they have? These questions echo those that were central to research leading to the instrumental approach (Artigue 2002; Lagrange 1999; Drijvers 2000; Guin et al. 2004). This particular theory showed the importance of instrumentation and its relation to conceptualisation within CAS environments, another type of tool, like spreadsheets, that was not initially created for teaching. These issues lead directly to the theoretical construct that is *instrumentation*, which allows us to understand more clearly the problems of technological integration, by showing the need to take account of the process of *instrumental geneses*.

The Instrumental Approach: Some Theoretical Elements

ICT use in mathematics education is a domain within the more general area of technology use in human activity, which has been studied within the field of cognitive ergonomics. A psychological and socio-cultural theory of instrumentation, developed in this field, provides a frame for tackling the issue of learning in complex technological environments (Vérillon and Rabardel 1995; Rabardel 1993, 2002). The instrumental approach in didactics took some elements of this frame, including two of its key ideas: the artefact/instrument distinction, and the fact that using a tool is not a one-way process; rather, there is dialectic between the subject acting on/her personal instrument and the instrument acting on the subject's thinking.[8] Within the

---

[8] Because of this dialectic "it is not possible to clearly distinguish between these two processes" (Trouche 2004).

activity of a subject, an artefact[9] becomes an instrument through a long individual process of instrumental genesis, which combines two interrelated processes: 'intrumentalisation' (the various functionalities of the artefact are progressively discovered, and may be transformed in personal ways) and 'instrumentation' (the progressive construction of cognitive schemes of instrumented actions).

The two processes also indicate that the instrumental geneses are not *neutral* for the subject: instruments have impact on *conceptualisation*. For example, using a graphic calculator to represent a function may play on pupils' conceptualisations of the notion of limit. This idea of non-neutral 'mediation' provides a way to report on the strong overlaps that exist, and have always existed, between mathematics and the instruments of the mathematical work. This idea has been used in several research studies on symbolic calculators in mathematics education (Artigue 2002; Lagrange 1999; Drijvers 2000; Guin et al. 2004, Trouche 2004).

In what follows I articulate in more detail the two notions that were used.

Instrumental distance (Haspekian 2005b), which will be used to analyse relations between spreadsheet and mathematics.

Instrumental genesis which will give more precisely a phenomenon of *double instrumental genesis* within the context of analysing teaching practices. Indeed, for students, the spreadsheet may become a mathematical instrument through an instrumental genesis. However, as a spreadsheet is not by definition a didactical tool to serve mathematics education, it also has to progressively become such an instrument during a professional genesis on the part of teachers (Haspekian 2006). These are two different instruments, which both exist for the teacher.

Instrumental Distance

In French curricula, dynamic geometry software is prescribed with as much emphasis as spreadsheets. However, the former find a better integration in mathematics classrooms than the second does. The notion of *distance* to the referential environment seems to play an important role in the explanation of this phenomenon (Haspekian 2005a). It intends to take into account, beyond the 'computer transposition' (Balacheff 1994), the set of changes (cultural, epistemological or institutional) introduced by the use of a specific tool in mathematics 'praxis'. For a given tool, if the distance to the 'current school habits' is too great, this acts as a constraint on its integration (Haspekian 2005b). On the other hand, the didactical potential of technology relies on the distance it introduces regards to paper-pencil mathematics as, for instance, by providing new representations, new problems, increasing calculation possibilities, etc. This is the case for the dynamic figures in geometry software, with respect to the static figures in paper-pencil geometry.

---

[9] We limit ourselves to the case of the material artefacts, but the ergonomic approach is extended to 'psychological' artefacts: symbols, signs, cards, etc.

The didactic potential of these dynamic objects and their benefits for students' learning have been evidenced by many research studies, (see for example Laborde 2001). For the concept of 'figure', a central object in geometry, the dynamic geometry does not only broaden the conception of such objects but it offers a representation that corresponds more closely to the abstract concept of 'figure' than its paper-pencil equivalent. The dynamic dimension helps to realise the famous distinction of *spacial drawing/geometrical figure* (Laborde 2001; Parzysz 1988; Laborde and Capponi 1994). One can also consider the interesting possibility of creating new types of geometrical problems for students by varying the different tools available in the toolbars of this software. Geometric construction problems can be completely different as a result of the suppression of traditional geometric tools or through the addition of new tools by the creation of macro-constructions.

Four types of elements have been brought out that can generate such instrumental distance (Haspekian 2005a). Some of these elements relate directly to the *computer transposition*, such as the representations and the associated symbolism. Some others are of different nature: *institutional*, or *didactical* (vocabulary, field of problems whose solution they allow, etc.), and *epistemological* (i.e. what gives a tool an epistemological legitimacy). For example, the vocabulary in spreadsheets is far from the mathematical one; teachers must even create it for themselves.[10] There is no official reference to help the mathematics teacher to relate this vocabulary (and the objects within spreadsheets) to their mathematical equivalents. Many questions arise for teachers, such as:

- What is a cell?
- Is it a variable?
- What is a column (or a row)?
- Is it a set of several variables, or another representation of a *unique* variable?
- What is a relative address? Is there an algebraic equivalent?
- What is 'filling/dragging down' (a gesture embodying the concept of formula?)
- Is the numeric feedback: a number? a result of a formula? the permanent appearance of the cell containing a formula whereas the formula itself would be its temporary appearance? etc.

In fact, beyond the computer transposition that modifies the mathematical objects, the modification, from an institutional point of view, actually concerns the whole ecology of these objects as the tasks, techniques, and theories can all be modified. The idea of *'distance'* reflects this gap between the praxeologies[11] associated to two different environments (considering paper-pencil as a peculiar environment of mathematical work). As for the epistemological aspect, distance relates to the teachers' personal component (their representations of mathematics, of teaching,

---

[10] This raises difficulties for teachers, see the experiment described in Haspekian 2005b.

[11] Mathematical objects are not isolated, in educational institutions they live through mathematical and didactical organisations that are praxeologies: a quadruplet composed of tasks, techniques, technologies (discourse about the techniques: explanations, justifications...) and theories. See (Chevallard 2007).

Fig. 1 A2 is the cell argument, B2 calculates the square of the value in A2

	A	B
1		
2	5	= A2^2

of the role this tool plays in the development of mathematics etc.). This idea is developed later in the chapter.

In what follows, I apply this instrumental approach to the spreadsheet for the teaching and learning of algebra in order to study the impact of the spreadsheet on algebra (the objects, techniques and symbolisations) through the notion of *distance* between paper-pencil algebra and algebra with spreadsheets. The relationship between spreadsheets and mathematics is not simple as mathematical knowledge is needed to achieve spreadsheet mastery.

Mathematics Within Spreadsheet Objects

Some computer characteristics within spreadsheets do not strictly correspond to mathematical knowledge transposed to a computer environment, or even to a computer transposition of school knowledge, however they are linked with mathematics. The basic principle of the spreadsheet, which consists of connecting cells by formulas, gives an example of these objects, linking spreadsheets to the domain of algebra. This particular relationship with mathematics is precisely the reason why many studies in didactics from different countries give spreadsheets a positive role in the learning of elementary algebra, identifying them as tools of an arithmetic-algebraic nature (Ainley (1999); Arzarello et al. (2001); Capponi (2000); Dettori et al. (1995) or Rojano and Sutherland (1997)). But, in spite of the apparent simplicity of use of speadsheets, it is not so evident for teachers to take advantage of their characteristics. In (Haspekian 2005a) I showed that the tool generates some complexity as it transforms the objects of learning and the solution strategies by creating new modalities of actions, new objects, and by modifying the usual objects, such as: variable, unknown, formula; and equation.

For example, in the paper and pencil environment, variables in formulae are written by means of symbols (generally a letter for the school levels concerned here). This 'letter variable' relates to a set of possible values (here numerical) and it exists in reference to this set. In a spreadsheet, let us take for example the formula for square numbers. Figure 1 shows a cell argument A2 and a cell B2 where the formula was edited, referring to this cell argument.

Here again the variable is written with symbols (those of the spreadsheet language) and exists, as with the paper and pencil environment, in reference to a set of possible values. But this referent set (abstract or materialised by a particular value, e.g. 5 in Fig. 1) appears here through an intermediary, the cell argument A2, which is simultaneously:

- An abstract, general reference: it represents the variable (indeed, the formula does refer to it, making it play the role of variable);

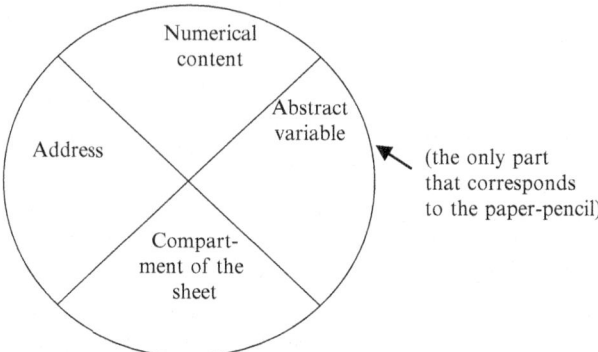

Fig. 2 The 'cell variable'

Table 1 The distance between different 'algebraic worlds'

'Values' of algebra	In paper-pencil environment	In the spreadsheet environment
Objects	Unknowns, equations	Variables, formulae
Pragmatic potential	Tool for resolution of problems (sometimes involving proof)	Tool of generalisation
Process of resolution	*Algorithmic* process, application of algebraic rules	Arithmetical process of trial and improvement
Nature of solutions	Exact solutions	Exact or approximate solutions

- A particular concrete reference: here, it is a number (in case nothing is edited, some spreadsheets attribute the value 0);
- A geographic reference (it is a spatial address on the sheet);
- A material reference (as a compartment of the grid, it can be seen as a box).

So, whereas in a paper and pencil environment, we would place a set of values, here we have an overlapping cell argument, bringing with it, besides the abstract/general representation, three other representations that do not have an equivalent representation in paper and pencil (Fig. 2). Other examples of spreadsheets' impact on algebra are given in Haspekian 2005a.

From an institutional point of view, these changes have different impacts depending on the range of ways that algebra is introduced. As one of the previous ICMI studies has showed (Stacey et al. 2004), different aspects of algebra can be focused on: a tool of generalisation; a tool of modelling; or a tool to solve arithmetical, geometrical or everyday life problems through the, so called, Cartesian *analytical method*. Depending on the focus, different mathematics are brought to the fore: variables, formulae and functions on one hand; unknowns, equations and inequalities on the other hand. The traditional French school culture adopts the analytic approach. The resolution of various problems through the solving of equations is emblematic of pupils' introduction to algebra. Table 1 provides a brief insight into the distance between the algebraic culture in the French secondary education and the algebraic world that is characteristic of spreadsheets.

Beyond the vocabulary, it is the whole set of the 'valued algebraic' objects that is modified in the spreadsheet environment. Within the paper and pencil algebra of junior high schools in France, the move is from algebra as a tool of resolution where equations and unknowns are valorised, towards the algebra of variables and formulae in their functional aspect, where algebra is more seen as tool of generalisation.

Overall, the mathematical culture sustained by spreadsheets is an 'experimental' one of approximations, conjectures, graphical and numerical resolutions, implementing everyday life/concrete problems, statistics, etc. Thus, this vision does not fit with the one usually attached to traditional mathematics in the secondary school of the French education system.

What Are the Consequences of Such Changes for the Teaching?

The idea of *distance* allows one of the conditions of viability of an instrument in teaching to be translated by considering the whole set of modifications that it introduces, not only at the level of computer transposition, but also through the cultural, epistemological and institutional aspects (Haspekian 2005b).

In the case of the spreadsheet for algebra, this distance seems to play a role in the teachers' resistances to its use because they have to grant to it a personal legitimacy, as the institutional legitimacy (the programs) or the social legitimacy (stemming from it as a modern tool that is used widely in industry) are not sufficient. Hence, the mediative, cognitive and personal components of the teachers (their history, perceptions of teaching, of algebra, etc.) come into play here. This also partly explains why not all instruments are treated alike in mathematics teaching and learning! Do teachers consider this distance 'legitimate' with regard to their epistemology of mathematics on the one hand, and to the didactic potential they foresee on the other hand? The interviews carried out with novice teachers (Haspekian 2005a) show that this is not self-evident. Furthermore, if a certain distance is necessary for the tool be seen to be interesting, this distance involves a mathematical and didactic reorganisation and thus an additional workload for the teacher. As we saw above, not only are there new praxeologies to create (that the programs and the resources, however many, are not enough to release) but additional tasks arise for teachers as they consider the management of pupils' instrumental geneses in a new environment. Last, but not least, this management should lead pupils to mathematical concepts (variable, formula, etc.) that remain relevant to the traditional paper-pencil environment.

Finally, the integration (or not) of a new tool requires equilibrium between the various elements. Do the teacher's own convictions about the expected benefits and/or the official directions to use the tool counterbalance the additional workload he/she can foresee in that task of integration? Moreover, a phenomenon of *double genesis* can come into play and add further complexities for teachers who are not very familiar with the tool, which is described later in the chapter. For the spreadsheet, one can assume that the praxeologies are far from the mathematical and didactic organisations currently practiced within early algebra in France.

This idea of *instrumental distance* prompts a number of questions concerning spreadsheet integration within mathematics education such as: do the many resources available to teachers consider it? and How do teachers who have integrated spreadsheets take advantage of this distance in their practices?

The next section reports on a case study involving an experienced teacher during the first two years of her integration of spreadsheets into her teaching, showing that the evolution during the second year moves precisely in the direction of reducing this distance.

Understanding Practices with ICT: A Case Study on Integrating Spreadsheets

Taking into account the idea of distance, I turn to the question of the teaching practices, with some additional tools to support the associated analysis.

In a study concerning teachers' initial training involving the integration of CAS calculators, Trouche (2004, p. 307) had already noticed the importance of two factors relative to the teachers themselves: their degree of mastery of the tool and the range of their positivity or negativity of the representation/conception of its integration.[12] In the same way, the numerous works analysing practices inspired by the *double approach* (Robert and Rogalski 2002) underline that teachers' activity is not only related to the mathematical content to be taught or the learning experiences of the students but also to a number of teacher-related factors such as individuals exercising a job which has its own constraints and freedoms. When considering ICT integration, it is relevant to take this personal component into account.

Additional Theoretical Elements to Analyse teachers' Practices

The didactic and ergonomic approach (Robert and Rogalski 2002) is an interesting theoretical support for the analysis of teachers' practices as it frames teacher's activity through different components, one of which is this important *personal component*. By turning the spotlight onto this personal component and because we want to take into account teachers' apprehension of the instrumental issues, I distinguish a professional instrument from a personal one (Haspekian 2006) and consider their corresponding instrumental geneses, professional and personal.

Didactic and Ergonomic Approach

The didactic and ergonomic approach analyses practices by means of five components: *cognitive, mediative, institutional, social* and *personal* (Robert and Rogalski 2002). The *cognitive* and *mediative* components relate to the choices made by the

---

[12] The words 'representation' and 'conception' are not problematised in this chapter and used in their common senses.

teacher in the spatial, temporal and mathematical organisation of the lessons. These choices are made according to the teacher's *personal component*. The personal component relates to the teacher as a singular subject with his/her own history, practices, vision of mathematics, way of conceiving mathematics learning, teaching, etc. Yet, the personal factor is not the only one to consider. Teachers are not completely free in their choices as they are more or less constrained by *institutional* and *social* dimensions. The institutional and social dimensions relate to the curricula, lesson duration, school social habits, mathematics teachers' habit, etc.

In the case of ICT practices, instrumental aspects seem to interfere with each of these components (Haspekian 2005a). In particular, the personal component plays a crucial role in determining whether ICT in mathematics teaching is supported. For example, teachers integrate ruler and compass without any problem as they are accepted as part of the mathematical culture. This might be because historically, the ruler and compass played an essential and epistemological role in the development of mathematics. (Chevallard 1992) This role and the number of mathematical problems generated by these traditional tools serve to legitimise their place in mathematics education. Is it the same for spreadsheets? How is their introduction in mathematics teaching justified? Do teachers feel this tool relevant to their mathematics and the ways they learned, learn, do and teach mathematics?

The consideration of these questions led to the use of the instrumental approach to analyse more locally some of the phenomena observed with ICT practices, in particular the teachers' professional instrumental genesis with the spreadsheet.

Professional Instrumental Genesis

This case study shows that, at the early stages, the way that teachers orchestrate and support pupils' instrumental geneses evolves year by year. Starting from the premise that the spreadsheet as an instrument for the teacher, which allows her to achieve some teaching goals, the process of instrumental genesis is considered *from the teacher's perspective* (Haspekian 2006). The same artefact, the spreadsheet, becomes an instrument for pupils' mathematical activity and an (other) instrument for teacher's didactical activity. Thus, when applying the instrumental approach to the spreadsheet as a *teaching* instrument created by the teacher through a professional genesis, two processes are highlighted:

- A process of instrumentalisation as teachers instrumentalised the tool in order to serve didactic objectives. It is transformed from its initial functions and its didactical potential is progressively created (or discovered and adapted in the case of an educational tool).
- A process of instrumentation in which the teacher, as a subject, is required to incorporate within her (already stable) teaching schemes some new schemes that integrate the use of the tool. Progressively, the teacher will specify the use of the tool for a particular class of situations (like, for example, "take advantage of spreadsheet for algebra learning") and organise her activity in a way progressively stable for this class of situation (Ann's case already shows some regularities from year 1 to year 2).

The instrument that is created as a result of this process of professional genesis (for instance the 'spreadsheet as a tool to teach algebra') is different from the instrument built through a *personal* genesis (the spreadsheet as a tool of personal work of calculation, plotting, data treatment, etc.). From the same artefact, two instrumental geneses (that may have interferences/interactions on each other) lead to two different instruments. The spreadsheet in these two situations is not at all *the same instrument*. The second one is close to the instrument we want pupils to build. The teacher's professional genesis with the tool is much more complicated as it includes the pupils' instrumental geneses. Here again, the phenomena are imbricate and interfering.

This notion of *double instrumental genesis* together with the *didactic and ergonomic approach* is used in the next section to analyse the observation of a teacher who is integrating the use of a spreadsheet in mathematics. The case of the spreadsheet provides a good amplification of the phenomena that play in the development of ICT practices for at least two reasons. Firstly, the spreadsheet is a professional tool without any *a priori* didactical functionality. In this case, the instrumental distance is not negligible and plays a considerable role in the difficulties surrounding the integration of spreadsheets. Secondly, the teacher has to turn this non-educational tool into a didactical instrument through a process of professional genesis, a process made more complex by this instrumental distance.

A Case Study: Ann's Practices and Evolution in ICT Integration

The next section reports the data and subsequent analyses of a study that observed how a very experienced teacher integrated spreadsheets within her practices for the first time and the evolution of this integration during the subsequent year.

The Data

Ann is not a trainee; she has taught mathematics for more than 10 years but is not an expert in the use of technology within mathematics teaching and learning. She has already some experience of dynamic geometry software and now she is beginning to integrate spreadsheets in her classroom. In this first year, Ann's choices were motivated by her participation in a 1-year research project that focused on spreadsheet use for learning *algebra* (Haspekian 2005a). The data that was collected included: observations of all of her spreadsheet lessons (6 sessions); teacher interviews before and after each session; and the students' spreadsheet files. At the end of the research, an interview collected Ann's thoughts and feelings about this experience.

After the completion of the research, Ann continued to use spreadsheets in the following year. During this second year, I observed and recorded her first spreadsheet session and the subsequent session in a paper and pencil environment. I collected the problems as they were given to the students and the associated homework, and I carried out some interviews concerning her intentions for this second year.

Table 2 Ann's approach to the introduction of spreadsheet in her teaching

Use of spreadsheet	Year 1 of the introduction of the spreadsheet	Year 2 of the introduction of the spreadsheet
Variations		
Class level	7th Grade (12 year old)	8th Grade (13 year old)
Old/new content	New	Old
Mathematical domain	Algebra	Statistics
Spreadsheet location	Limited to computer classroom	Computer/ordinary classroom
Synthesis	No	Yes
Interactions teacher-students	Mostly individual	Individual and collective
Use of the video and collective presentation	Piloted by teacher, limited role	Teacher and student. Important role
Students configuration	Work by pairs	Work by pairs + collective work: one student at the board
Regularities		
Maths objectives, teacher aims	Algebra	
Additional material	Worksheet for pupils and pre-organised spreadsheet file	
Institutionalisation	In an ulterior lesson, in ordinary classroom	

The resulting analyses showed an evolution of her practice. This evolution converges towards the characteristics of experts' practices described in the next section.

During the second year, Ann introduced the spreadsheet not within algebra but within statistics (headcounts, frequencies and cumulative frequencies), after having seen these notions in paper and pencil environment. In this context, some of the observed elements were surprising as the lesson revealed very little statistical content and mostly centred on the tool use and functionalities, revealing unexpected mathematics such as notions of variable, formula and the distinction between numeric and algebraic functions. Of course, this reflects the influence of the first year of her experience, centred on algebra, but this does not explain the complete evolution (variations and regularities) summarised in Table 2 of Ann's choices for introducing spreadsheets.

In both years, Ann met the institutional demand to integrate spreadsheets within her mathematics teaching but the way that she did this was different in each year. Table 2 shows an evolution of two components. The mediative and cognitive components have evolved with respect to the chosen mathematical domain, the way that the spreadsheet was introduced and the level of the class that was chosen. This prompts the questions: Why did she evolve, and how can we state more specifically her professional genesis with the tool?

Ann's Professional Genesis with the Spreadsheet as a Didactical Tool

In both years, Ann's activity with the spreadsheet is oriented by the goal of using it to teach algebraic concepts such as variables and formulae, for example, by using the copy function, or by profiting from the numerical feedback to infer the equivalence of two formulae.

This brings into play some usage schemes[13] concerning the material and organisational aspects that are being developed from one session to another towards a more stable set of practices that concern: integrating the tool within a larger set of instruments (with the data projector); using the data projector at the beginning of the lesson to make collective explanations; requiring the pupils to communicate and work in pairs; giving an instruction sheet and a pre-built file to save time and regularly clicking on a cell to check whether pupils have edited a formula or numerical operation, or the numerical result.

In Ann's case, this professional genesis was not independent from her personal genesis with spreadsheet as the observations show how these interfered (i.e. they interacted in a relational sense) with each other.[14] These interferences were made more complex by the fact that she wanted her pupils to manipulate the spreadsheet for themselves (one could imagine a spreadsheet usage only under a teacher's control) and learn mathematics as a result of this activity. As already stated, as the pupils' instrumental geneses forms part of the teacher's professional genesis with the tool this leads to another interference.

Observation of some of Ann's activity in these first two lessons in her second year result from these interferences and an example of this now follows.

The Interferences Between the Teachers' Double Instrumental Genesis and the Pupils' Instrumental Geneses

As already mentioned, Ann chose to introduce the spreadsheet to a different class within the domain of statistics. Figure 3 is an abstract of the task she developed for her pupils that shows the corresponding spreadsheet file with the pre-edited formula built by Ann:

It is interesting to notice that Ann modified this file three times. In its first version, the formula calculating the frequency (in B7) was =**B6*100/50**. This formula, if copied along row 7 calculates the correct frequencies for the corresponding data of row 6. But it is not adequate regarding the question b.[15]

The day before the lesson, Ann realised the mistake and changed the formula to =**B6/F6*100**. She confided that she did not yet feel very comfortable with spreadsheets. Her own instrumental genesis with spreadsheets as a mathematical instrument probably plays a role here as we also see that the key point of the problem comes from the spreadsheet as a *didactic-oriented* instrument. From the point of view of the spreadsheet as a *calculation-oriented* instrument, the formula was adequate. The didactical aim (showing the mathematical dependency between

---

[13] Rabardel (2002) distinguishes the *usage schemes* (related to the *material* dimension of the tool) from the *schemes of instrumented action* (related to the global achievement of the task, with goals and intentions).

[14] It may not be the case for all teachers: unlike Ann's case, the first instrument can be already constituted in a more advanced way, long before trying to make it a didactical instrument.

[15] The formula refers to the value 50 for the total. If one changes the value of any headcount, then the total will change and the formula becomes wrong.

| Step 2: use of formulae and the fill command |

distance (km)	0<d ≤5	5<d ≤10	10<d ≤15	15<d ≤20	total
headcounts	16	14	12	8	50
frequencies (%)	32				100

1) a) What is the total number of items? _____
Where is this number located? What is the formula to calculate it? _____
b) If one changes the headcounts for 0<d ≤5, does the frequency change? _____

	A	B	C	D	E	F
1						
2						
3	**Etape 2 :**		**calcul de fréquences**			
4						
5	distance (km)	0<d ≤5	5<d ≤10	10<d ≤15	15<d ≤20	total
6	effectif	16	14	12	8	=SOMME(B6:E6)
7	fréquence (%)	=B6/F6*100				100

Fig. 3 Ann's final version of the formulae

numbers and frequencies) led Ann to ask the question b which resulted in an incorrect formula. She did not realise this when she first constructed her formula. At that moment, the personal instrument stands at the front of the scene, and obscures the professional instrument and its associated didactical aims (the question b.).

Interference between the personal and the professional instrument can be seen again within the continuation of the story. The new formula, =B6/F6*100, is now adequate for question b, but still not convenient if we consider the next question (Fig. 4) for inverse reasons! Ann wants pupils to copy the formula in order to fill row 7 and meet this filling functionality with the automatic increasing of cell references (B6 becomes C6...). This time, this is part of her goals for students' instrumental geneses.

The formula above, if copied along row 7, is no longer valid, as the cell referring to the total, F6, will change into G6, H6... along the row. A solution to this problem is to fix the cell F6 in the recopy by using the $ symbol. But Ann did not want this functionality to appear in the first spreadsheet session as it was above the level of instrumentation she wanted for her pupils at that moment. When she built her new formula for question b, the $ symbol was not in her mind and she did not include it, forgetting that it would create false results at question 3. The day before the session, we had a phone call to finalise our meeting during which she realised the new issue and included the $ symbol as a last-minute decision.

Thus, this time the formula was 'wrong' with regards to an instrumental goal, that is the use of the $ symbol was above Ann's instrumental objectives and she did not have it in her mind. It is neither easy nor trivial to adapt to meet all of the constraints,

> 3) Complete the table using the formula in B7:
> Recopy the formula on the right. (see instructions below for the "cell recopy")
> What is the formula contained in C7? D7? E7?

Fig. 4 The next stages of the task

particularly as she had already changed her very first version of formula for a mathematical aim and now she had to change it again for an instrumental aim. This time, the professional-oriented instrument overrode the personal one, by taking into account pupils' geneses and the level of instrumentation that she wanted them to reach.

These successive formulae disrupted the session and finally Ann put the $ sign into the formula but expected to avoid speaking about it with the pupils. Unfortunately, it arose of course during the session! Being compelled by pupils' questions to explain, she only said that it is not important to write it with a paper and pencil environment. Then, when a pupil came to the board to write the spreadsheet formula, he forgot the $, the 'division by zero Error' appeared after filling and Ann said *"now you're happy?"* but did not explain the message nor the division by zero.[16] In that sense, the perturbation due to the '$' sign appears as one of Clark-Wilson's lesson hiccups (Clark-Wilson 2010b) defined as:

> These were the perturbations experienced by the teachers during the lesson, triggered by the use of the technology that seemed to illuminate discontinuities in their knowledge and offer opportunities for the teachers' epistemological development within the domain of the study (Clark-Wilson 2010b, p. 138).

Interpretation of the Complex and Divided Geneses on the Part of the Teacher

The example above shows how the double genesis on the teacher side may interfere with pupils' geneses. The spreadsheet's constraints interacted with the teacher's goals and didactical expectations (she wanted to introduce a basic level of spreadsheet functionalities but did not want to go any further). This is evidence that she has not yet turned her personal instrument into a mathematics-teaching instrument. This process is made more complex by the different geneses at stake. As we saw in the example, it is constrained by:

- The teachers' aims for the mathematical learning, i.e. concerning statistics and algebra.
- The pupils' instrumentation that is, how to support pupils' mathematical work through their interactions with the spreadsheet i.e. the mathematical headcount-frequency dependence through the change of the frequency cell after changing the value of the headcount cell.

---

[16] Increment of references after filling makes the formula refer to empty cells. By default, empty cells are treated in formulas as if they contain the value 0, this option that can be changed.

- The pupils' instrumentalisation, that is the choice of functionalities to be used, the desired schemes of use, i.e. relative references and the automated increments for cell references using the copy function, but not yet the absolute references, the $ sign and its specificity in the filling of formulae.

The simultaneous management of these constraints is not easy as the spreadsheet is not *a priori* a didactical instrument. According to Rabardel's theory (Rabardel 2002) Ann's case study on making the spreadsheet a didactic instrument shows that such an instrument is, as any instrumental genesis, only developed progressively in a long-term and complex process. Here, the teacher's and the students' personal instrumental geneses are elements that are adding complexity to this professionally oriented genesis.

How to Understand Ann's Evolutions?

The way that Ann evolved from the first year to the second is related to this professional instrumental genesis.

In the previous section, using both the notions of distance and double instrumental genesis, I have described the beginning of such a genesis and analysed locally the associated complexity through the case of Ann's use of spreadsheet. In particular, the way that teachers orchestrate and support pupils' instrumental geneses evolves year after year.

Ann's goal is to use the spreadsheet to teach algebraic concepts and she develops some instrumented schemes of action for this that concern the material aspects, the organisation of the sessions and the orchestration of pupils' instrumental geneses. Ann's practice with speadsheets includes, for instance, the following elements that emerged during the first year (not necessarily since the beginning) and seemed to stabilise in the second year:

- Using a data projector at the beginning of the session to make collective explanations;
- Requiring pupils to communicate and work in pairs;
- Giving pupils a sheet of instructions and a pre-built computer file to save time;
- Regularly 'click' on individual cells to check whether pupils have edited a formula or numerical operation, or even directly the numerical result.

Some other elements of her orchestrations were modified during the second year:

- The use of the spreadsheet with a higher level of class, i.e. with Grade 8 instead of Grade 7;
- Fewer 'new' concepts were introduced at one time, i.e. the introduction of the spreadsheet and the introduction of new mathematical notions;
- She changed the mathematical domain, i.e. it was introduced within with statistics, which seemed to Ann to be more appropriate than algebra;
- A deeper articulation was made between social and individual schemes, something that Trouche (2005) has stressed the importance of within the process of instrumental geneses. In the interview, Ann said she had not organised enough moments of 'mutualisation' (whole class discussions) and she explicitly wished to take care of this point in the second year.

The next section observes these evolutions more closely, and shows that they all appear to converge in the direction of *reducing* the instrumental distance.

Changing the Class Level: Higher Level of Class

This modification comes with the change of the mathematical domain. In the French curriculum the spreadsheet is explicitly mentioned for the teaching and learning of statistics for Grade 8 pupils. In the Grade 7 curriculum the spreadsheet appears in a more general and vague way and teachers are required to reflect more deeply to define its potential for the learning of mathematical notions. These notions appear more distant from spreadsheet mathematics than within the Grade 8 curriculum, where the spreadsheet is more clearly specified with respect to precise mathematical notions. Thus, by choosing this level Ann was able to reduce the distance and match the official prescriptions more easily. In addition, during year 1 Ann did not find the Grade 7 pupils' instrumentalisation process easy. The pupils had difficulty in filling cells, selecting a single cell and editing a formula. Older pupils seemed to be more skilful and problems that were linked to instrumentalisation should interfere less with the mathematical work. With Grade 7, the manipulations of the tool seemed more difficult and the tool appeared less transparent.

The 'Old/New' Knowledge Game with Respect to the Mathematical and Instrumental Content

During year 1, Ann introduced a new instrument at the same time as she introduced some new mathematical content (algebraic notions). The relationship between the old knowledge and the new knowledge is different in year 2, which tends to reduce the instrumental distance by lessening the amount of newness. For example, all of the mathematical notions at stake in the spreadsheet session (headcounts, frequency, cumulative frequency) had already be seen previously by the pupils in the paper and pencil environment. This experience (new environment with 'already-seen' concepts) will then serve Ann as a base to introduce algebraic notions (new concepts in an 'already-seen' instrument).

Domain Changing

There are at least three reasons why the mathematical domain chosen by Ann in year 2 also reduces the distance with respect to algebra. The domain of statistics is usually seen to conform more closely to the representations within a spreadsheet than the domain of algebra. Furthermore, institutional pressure is less important in statistics than algebra, which is a more classic and traditional domain that is strongly linked to paper and pencil mathematics. On the contrary, nowadays statistics is more aligned to the use of technology. Finally, within the language of the spreadsheet, one can find terms that are more commonly used within statistics whereas the distance to the traditional vocabulary of algebra is wider (and more important) (Haspekian 2005b).

Moments of Mutualisation and Articulation with Paper and Pencil Mathematics

In her second year, Ann introduced some moments of mutualisation during her spreadsheet sessions. In the interview, she affirmed her will to increase the similarity between these sessions and the traditional ones. She felt that it was necessary to increase the links to the paper and pencil mathematics. For example, she started the sequence with a paper and pencil session, then revisited the same notions in a spreadsheet session, and then returned to the work done with spreadsheet within a subsequent paper and pencil session.

Thus, at a range of different levels, Ann's modifications tended to minimise the spreadsheet's instrumental distance. All of these actions contributed to reduce the distance with paper-pencil and to mix in a greater proximity the mathematics within these two environments.

Another notable development is that Ann's evolution gains some characteristics of experts' practices, as evidenced in the research. This is explored in the next section.

Bringing Together the Results from Different Research

In this section, I am bringing together Ann's case study with the results of a second research study. This latter research studied the practices of what we have called *expert* teachers, that is, non-novice teachers who have been integrating ICT and spreadsheet for a long time and who are also 'ICT trainers' and 'spreadsheet trainers' within the context of mathematics teacher training. By comparing the practices of these expert teachers alongside the practices of pre-service teachers, I have highlighted some overarching characteristics of practices with ICT.

An interesting outcome of this cross analysis is that Ann's evolution with the spreadsheet converges towards the characteristics of experts' practices. The next section presents this in more detail by first giving some results and regularities found in the data collected with *expert* and novice teachers.

Some Characteristics of Experts' Practices with ICT

Are there regularities of practice amongst teachers who have successfully integrated the spreadsheet? In making a comparison with novice teachers, what are the characteristics of the expert teachers' practices that seem to contribute fundamentally to their success? How do they manage the pupils' instrumental geneses? And how do they take into account the instrumental distance generated by the spreadsheet? In order to answer these questions, I looked for regularities at the following levels: in teachers' conceptions; in the evolution of their practices; and in the changes that resulted from this evolution. The notions of *coherence* and *stability* as defined by Robert & Rogalski can enlighten these questions:

> the coherence of the system of the practices of a teacher (...) would prevent the introduction of inconsistent elements with this system (Robert and Rogalski 2002, p. 521).

Within an alternative theoretical framework, the considerations of Lagrange are in the same direction. Lagrange (2000) underlines that the introduction of a tool into mathematics lessons generates an upheaval of the *praxeologies*, which may hinder its integration into the practices. How did expert teachers deal with these obstacles?

As said in the introduction, I carried out questionnaires and interviews with trainees and expert teachers. The questionnaire for trainees contained 41 questions divided in three parts (see Appendix). The first was general information about the teacher (age, training, etc.), the second concerned their general opinions about the use of technology and the third concerned their use of spreadsheet in mathematics classroom and their opinions about this. There were 23 questionnaires returned by the trainees and four additional group discussions (in groups of 3 or 4) were held in which we allowed the trainees to discuss their answers to parts 2 and 3 of the questionnaire in order to gain a better understanding of their opinions. The questionnaire given to the expert teachers was an identical one and six individual interviews lasting 2–3 h were conducted about their effective practices with ICT and spreadsheets. We also collected all of their teaching materials, which evidenced their progression in use of the spreadsheet, examples of tasks, frequency of use, etc.

The research study compares the trainees with the experts (Haspekian 2005a) and outlines some common findings about the novices, such as their obvious difficulties in perceiving the tool's potential and to conceive mathematical and classroom organisations, which as yet they had not seen or experienced. It also suggested some convergence of practice amongst the experts that can be connected to their successful integration of spreadsheets.

The first result concerns the nature of the tasks chosen for a spreadsheet use. Parts 2 and 3 of the questionnaire included a set of different spreadsheet tasks that included very basic use of the spreadsheet as a calculator to a more interesting use that took greater advantage of the spreadsheet's potential. These latter tasks were based on research situations mentioned in Capponi 2000, Arzarello et al. 2001, and Rojano and Sutherland 1997, and they had been analysed by their authors as being positive for mathematics learning. In the questionnaire we presented different ways of using spreadsheets and asked the teachers to choose which of these situations they found interesting for mathematics teaching and learning. The results of this study concurred with those from other research (Laborde 2001; Monaghan 2004), that is novice teachers who are non-expert in the use of the spreadsheet have difficulty in realising the potential of the tool and in identifying interesting situations for its use. The choices and underlying rationales of the beginner teachers were *systematically opposed* to those of the expert teachers, which corresponded to the interesting situations. Thus, the teachers' first approach to the use of spreadsheets did not take advantage of the tool's potential. As Artigue recalls, the observed (and quite understandable) tendency amongst novice users is to use technological tools not for their epistemic value (as a support to understand mathematical objects) but only for their pragmatic value (to produce results quickly and easily) within tasks that are very similar to those given in traditional paper and pencil tasks (Artigue 2002).

In the analysis of the expert teachers' practices and the subsequent comparison of these findings with the novice practices, a set of common characteristics appears (for more detail on this see Haspekian, (2005a)). This prompts the question as to whether there are fundamental elements contributing to teachers' success in the integration of spreadsheets. The first element is the importance of taking into account not a single tool but a system of instruments. This confirms the importance of the *instrumental distance* as these characteristics are a way to minimise the distance imposed by the spreadsheet. Another common characteristic was the fact that, using this system of instruments, these teachers play an *old/new game* concerning the mathematical content with equal attention to the various technological tools that they integrate. This means that they alternate new/old instruments with new/old content and do not try to introduce, for example, a new instrument with new concepts. This game also helps to articulate the work involving the technology with the paper and pencil work.

These two characteristics provide an economic way to both manage the class in ICT sessions, and to manage the pupils' instrumental geneses. For example, concerning the mathematical content, one teacher said that it offered "a way of making revisions by bringing something more". Another said that he had "the same notions presented in two different environments". A third *expert* teacher who was interviewed said that she systematically works on the same notion using by hand methods after an ICT session, and combines paper-calculator-spreadsheet and so on: "I make links non-stop, again and again…"

For all of these expert teachers, the integration of the spreadsheet is based upon this orchestration of a whole system of instruments. As they perceive the spreadsheet as more complex, they introduce it to their pupils after other software. This allows:

- **Time saving** on the management of the class in ICT sessions (introduce the classroom, organise the didactic contract, etc.);
- **Time saving** with respect to the instrumental geneses with the spreadsheet as some aspects have been addressed through other technological tools (physical manipulation of the materials, the computer room, virtual manipulation of files, etc.).

Within the common characteristics, we also found an increased attention paid to the questions of *mutualisation* and *socialisation*, which was accomplished in two ways. Firstly, the expert teachers all organised their sessions with the pupils working in pairs and secondly, the teachers have developed the habit to use the data projector in order to mutualise or bring together the scattered knowledge of the pupils leading to more homogenous mathematical and instrumental knowledge.

Table 3 resumes the common characteristics that appear to contribute fundamentally to the expert teachers' successful integration of the spreadsheet:

> → the taking into account of a system of instruments, including the articulation with the paper and pencil environment;
> → a game of *old/new*, which is played at both the level of the mathematical content and at the level of the instrument;
> → a certain art/skill to know how to mix these two games,
> → the use of *mutualisation* and *socialisation* (students work in pairs, use of the data projector).

Table 3 Some common elements found in experts' practices

What is noticeable is that some connections can be seen then between these characteristics and Ann's evolution of practice as a result of the changes she introduced in the second year.

Reducing Instrumental Distance: Towards Experts' Practices

In the analysis of the expert teachers, there were some common characteristics in their successful integration of ICT, in particular concerning spreadsheets. In this section, I will show that Ann's evolution, as analysed previously, *tends towards* some of these characteristics and gives an indication of the importance of these characteristics.

First, as seen in both cases, we find the tendency to minimise the instrumental distance. Actually, some of Ann's evolutions can be explained in terms of a *reduction* of the distance, either by making this distance more explicit or by increasing the times when she alternated the work in both the spreadsheet and paper and pencil environments, which enriched both of them. This mixing of different environments and, in particular, the articulation within the paper and pencil environment, appeared precisely as a common characteristic of the teachers who have integrated the spreadsheet successfully. Thus, it is interesting to notice that Ann's professional genesis follows the same path (even though she did not achieve a level of expert practice with respect to all characteristics). For instance, the moments of mutualisation and articulation with paper and pencil mathematics by Ann are more successful in the second year, whereas she did not pay much attention to this in the first year.

The *old/new* game mentioned above is another characteristic found in the expert teachers' practices. They manage ICT integration by adjusting and adapting the degree of novelty to incorporate a degree of complexity of the tool. When introducing a complex artefact such as the spreadsheet, they choose familiar content, which has already been introduced within the paper and pencil environment. Once the students have more familiarity with the spreadsheet with more familiar mathematical content, they use it subsequently to develop new mathematical knowledge.

Again, it can be noted that Ann's evolution is moving in that direction. In the first year, she introduced both the spreadsheet and a *new* mathematical domain (algebra), whereas in the second year she changed her approach to introduce spreadsheets by choosing an *old* mathematical domain, statistics. The pupils, having already seen statistics in a paper and pencil environment, then meet the new instrument, a spreadsheet, in the context of old content. Ann's long term intention, as stated in her interview, is to use the spreadsheet within the context of algebra, but now she intended to do this after the pupils have seen spreadsheets in another area of mathematics (an *old* one) to avoid introducing both new artefact and new contents.

Of course, when I observed Ann at the beginning of the second year, she had not achieved all of the common characteristics of the expert teachers as listed in Table 3, but this is not surprising. She was at a stage within her professional genesis with the spreadsheet where she was integrating it for the second time in her career. It is predictable that her practices are not completely stabilised and that these will continue to evolve. For instance, for the expert teachers, the game 'old/new' concerns not only the mathematical content and not only one tool, but a complex system of instruments that incorporate paper and pencil articulations. Expert teachers do not expect pupils to first meet computers through the use of spreadsheets but with other software, such as dynamic geometry software, which presents a smaller instrumental distance than the spreadsheet. In that way, pupils meet the computer classroom, the basic instructions about the use of the computers, the files, the opening and closing sessions, the articulation within the paper and pencil environment, the work in pairs, and so on, with a software that seems easier to integrate than the spreadsheet. Once they are used to these basic manipulations and orchestrations on a more familiar *old* instrument, they are ready to meet a new, more difficult one, such as the spreadsheet.

Discussion and Perspectives

In the section, I will come back to the general purpose of this work, which was to gain a better understand of teachers' practices with technology and the process of their instrumental geneses. To this aim, the previous sections have introduced some important elements and lead me to draw conclusions on their instrumental professional geneses with ICT, which I will discuss here.

I have analysed Ann's evolutions in terms of a *reduction* of the instrumental distance, either by making this distance more explicit, or by multiplying the opportunities to alternate work in the two environments, enriching both of them. This distance is more or less important, depending upon the tool. The integration of spreadsheets in the teaching and learning mathematics constitutes a significant creative task for teachers as the tool is not given with any didactical functionality. It requires a professional instrumental genesis on the teacher's side that differs from the teacher's personal genesis with the tool (even if they interfere) and different again from that of the pupils. Here again, one can hypothesise that a professional instrumental genesis with dynamic geometry software is easier.

These combined considerations helped the analysis of Ann's genesis and the conclusion that Ann tended to acquire in her evolution some of the characteristics found as commonalities among the expert teachers as follows:

- Articulation with paper-pencil mathematics;
- Moments of mutualisation and socialisation;

- The game old/new, concerning the mathematical content (not yet on the instruments for Ann).

These are all included in the experts' characteristics Table 3. The inverse is not true because Ann did not demonstrate all the characteristics of the experts. For example, in her evolution, this exploitation of different instruments to facilitate the introduction of spreadsheets does not appear yet, but it seems reasonable to think that one does not gain all of the characteristics of the expert teachers after only 1 year. This instrumental professional genesis is a long process, as is any instrumental genesis. This raises questions for the professional training of teachers such as: How to take into account the importance of working within a system of instruments instead of the isolated tools? How to take into account the *socialisation* dimension? Is it possible through these improvements to shorten the time needed for the instrumental professional genesis?

I conclude on the fact that these results are at the stage of hypotheses, as key issues in ICT integration. To extend this result, a larger scale study is needed with more than six expert teachers, and with some observations of their actual practices in the classrooms. The fact that Ann's evolution tends towards some of their common characteristics is a simple indication that these elements may constitute good *candidates* of ICT practices, but this hypothesis does requires further research.

Other questions remain for research. For example, concerning ICT integration and evolutions of teachers' practices, a criterion which we have seen as important in this chapter is the notion of instrumental *distance*. If it does reveal itself as a source of difficulty for teachers, then it is crucial to advance in the comprehension of ICT impact on mathematics and the way teachers take into account instrumental distance, drawing some important characteristics from experts' practices. However, it is also necessary to determine which elements may counterbalance this distance and may support the process of tool integration, such as institutional injunctions, or the tool's epistemic value and its didactical design. As technology evolves, the instrumental distance can thus be important for educational tool designers. As for the epistemological legitimacy, it also relates to teachers' representations and beliefs about ICT and mathematics. This dimension has been investigated in other research, see for instance Norton et al. (2000), who conclude that teachers' resistance is related to their beliefs about mathematics teaching and learning. If knowledge and beliefs about teaching mathematics with ICT are actual barriers, can this dimension be considered in teachers' training and how?

Finally, the issue of 'isolated' potential of technology for mathematics education does not solve the problem of their integration in teaching practices (for example in teaching algebra in the case of the spreadsheet), due to this instrumental distance. Several questions remain and a better understanding of the characteristics of experts' practices and of course the way to develop these, may be important also in a training perspective. This remains an open field for further research.

Acknowledgments I would like to thank Rebecca Freund, and the anonymous second reviewer, who very carefully reviewed the English of the text.

Appendice-Extract of the Questionnaire Trainees and Experts

When several answers are possible, please number them according to you preferences:1 for the 1rst, 2 etc.

PART I:

14. A **priori**, a) Which maths content do you think is more appropriate for using spreadsheet in the mathematics class? *(number if several answers)* [ ]Algebra   [ ]Arithmetics   [ ]Statistics   [ ]Calculus   [ ]Probabilities   [ ]Others :

b) For which topic *(1choice or more)*? Simulations [ ]   Fonctions [ ]   Introducing Algebra [ ]   Series [ ]
   Implementing algorithms [ ]   Problems of approximations [ ]   Resolution of algebraic problems [ ]   Others [ ] :

19. How do you envisage the use of ICT in your current and future teaching ? *(number if several answers)*
   [ ] rather ponctually, as a outil to free oneself from tedious calculations so that pupils focus their work on the concepts
   [ ] rather for individual help and remediation for pupils who have difficulties in mathematics
   [ ] rather well integrated in my year progression, as a new tool to create learning situations

20. A priori, do some tools appear to you easier to integrate than others? *(number from the most the the less)*:
   [ ] dynamic geometry software (as Cabri or Géoplan)   [ ]spreadsheet (as Excel)   [ ]internet   [ ]calculators

21. Vos professeurs avaient d'autres programmes et manières d'enseigner. Auriez-vous aimé procéder comme eux ?

22. Your teachers had other programs and ways of teaching. Would you have liked to proceed like them? [] yes, I would have liked much   [] that is equal for me   [] no, especially not.   Explain:

PART II: "Fictitious Teachers": A, B, C, D, E and F are 6 fictitious maths teachers. The first five ones have never really integrated ICT in their teaching, hjere are their arguments:

<u>**A:**</u> *This year, I was firmly decided to use ICT with my pupils, but the key of the computer classroom seemed so difficult to obtain that I finally gave up*

<u>**B:**</u> *I have already thought about having a computer session for my pupils. But, I felt afraid not to know well how to manage technocally (especially that some pupils manage surely better than me) so I gave up*

<u>**C:**</u> *I wish I could use ICT to teach mathematics, but this requires too much time of preparation!!*

<u>**D:**</u> *I do not have anything against using ICT with my pupils, and I've never been trained to integrate it in my teaching. If I were suitably prepared for that, I would do it readily "*

<u>**E:**</u> *Honestly, I feel as abnormal incorporating ICT in our programs because our role is to teach mathematics. It is obvious that if I had the choice,, I would teach mathematics without using any ICT*

23. *Have you already felt or thought like one of these teachers? (number if several answers)* A[ ]   B[ ]   C[ ]   D[ ]   E[ ]
   Was that concerning a particular tool ?

24. According to you, which one could concretely happen to you? A[ ]   B[ ]   C[ ]   D[ ]   E[ ] *(number if several answers)*

25. Que pensez-vous plus précisément de chacune des déclarations ? *A – B –C – D et E*

26. According to their declaration, **which** of these 5 teachers appears to you as beeing: *(number if several answers)* :
   a) the least fictitious: A[ ]   B[ ]   C[ ]   D[ ]   E[ ]   b) the most representative: A[ ]   B[ ]   C[ ]   D[ ]   E[ ]

27. Which of them do you feel the most resembling?   A[ ]   B[ ]   C[ ]   D[ ]   E[ ] *(number if several answers)*

28. Which of them do you feel the less resembling?   A[ ]   B[ ]   C[ ]   D[ ]   E[ ] *(number if several answers))*

29. And you? What is your position concerning the use of ICT in mathematics teaching?

30. In general, how do you see the introduction of computer tools into mathematics teaching? (use 5 or more adjectives to describe your feeling) (if needed, put at the back any other comment)

31. Lastly, teacher F says: *"I do see well the value of teaching mathematics with ICT, I would like, for example, use the spreadsheet with my 8th grade pupils, who have already an experience of it"*

A colleague proposes 2 activities to him, each of them having an instruction sheet to guide pupils and a spreadsheet file prepared by the teacher:

<u>1 :</u> "Formulas" : *A ready-made worksheet contains a formula in B3, the pupils must initially identify that the value in B3 depends on the values in A1 and C1, then identify the formula and use it to answer the question 2.*

Instructions:	Spreadsheet file ready made		
		A	B	C
1. Replace 8 and 9 by other integers and observe what happens.	1	8		9
2. What numbers must the cells A1 and C1 contain for having 50 in cell B3? And 100? and 300? Can we get all the integers?	2			
Explain.	3		43	

(the formula in B3 is here "=2*A1+3*C1")

2: "Theorem of Pythagore" : *Students draw on paper 5 right-angled triangles, then use a spreadsheet file ready-made by the teacher to calculate automatically the squares of the three sides and the sums of the numbers*

Instructions:	Spreadsheet (**formulas in E, F, G et H are ready made**):
1. Draw 5 right-angled triangles of different measures and fill the columns B, C D *the spreadsheet calculates automatically the results in columns E, F, G and H* 2. What do you notice ?	

	A	B	C	D	E	F	G	H
1		Mesures des 2 côtés de l'angles droit		Mesure de l'hypoténuse	Carrés des côtés de l'angle droit		Somme des carrés	Carré de l'hypoténuse
2								
3	Triangle 1							
4	Triangle 2							
5	Triangle 3							
6	Triangle 4							
7	Triangle 5							
8								

Which one would you advise him according to the different hem ay have: mathematical interest, classroom management, easyness to integrate in a progression (*explain your choice for each criterion*)

32. What about you ? If you had to, which one of these 2 activities would you be ready to implement in the classroom? rather 1 [] rather 2 [] Why?

33. Which one would you rather have as a student? rather 1 [] rather 2 [] Why.

PART III: "A supposition..."

Suppose that tomorrow you want or need (for whatever reason) to make your Grade 8 pupils use the spreadsheet to write algebraic formulas (they have experience in spreadsheets but little knowledge in algebra). Your objective is that the pupils find, from the given spreadsheet-file (a sequence of consecutive integers), the formula "2n+1" as the general expression of an odd number. A colleague proposes to you the two following statements:

	A	B
1		
2	0	
3	1	
4	2	
5	3	
6	4	

Statement A: Enter in the cell B2 "=2*A2+1", pull down the filling handle. What do you notice? Can you explain it?
Statement B: From the numbers of the column A, find a general formula that makes in column B a sequence of odd numbers?

34. According to you, which are the advantages and disadvantages of these 2 statements?

35. With the same file and the same objective, which of the 2 statements would you use unchanged (without any modification)? [ ]rather **A**   [ ] rather **B**   [ ]**None**: : I would have modifications to be made, here is my statement:

Topics: writing of formulas.				**Personal statement :**
Objective: Find, through the use of the spreadsheet, the formula "2n+1" as a general expression of an odd number **File given** :	<table><tr><td></td><td>A</td><td>B</td></tr><tr><td>1</td><td></td><td></td></tr><tr><td>2</td><td>0</td><td></td></tr><tr><td>3</td><td>1</td><td></td></tr><tr><td>4</td><td>2</td><td></td></tr><tr><td>5</td><td>3</td><td></td></tr><tr><td>6</td><td>4</td><td></td></tr></table>			

36. Let's try to build up a sequence:

a) How do you introduce the selected activity? With what set of instructions?

b) A priori which are the possible strategies of the students, and with which functionalities of the spreadsheet?

c) What are the foreseeable difficulties and which help can you bring so that the objective is achieved?

d) In view of b) and c) (foreseeable strategies, difficulties and assistances) describe the way you would management the chosen activity (your role, that of the students, the different phases of the sequence...)

37. What would you write in the copybook about this activity?

38. Which precise continuation would you give to this activity?

39. On the topic "introduce students to algebraic work", if your objective was to make students comfortable **in the writing of algebraic expressions to solve problems by equations and handle variables in formulas**, which progression would you build around this activity (notions, concepts that you plan before, after)?

Progression		
Before:	Odd numbers activity	After:
- ...... -		- ...... - ......

40. Let us go back to current reality: today, would you use the activity you have chosen? yes [ ]   non [ ]

41. Why?

Annexe A- Le questionnaire initial intégral

Consigne : quand plusieurs réponses sont permises, les classer par ordre de préférence:1 pourla 1ère, 2etc...

PREMIERE PARTIE : Vous connaître : Madame [ ]   Monsieur [ ]   Année de naissance :..........

1. Dans quel type d'établissement enseignez-vous ?   ZEP [ ]   Difficile [ ]   Sensible [ ]   Normal [ ]   Bon [ ]
2. Avec quelle(s) classe(s) ? Pour combien d'heures de cours/ soutien/ aide individualisée etc ?
   Classe : .... pour ....h......de cours et ......h........de.........................
   Classe : .... pour ....h...... de cours et ......h........de.........................
3. Vous êtes :   Certifié [ ]   Agrégé [ ]   Dernier diplôme obtenu :................................
4. Avez-vous exercé une autre profession avant d'être enseignant ?   oui [ ]   non [ ]
   Si oui, laquelle ?..
5. Avez-vous déjà enseigné auparavant ? oui[ ] non[ ] Si oui, combien d'années ?........Dans quelles classes ?..............
6. Vous intéressez-vous à :   L'histoire des mathématiques [ ]   L'informatique [ ]   La didactique [ ]
   Autre(s)[ ] :..
7. a)Connaissez-vous la littérature enseignante ?   oui [   ]laquelle :.......................... non [
   b)Consultez-vous régulièrement des sites enseignants ? oui[   ]lesquels :........................... non [ ]
8. Lisez-vous des revues en lien avec les mathématiques ?   Jamais [ ]   Parfois [ ]   Régulièrement [ ]
9. Au cours de votre scolarité au collège ou au lycée, avez-vous eu l'occasion ?
   - d'utiliser en classe de mathématiques un logiciel d'enseignement ?   oui [ ]Préciser :........................   non [ ]
   - de consulter des sites web concernant les mathématiques ?   oui [ ]   non [ ]
10. Au cours de vos études (hors IUFM) avez-vous eu des formations en informatique ou suivi personnellement des cours d'informatique (à l'université ou avec des organismes privés) ?   oui [ ]   non [ ]
    Préciser :..
11. **Equipement** : possédez-vous un ordinateur personnel ? oui [ ] non [ ] une adresse électronique ? oui [ ] non [ ]

12. **Connaissances informatiques :** Savez-vous	oui	un peu	non
déplacer, copier, supprimer un fichier ?			
créer un document texte (ex : avec Word) ?			
créer un tableau dans un document texte ?			
mettre en forme (styles…) un document texte ?			
créer automatiquement une table des matières ?			
utiliser un éditeur d'équation pour écrire des formules mathématiques dans un texte ?			
composer des pages html ?			

13. **Tableur :** Avant l'IUFM saviez-vous	Avant			Et maintenant ?		
	oui	un peu	non	oui	un peu	non
- saisir et utiliser des formules dans un tableur ?						
- créer un graphique à partir de données saisies dans un tableur ?						

14. A priori, a) à quel domaine des mathématiques l'usage du tableur vous semble-t-il le plus approprié ? *(1 choix ou plus, dans ce cas numéroter)* [ ]Algèbre   [ ]Arithmétique   [ ]Statistiques
    [ ]Analyse   [ ]Probabilités   [ ]Autres :...............................

b) pour quelle partie *(1choix ou plus)*?   Simulations[ ]   Fonctions[ ]   Introduction de l'algèbre[ ]
   Mise en oeuvre d'algorithmes[ ]   Suites[ ]   Problèmes d'approximations[ ]
   Résolution de problèmes d'algèbre[ ]   Autres[ ] :...

15. **Enseignement** : Connaissez-vous	oui, très précisément	un peu, vaguement	non, pas du tout
16. les logiciels installés dans votre établissement ?			
17. le matériel informatique (salles, postes…) dont il est équipé ?			

18. Votre tuteur utilise-t-il l'outil informatique ?   très régulièrement [ ]   assez souvent [ ]   jamais [ ]
19. Savez-vous s'il y a des logiciels de mathématiques accessibles aux élèves au CDI ?   je sais [ ]   je ne sais pas [ ]

20. Comment envisagez-vous l'usage de l'ordinateur dans votre enseignement actuel et futur ? *(numéroter si plusieurs réponses)*
[ ] plutôt ponctuellement, comme outil pour se dégager des calculs fastidieux et concentrer le travail sur les concepts
[ ] plutôt pour l'aide individualisée et la remédiation avec les élèves en difficultés
[ ] plutôt bien intégré à ma progression annuelle, comme nouvel environnement pour créer des situations d'apprentissage

21. A priori, certains outils vous paraissent-ils plus faciles à intégrer que d'autres ? *classer du plus facile(1) au moins facile(4)* :
[ ] logiciel de géométrie dynamique (tels Cabri ou Géoplan)   [ ] tableur (tel Excel)   [ ] internet   [ ] calculatrice

22. Vos professeurs avaient d'autres programmes et manières d'enseigner. Auriez-vous aimé procéder comme eux ?
[ ] oui, j'aurais beaucoup aimé   [ ] ça m'est égal   [ ] non, surtout pas. Expliquer :...
....

DEUXIEME PARTIE : « Des professeurs fictifs »

A, B, C, D, E et F sont 6 professeurs de mathématiques fictifs du second degré. Les 5 premiers n'ont jamais réellement intégré l'informatique dans leur enseignement, voici leurs arguments :

A: « *Cette année, j'étais fermement décidé à utiliser l'informatique avec mes élèves, mais la clé de la salle semble si difficile à obtenir que j'ai finalement abandonné* »

B: « *J'ai déjà envisagé de préparer une séance informatique pour mes élèves. Mais, j'ai eu peur de ne pas bien savoir me débrouiller techniquement (surtout que certains élèves se débrouillent sûrement mieux que moi) alors j'ai renoncé* »

C: « *Je voudrais bien utiliser l'informatique pour enseigner les mathématiques à mes élèves, mais ça demande trop de temps de préparation !!* »

D: « *Je n'ai rien contre utiliser l'informatique avec mes élèves, mais on ne m'a jamais formé à l'intégrer dans mon enseignement. Si j'étais convenablement formé, je le ferais volontiers* »

E: « *Franchement, je ressens comme anormale la présence de l'informatique dans nos programmes car notre rôle est d'enseigner les mathématiques. Il est évident que si on me donnait le choix, j'enseignerais à mes élèves les mathématiques sans faire intervenir l'informatique* »

23. Avez-vous déjà ressenti ce qu'exprime l'un d'eux ? *(numéroter si plusieurs réponses)* A[ ]   B[ ]   C[ ]   D[ ]   E[ ]
Envisagez-vous un outil informatique en particulier ?...

24. Selon vous, cela pourrait-il concrètement vous arriver ? A[ ]   B[ ]   C[ ]   D[ ]   E[ ]   *(numéroter si plusieurs réponses)*

25. Que pensez-vous plus précisément de chacune des déclarations ? *(compléter au dos si besoin)*
A :...
....
B :...
....
C :...
....
D :...
....
E :...
....

26. D'après leur déclaration, **lequel** des 5 enseignants vous paraît être *(numéroter si plusieurs réponses)* :
a) le moins fictif : A[ ]   B[ ]   C[ ]   D[ ]   E[ ]   b) le plus représentatif : A[ ]   B[ ]   C[ ]   D[ ]   E[ ]

27. Duquel vous sentez-vous le plus proche ?   A[ ]   B[ ]   C[ ]   D[ ]   E[ ]   *(numéroter si plusieurs réponses)*

28. Duquel vous sentez-vous le moins proche ?   A[ ]   B[ ]   C[ ]   D[ ]   E[ ]   *(numéroter si plusieurs réponses)*

29. Et vous ? Quelle est votre position concernant l'utilisation d'un ordinateur dans les cours de mathématiques ?
....

30. En général, comment ressentez-vous l'introduction de l'outil informatique dans les programmes de mathématiques ? (utiliser 5 adjectifs, ou plus, pour décrire votre sentiment) (si besoin, mettre au dos tout autre commentaire)
....

31. Enfin, **le professeur F dit :** « *Je vois bien à quoi l'informatique peut servir, je voudrais, par exemple, utiliser le tableur avec mes 4è qui en ont déjà une expérience* »

Un collègue lui propose 2 activités que voici : (*Il s'agit à chaque fois d'une séance en salle informatique où les élèves ont une fiche élève guidant leur travail et un fichier-tableur préparé par le professeur*)

1 : "Formules" : *Une feuille de calculs toute prête contient une formule en B3, les élèves doivent dans un premier temps repérer que la valeur en B3 dépend des valeurs en A1 et C1 puis identifier la formule et l'exploiter pour répondre au 2.*

Fiche Elève :	Tableur (**feuille prête, formule déjà créée**) :
1. Remplacer 8 et 9 par d'autres nombres entiers et observer ce qui se passe. 2. Que placer dans les cellules A1 et C1 pour obtenir 50 dans la cellule B3? Et 100 ? et 300 ? Peut-on obtenir tous les nombres entiers ? Expliquer	

	A	B	C
1	8		9
2			
3		43	

(la formule entrée en B3 est ici"=2*A1+3*C1')

2 : "Théorème de Pythagore" : *Les élèves tracent d'abord sur papier 5 triangles rectangles puis calculent les carrés des 3 côtés grâce à une feuille de calculs déjà prête qui calcule des carrés et des sommes de nombres.*

Fiche Elève :	Tableur (**feuille prête, formules dans E, F, G et H déjà créées**) :
1. Tracer 5 triangles rectangles de mesures différentes et compléter les colonnes B, C et D *le tableur calcule automatiquement les résultats en E, F, G et H* 2. Que remarque-t-on ?	

	A	B	C	D	E	F	G	H
1								
2		Mesures des 2 côtés de l'angle droit		Mesure de l'hypoténuse	Carrés des côtés de l'angle droit		Somme de ces carrés	Carré de l'hypoténuse
3	Triangle 1							
4	Triangle 2							
5	Triangle 3							
6	Triangle 4							
7	Triangle 5							
8								

Laquelle lui conseilleriez-vous en fonction des différents critères qu'il peut avoir : intérêt mathématique, gestion de classe, facilité à l'intégrer dans une progression *(expliquer votre choix pour chaque critère)* :

....
....
....
....
....

32. Et vous, s'il le fallait, laquelle seriez-vous prêt à mettre en œuvre en classe ?   plutôt la 1[ ]   plutôt la 2[ ]
Pourquoi ?...
....
....
....
....

33. Laquelle auriez-vous préféré avoir en tant qu'élève ?   plutôt la 1[ ]   plutôt la 2[ ]   aucune [ ]
Pourquoi ?...

TROISIEME PARTIE : « Une supposition… »

Supposons que, demain, vous vouliez ou deviez (pour une raison quelconque) utiliser le tableur pour faire travailler vos élèves de 4e (ayant déjà une expérience du tableur mais peu de connaissances en algèbre) sur l'écriture de formules algébriques. Votre objectif est que les élèves trouvent, à partir du fichier-tableur ci-contre (une suite d'entiers consécutifs), la formule **« 2n+1 »** comme expression générale d'un nombre impair. Un collègue vous propose les 2 énoncés suivants :

	A	B
1		
2	0	
3	1	
4	2	
5	3	
6	4	
	…	

Enoncé A :   Dans la cellule B2, tape : « =2*A2+1 », tire la poignée de recopie vers le bas. Que remarques-tu ? Peux-tu l'expliquer ?
Enoncé B :   A partir des nombres de la colonne A, trouver une formule générale qui donne, dans la colonne B, des nombres impairs.

34. Quels sont, pour vous, les avantages et inconvénients de ces 2 énoncés ?

	Avantages	Inconvénients
Enoncé A		
Enoncé B		

35. En gardant le même fichier et le même objectif, lequel des 2 énoncés donneriez-vous à vos élèves tel quel (sans aucune modification) ? [ ]Plutôt **A**   [ ]Plutôt **B**   [ ]**Aucun** : j'aurais des modifications à apporter, voici mon énoncé :

Thème : écriture de formules. **Objectif** : Trouver, grâce au tableur, la formule « 2n+1 » comme expression générale d'un nb impair **Fichier donné** :	**Enoncé personnel :**
<table><tr><td></td><td>A</td><td>B</td></tr><tr><td>1</td><td></td><td></td></tr><tr><td>2</td><td>0</td><td></td></tr><tr><td>3</td><td>1</td><td></td></tr><tr><td>4</td><td>2</td><td></td></tr><tr><td>5</td><td>3</td><td></td></tr><tr><td>6</td><td>4</td><td></td></tr><tr><td></td><td>...</td><td></td></tr></table>	

36. Essayons de construire un scénario… :
a) Comment introduisez-vous l'activité choisie ? Avec quelle (s) consigne(s) ou contrat de travail ?.............................
…….

b) A priori quelles sont les stratégies possibles des élèves, et avec quelles fonctionnalités du tableur ?
…….
…….
…….

c) Quelles sont les difficultés prévisibles et quelle aide pouvez-vous apporter pour que l'objectif soit atteint ?
…….
…….
…….

d) Au vu du b) et c) (stratégies, difficultés et aides prévisibles) décrire la gestion que vous feriez de l'activité choisie à travers les grands moments a priori du scénario : votre rôle, celui des élèves, les différentes phases de la séance suivant les changements de prise de parole (professeur/élève)
…….
…….
…….

37. Que feriez-vous écrire dans le cahier de cours concernant cette activité ?...
…….
…….
…….

38. Quelle suite précise donneriez-vous à cette activité ?...
…….
…….
…….

39. Sur le thème : « faire entrer les élèves dans un travail algébrique », quelle progression construiriez-vous autour de cette activité (notions, concepts que vous planifiez avant, après) si votre objectif final est que **les élèves soient à l'aise dans l'écriture d'expressions algébriques pour résoudre des problèmes par équations ou manipuler des variables dans des formules ?** *(compléter au dos un tableau suivant ce modèle)*

Progression		
Avant : - …… - ……	L'activité des nombres impairs	Après : - …… - ……

40. Retournons à la réalité actuelle : aujourd'hui, utiliseriez-vous l'activité que vous avez choisie ? oui [ ]   non [ ]
41. Pourquoi ?...
…….
…….
…….

Merci de votre collaboration
Si besoin, seriez-vous d'accord pour que l'on vous contacte pour un entretien ?(téléphone ou mail)............................

Noter ici et au dos toute remarque supplémentaire que vous avez envie d'exprimer :

References

Ainley, J. (1999). Doing algebra-type stuff: Emergent algebra in the primary school. In O. Zaslavsky (Ed.), *Proceedings of the twenty third annual conference of the International Group for the Psychology of Mathematics*, Haifa, Israel.

Ainley, J., Bills, L., & Wilson, K. (2003). Designing tasks for purposeful algebra. *Proceedings of the third conference of the European Society for Research in Mathematics Education*, Bellaria, Italy.

Artigue, M. (2002). Learning mathematics in a CAS environment: The genesis of a reflection about instrumentation and the dialectics between technical and conceptual work. *International Journal of Computers for Mathematical Learning, 7*, 245–274.

Arzarello, F., Bazzini, L., & Chiappini, G. (2001). A model for analysing algebraic processes of thinking. In R. Sutherland, T. Assude, A. Bell, & R. Lins (Eds.), *Perspectives on school algebra* (Vol. 22, pp. 61–81). Dordrecht: Kluwer.

Baker, J., & Sugden, S. (2003). Spreadsheets in Education –The First 25 Years. *Spreadsheets in Education (eJSiE)*: Vol. 1–1, Article 2.

Balacheff, N. (1994). La transposition informatique. Note sur un nouveau problème pour la didactique. In M. Artigue (Ed.), *Vingt ans de Didactique des Mathématiques en France* (pp. 364–370). Grenoble: La pensée sauvage.

Balanskat, B., Blamire, R., & Kefala, S. (2006). The ICT impact report. A review of studies of ICT impact on schools in Europe. *Report written by European Schoolnet in the framework of the European Commission's ICT cluster.* http://ec.europa.eu/education/pdf/doc254_en.pdf

Bruillard, E. & Blondel, F.-M. (2007). *Histoire de la construction de l'objet tableur*, document de pré-publication, version 1 du 22 octobre 2007. http://hal.archives-ouvertes.fr/docs/00/18/09/12/PDF/histoire_construction_objet_tableur_v1.pdf

Bruillard, E., Blondel, F.-M., & Tort, F. (2008). DidaTab project main results: Implications for education and teacher development. In K. McFerrin, R. Weber, R. Carlsen, & D. A. Willis (Eds.), *Proceedings of the International Conference, SITE 2008* (pp. 2014–2021). Chesapeake, USA: AACE.

Capponi, B. (2000). Tableur, arithmétique et algèbre. L'algèbre au lycée et au collège, *Actes des journées de formation de formateurs 1999* (pp. 58–66). IREM de Montpellier.

Capponi, B., & Balacheff, N. (1989). Tableur et Calcul Algébrique. *Educational Studies in Mathematics, 20*, 179–210.

Chevallard, Y. (1992). Intégration et viabilité des objets informatiques dans l'enseignement des mathématiques. In B. Cornu (Ed.), *L'ordinateur pour enseigner les mathématiques* (pp. 183–203). Paris: Presses Universitaires de France.

Chevallard, Y. (2007). Readjusting didactics to a changing epistemology. *European Educational Research Journal, 6*(2), 131–134.

Clark-Wilson, A. (2010a). Connecting mathematics in a connected classroom: Teachers emergent practices within a collaborative learning environment. *British Congress on Mathematics Education, BSRLM Proceedings, 30*(1). University of Manchester.

Clark-Wilson, A. (2010b). *How does a multi-representational mathematical ICT tool mediate teachers' mathematical and pedagogical knowledge concerning variance and invariance?* Ph.D. thesis, Institute of Education, University of London.

Coulange, L. (1998). Les problèmes "concrets à mettre en équation" dans l'enseignement. Petit x n 47. 33–58.

Dettori, G., Garuti, R., Lemut, E., & Netchitailova, L. (1995). An analysis of the relationship between spreadsheet and algebra. In L. Burton & B. Jaworski (Eds.), *Technology in mathematics teaching: A bridge between teaching and learning* (pp. 261–274). Bromley: Chartwell-Bratt.

Dettori, G., Garuti, R., & Lemut, E. (2001). From arithmetic to algebraic thinking by using a spreadsheet. In R. Sutherland, T. Rojano, A. Bell, & R. Lins (Eds.), *Perspectives on school algebra* (pp. 191–207). Dordrecht: Kluwer.

Drijvers, P. (2000). Students encountering obstacles using CAS. *International Journal of Computers for Mathematical Learning, 5*(3), 189–209.

Eurydice. (2004). Keydata on information and communication technology in schools in 2004. [On line] http://www.eurydice.org/ressources/eurydice/pdf/048EN/004_chapB_048EN.pdf

Eurydice. (2005, October 2005). How boys and girls in Europe are finding their way with information and communication technology? Eurydice in brief. Brussels. [On line] http://bookshop.europa.eu/en/how-boys-and-girls-in-europe-are-finding-their-way-with-information-and-communication-technology--pbEC3212339/

Guin, D., & Trouche, L. (1999). The complex process of converting tools into mathematical instruments: The case of calculators. *International Journal for Computer Algebra in Mathematics Education, 3*(3), 195–227.

Guin, D., Ruthven, K., & Trouche, L. (2004). *The didactical challenge of symbolic calculators: Turning a computational device into a mathematical instrument*. New York: Springer.

Haspekian, M. (2005a). *Intégration d'outils informatiques dans l'enseignement des mathématiques, Etude du cas des tableurs*. Ph.D. thesis, University Paris 7. tel.archives-ouvertes.fr/tel-00011388/en/.

Haspekian, M. (2005b). An "Instrumental Approach" to study the integration of a computer tool into mathematics teaching: The case of spreadsheets. *International Journal of Computers for Mathematical Learning, 10*(2), 109–141.

Haspekian, M. (2006). Evolution des usages du tableur. In Rapport intermédiaire de l'ACI-EF *Genèses d'usages professionnels des technologies chez les enseignants*. http://gupten.free.fr/ftp/GUPTEn-RapportIntermediaire.pdf

Haspekian, M. (2011). The co-construction of a mathematical and a didactical instrument. In M. Pytlak, E. Swoboda & T. Rowland (Eds.), *Proceedings of the seventh Congress of the European Society for Research in Mathematics Education*. CERME 7, Rzesvow.

Laborde, C. (2001). Integration of technology in the design of geometry tasks with Cabri geometry. *International Journal of Computers for Mathematical Learning, 6*(3), 283–317.

Laborde, C., & Capponi, B. (1994). Cabri-géomètre constituant d'un milieu pour l'apprentissage de la notion de figure géométrique, *Recherche en didactique des mathématiques, 14*(1-2).

Lagrange, J. B. (1999). Complex calculators in the classroom: Theoretical and practical reflections on teaching pre-calculus. *International Journal of Computers for Mathematical Learning, 4*(1), 51–81.

Lagrange, J. B. (2000). Lintegration d'instruments informatiques dans l'enseignement : une approche par les *techniques*. *Educational Studies in Mathematics, 43*, 1–30.

Monaghan, J. (2004). Teacher's activities in technology-based mathematics lessons. *International Journal of Computers for Mathematics Learning, 9*, 327–357.

Norton, S., McRobbier, C. J., & Cooper, T. J. (2000). Exploring secondary mathematics teachers' reasons for not using computers in their teaching: Five case studies. *Journal of Research on Computing in Education, 33*(1).

Parzysz, B. (1988). Voir et savoir – la représentation du "perçu" et du "su" dans les dessins de la géométrie de l'espace. *Bulletin de l'APMEP, 364*.

Pelgrum, W. J., & Anderson, R. E. (Eds.). (2001). *ICT and the emerging paradigm for life-long learning*. Amsterdam: IEA.

Rabardel, P. (1993). Représentations pour l'action dans les situations d'activité instrumentée. In A. Weill-Fassina, P. Rabardel, & D. Dubois (Eds.), *Représentations pour l'action*. Octares: Toulouse.

Rabardel, P. (2002). *People and technology -a cognitive approach to contemporary instruments*. http://ergoserv.psy.univ-paris8.fr

Robert, A., & Rogalski, J. (2002). Le système complexe et cohérent des pratiques des enseignants de mathématiques: une double approche. *Revue canadienne de l'enseignement des sciences, des mathématiques et des technologies, 2*, 505–528.

Rojano, T., & Sutherland R. (1997). Pupils' strategies and the Cartesian method for solving problems: The role of spreadsheets. *Proceedings of the 21st international PME conference* (Vol. 4, pp. 72–79).

Ruthven, K. (2007). Teachers, technologies and the structures of schooling. In D. Pitta-Pantazi & G. Philippou (Eds.), *Proceedings of the fith congress of the European Society for Research in Mathematics Education* (pp. 52–68). Larnaca.

Stacey, K., Chick, H., & Kendal, M. (2004). *The future of the teaching and learning of algebra* (The 12th ICMI study). Kluwer, Dordrecht.

Trouche, L. (2004). Managing complexity of human/machine interactions in computerized learning environments: Guiding student's command process through instrumental orchestrations. *International Journal of Computers for Mathematical Learning, 9*(3), 281–307.

Trouche, L. (2005). Instrumental genesis, individual and social aspects. In D. Guin, K. Ruthven, & L. Trouche (Eds.), *The didactical challenge of symbolic calculators: Turning a computational device into a mathematical instrument* (pp. 97–230). New York: Springer.

Vérillon, P., & Rabardel, P. (1995). Cognition and artifacts: A contribution to the study of though in relation to instrumented activity. *European Journal of Education, X*(1), 77–101.

A Methodological Approach to Researching the Development of Teachers' Knowledge in a Multi-Representational Technological Setting

Alison Clark-Wilson

Abstract This chapter details the methodological approach adopted within a doctoral study that sought to apply and expand Verillon and Rabardel's (*European Journal of Psychology of Education*, *10*, 77–102, 1995) triad of instrumented activity as a means to understand the longitudinal epistemological development of a group of secondary mathematics teachers as they began to integrate a complex new multi-representational technology (Clark-Wilson, *How does a multi-representational mathematical ICT tool mediate teachers' mathematical and pedagogical knowledge concerning variance and invariance?* Ph.D. thesis, Institute of Education, University of London, 2010a). The research was carried out in two phases. The initial phase involved fifteen teachers who contributed a total of sixty-six technology-mediated classroom activities to the study. The second phase adopted a case study methodology during which the two selected teachers contributed a further fourteen activities. The chapter provides insight into the methodological tools and processes that were developed to support an objective, systematic and robust analysis of a complex set of qualitative classroom data. The subsequent analysis of this data, supported by questionnaires and interviews, led to a number of conclusions relating to the nature of the teachers' individual technology-mediated learning.

Keywords Hiccup • Instrumented activity • Instrument utilisation scheme • Multi-representational technology • Social utilisation scheme • TI-Nspire • Mathematical variance and invariance

---

A. Clark-Wilson (✉)
London Knowledge Lab, Institute of Education, University of London,
23-29 Emerald Street, London WC1N 3QS, UK
e-mail: a.clark-wilson@ioe.ac.uk

Introduction

The research that is reported in this chapter had the broad aim to articulate the nature of secondary mathematics teachers' epistemological development as they began to use a complex new multi-representational technological tool with students in their classrooms. The chosen technology was new in the sense that it offered linked multiple representations between numeric, syntactic and geometric domains (See Arzarello and Robutti (2010) for a more in-depth description). I defined a teacher's epistemological development as the trajectory of their growth in mathematical, pedagogic and technological knowledge within the context of the design and teaching of activities that privileged their students' explorations of variance and invariance. The research was carried out in two phases, July 2007 – Nov 2008 and April 2009 – December 2009, when groups of teachers were selected, and a series of methodological tools developed, to capture rich evidence of the teachers' uses of the technology in their classrooms to enable the aims of the study to be realised. The first phase of the project was located within a professional development setting, which blended opportunities for the teachers to learn about the affordances of the technology alongside time for the teachers to design activities and give subsequent feedback about the outcomes of their lessons. The second phase of the study was wholly situated within the participating teachers' mathematics classrooms.

Theoretical Background

The theoretical foundations for the study concerned three domains: coming to know new technologies and the role of technology in developing subject and pedagogic knowledge; the concept of variance and invariance in a multi-representational technological setting; and making sense of the process of teacher learning.

The theoretical framework that was developed for the study was rooted in Verillon and Rabardel's (1995) theory of instrumented activity systems as a model to describe the processes involved in human-instrument interactions. In this framework a distinction between artefact and instrument is introduced in order to distinguish between the object itself (as an independent artefact), and the same object as used by a subject. The object is referred to as an artefact when it is used by a person during an activity. The same object is referred to as an instrument when it has been endowed with specific utilisation schemes that have been introduced by the subject. Consequently, as these schemes of use are introduced by the subject, the relation between the artefact and its uses evolve, giving rise to the process of instrumental genesis. While the artefact is an object that can be considered statically, in the sense that it does not change its features over time, the instrument can be conceived dynamically, in the sense that it can change its features, according to the schemes of use that are activated by the user. Therefore, the same artefact can become different instruments, related to the purpose of the subject's actions. In their original model, the Subject-Instrument-Object triad assumed that the subject's primary consideration was to evolve uses of the instrument for some clear purpose, which is to carry out a particular, specified task. This model has been applied to a number of situations within mathematics education research where the lens has been trained on *students* of mathematics who were beginning to use chosen technologies for the purpose of solving mathematical problems (Guin and Trouche 1999; Artigue 2001; Ruthven 2002). However, the context for my own study brought another consideration to the fore. As the subjects within my study were *teachers*, there were two facets to the object for their subsequent use for the technology. It was obviously necessary for them to become familiar with the affordances of the technology but also, a simultaneous consideration for them was whether and how these affordances could be integrated into educationally legitimate classroom activities for mathematics.

Within my study, subjects were 'teachers as learners' and the objective for their technology-related activity concerned the processes of designing, teaching and evaluating explorations of mathematical variance and invariance. My research was interested in the teachers' epistemological development over several years as they were engaged in these processes. By epistemological development, I mean the development of their personal knowledge, which would incorporate mathematical, technological and pedagogic aspects. For my context, the instrument incorporated the mediating artefact, that is, the TI-Nspire handheld and software alongside the emergent utilisation schemes developed individually by each teacher or socially, where collaboration was involved. Hence the study sought to gain deeper insight into the mediating role of the technology. This sense of *double instrumentation* resonates with the findings of Haspekian's (2005 and Chap. 9 in this volume) research within the context of a spreadsheet environment in which she concludes that the spreadsheet is one instrument for teacher's personal mathematical work and *another* instrument for the teacher's professional didactical work (Haspekian 2006). This led to the notion of *double instrumental genesis* from the teacher's perspective.

The mathematical focus for the study concerned activities that privileged the students' *explorations of variance and invariance*. This is the approach whereby the technology is being used in an exploratory way, with the intention that the students will *discover* some mathematical generalisation(s) by varying some sort of input and observing the output provided by the technology. Essentially, this meant that the teachers were privileging explorations of variant and invariant properties within a chosen mathematical context. This focus was a constraint of the project's methodology in response to the teachers favouring the design of tasks that encouraged student autonomy by requiring them to make inputs to the technology and draw conclusions in relation to the resulting outputs.

The multi-representational features of TI-Nspire (Arzarello and Robutti 2010) prompted a review of key texts and research that had considered both the mediating role of technology in supporting such explorations alongside a review of literature on the nature of a mathematical variable (Bednarz et al. 1996; Moreno-Armella et al. 2008; Sutherland and Mason 1995; Kaput 1986; Kaput 1998; Kieran and

Wagner 1989). This review led me to define mathematical learning as being predominantly concerned with the privileging of students' opportunities to generalise and specialise as a means to constructing their own mathematical meanings.

Within the context of this study, the teacher's role was to design and orchestrate classroom activities and approaches, using the various functionality of the multi-representational technology to achieve this. However, as teachers' individual belief systems (in the usual sense) about mathematical learning (and the role of technology within this) would undoubtedly influence their decisions and actions, the trajectory of teacher development to which I refer also revealed evidence of these preconceptions.

Finally, as the study was concerned with the nature and processes of mathematics teachers' epistemological development, two areas of related literature were reviewed. The first area concerned definitions and interpretations of mathematics teachers' personal knowledge, subject knowledge for teaching and pedagogic knowledge (Shulman 1986; Rowland et al. 2005; Zodik and Zaslavsky 2008; Polanyi 1962, 1966). The second area examined constructs concerning the process of teacher learning (Schön 1984; Thompson 1992; Mason 2002; Jaworski 1994; Ahmed and Williams 1997). The review of literature referring to the content, nature and process of teacher learning led me to adopt a broad interpretation of knowledge as proposed by Shulman's *knowledge for teaching*. It also highlighted the complexities of the process of teacher learning and supported the development of methodological tools that would capture the evidence of this learning in line with my desire to describe teachers' trajectories of epistemological development. I use the word epistemology in a deliberate sense to indicate that I was most concerned with how their knowledge developed over time. This had implications for the methodological approach that was adopted as, although some of these theoretical ideas gave a framework for describing teachers' knowledge, they did not necessarily lend themselves to the development of a useful set of methodological tools and techniques.

Methodology

An extensive data collection period between July 2007 and November 2009 resulted in the participating teachers contributing eighty *lesson bundles* to the study. During the first phase of the study, a lesson bundle comprised all or some of the following:

- A compulsory lesson evaluation questionnaire – (see Clark-Wilson 2008b);
- An activity plan in the form of a school lesson planning proforma or a handwritten set of personal notes;
- A lesson structure for use in the classroom (for example a Smart NoteBook or PowerPoint file);
- A software file developed by the teacher for use by the teacher (to introduce the activity or demonstrate an aspect of the activity);
- A software file developed by the teacher for use by the students, which would normally need to be transferred to the students' handhelds in advance or at the beginning of the lesson;

- An activity or instruction sheet developed by the teacher for students' use;
- Students' written work resulting from the activity;
- Students' software files captured during and/or at the end of the activity;
- Audio or video clips of the activity;
- Notes or slides from presentations made by the teachers about the activity.

These lesson bundles resonate with the idea of the teachers' *documentation system* (See Aldon Chap. 12 this volume) that capture the complete set of resources developed (or made use of) such that teachers can make use of technologies for mathematics within classroom settings (Gueudet and Trouche 2009).

Summarising Lessons

The sets of raw data were imported to the qualitative data analysis software package, Nvivo8 (QSR International 2008), where they were subsequently scrutinised and coded to elicit three elements: a broad description of the lesson; an inference concerning the teacher's interpretation of variance and invariance within the designed activity; and the implied instrument utilisation scheme that the students were expected to use.

An example of this for a lesson 'Prime factorisation', submitted by one of the teachers early at the beginning of the first phase of the study is shown in Table 1.

The subsequent cross-case analysis of these individual lesson data led to the development of nine *instrument utilisation schemes*, which sought to generalise the flow of an activity in relation to the intended interactions by the student as they used the technology, using a constant comparison method. The resulting instrument utilisation schemes considered the broad representational input or output as being either numeric, syntactic or graphic. For example, the lesson Prime factorisation described in Table 1, would lead to the instrument utilisation scheme in Fig. 1 below.

In this activity the input was a combination of a syntactic entry (i.e. factor(n)) and a numeric entry (i.e. n) and the output was syntactic in that the representation $2^2 \bullet 5$ implies a mathematical syntax that is adopted by the technology.

A numeric input might involve entering numeric values into a spreadsheet or changing an input for a numeric variable. A syntactic input is considered to encompass both the syntactic forms of conventional mathematical notation in addition to the syntax required when using specific functionalities of the technology such as the need to use the specific syntax of the built-in 'Factor' command. In this respect, the word syntactic is not being interpreted in a wholly linguistic sense but it does embrace Shulman's sense of *syntactic structures* (Shulman 1986). As I began to classify the nature of the 'outputs' I initially used the same three categories. However, it quickly became apparent that the analysis became more informative if some sub-divisions of the initial three categories were made. Hence the *numeric* category was subdivided into *measured, calculated* and *tabulated*; the *geometric* category was subdivided into *graphical (data points), graphical (function graphs)* and *geometric (positional)*.

Table 1 The summary data for the lesson 'Prime factorisation'

Lesson code, title, activity description and age of students.	Relevant screenshots and implied evidence of teacher learning	Interpretation of variance and invariance	Implied Instrument Utilisation Scheme (with respect to the students)
| Lesson code: SJK1 12–13 years Prime factorisation Students created a new file and used the *factor()* command within the Calculator application to explore different inputs and outputs to encourage generalisation. Students recorded their work on a worksheet prepared by the teacher. | ```
1.1          DEG AUTO REAL
factor(15)              3·5
factor(21)              3·7
factor(10)              2·5
factor(12)             2²·3
factor(8)                2³
factor(16)               2⁴
                          6/6
```  *Definitely the worksheet idea as this enables pupils to work at their own pace. I needed the worksheet for me as well as for them. I was able to refer to the sheet and that helped my confidence. The sheet also allowed pupils to continue with the work whilst I went round to help students with a problem. BUT – the sheet could have been more structured i.e. not jump around haphazardly but be more systematic. Factor (1), Factor (2), Factor (3), etc....*  *I was nervous to use the device even though I am a very experienced teacher of maths. I needed the worksheet as support. Having done one lesson I would now be confident to try again. The worksheet could have been more interesting. Pupils seemed to enjoy the lesson.* [SJK1(Quest2)] | The teacher had constructed a worksheet (with support from her mentor) that did lead students through a set of suggested input numbers that progressed in their level of complexity. Variance = changing the input number (a manual text input to calculator application using factor() syntax) Invariance = all prime numbers had only two factors (by definition). | IUS1: *Vary a numeric or syntactic input and use the instrument's functionality to observe the resulting output in numeric, syntactic, tabular or graphical form.* Private utilisation scheme: Students worked with a numerical input and output within one application responding to the same set of numbers, as provided by the teacher. No use of multiple representations using the technology. Social utilisation scheme: Whilst JK developed this lesson in close collaboration with TP, who taught a similar lesson. The lessons had very different utilisation schemes. JK provided a structured worksheet for the students that contained a variety of questions that did not appear to follow any conceptual progression, although the factorised forms became increasingly complex. |
```

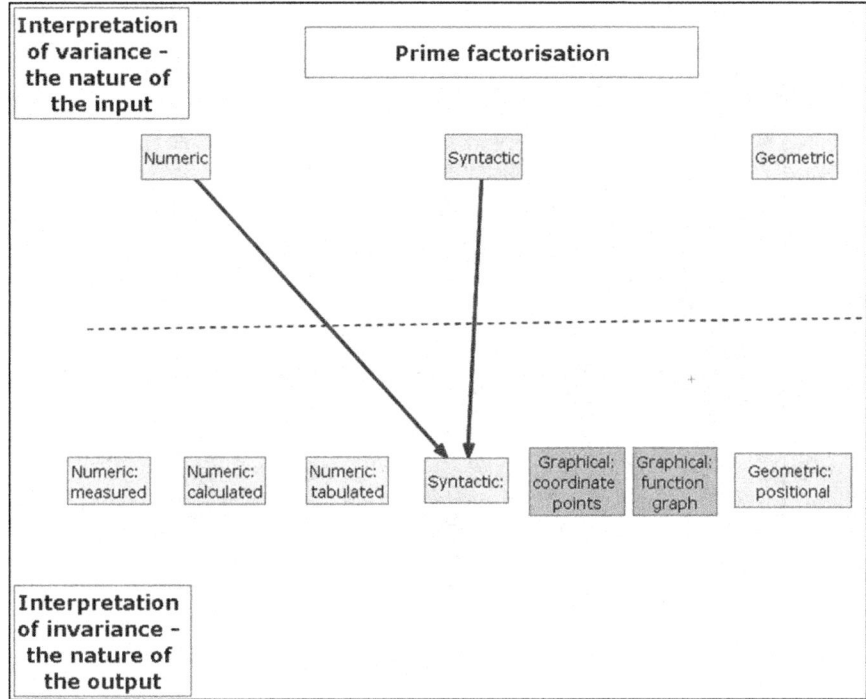

Fig. 1 The Instrument Utilisation Scheme (IUS1) for the lesson 'Prime Factorisation'

Instrument utilisation scheme type one (IUS1) was the simplest of all of the schemes, and it was also the most frequently used scheme by the teachers in the first phase of the study, with over half of the reported lessons being classified as IUS1.

By contrast, as the project progressed, there were three teachers who developed a diverse set of IUS. As the nature of the activities that the teachers created were all exploratory, they all had an initial input and output phase. However, a more diverse set of IUSs developed as teachers began to design tasks that elaborated on this initial phase by requiring different forms of interaction with the technology such as dragging or the inclusion of an additional representational form. One such example was the lesson activity developed by Eleanor, 'Perpendicular functions' which is described in detail in Table 2.

The instrument utilisation scheme for this lesson (IUS7) is shown in Fig. 2.

The second phase of the study still required the teachers to design, teach and evaluate lesson activities using the technology and, additionally, it involved lesson observations, which were all audio-recorded (with key sequences also video-recorded). The two case study teachers (Eleanor and Tim) were also interviewed before and after the classroom observations. This more substantive data was initially used to write a detailed description of the lesson (8–10 pages), interspersed with mediating screen shots from the teacher's and students' files. This process was greatly supported through the use of the handheld classroom network system

284 A. Clark-Wilson

Table 2 The summary data for the lesson 'Perpendicular functions' (Students aged 14–15 years)

Lesson code, title and activity description	Relevant screenshots and implied evidence of teacher learning	Interpretation of variance and invariance	Implied IUS
CEL5 Perpendicular functions Students created a new file and used the Graphs and Geometry application to define a linear function and draw a freehand 'perpendicular line'. They used the angle measure facility to check their accuracy, dragged the geometric line until it was perpendicular and then generated parallel functions to this geometric line. They used a Notes page to record their findings. Students saved their work.	[screenshot: DEG AUTO REAL, f1(x)=2x, 90°] *If the perpendicular value did not appear to fit the rule we checked angles were 90°.* Q. What changes would you make? *To reinforce looking for connections between perpendicular gradients as written for the value of m. To encourage more formal methods of recording results. To introduce a spreadsheet page to establish that the product of perpendicular gradients is −1. Slope can be measured. Ensuring all hand-held are in degree mode and all with a float of 3.* *Students were very intuitive with using the new technology. Would need to reinforce girls to do more mathematical thinking and reflection and to view the technology as a means to do this.*	Variance = initial definition (position-ing?) of linear function, 'by eye' positioning of geometric line, value of measured angle, the gradients of the two lines (by measurement or by definition). Invariance = the condition that, when the lines were perpendicular, the products of their gradients would equal −1.	IUS7: *Construct a graphical and geometric scenario and then vary the position of geometric objects by dragging to satisfy a specified mathematical condition. Input functions syntactically to observe invariant properties.* New possibilities for the action: This was a very innovative use of the technology – it combined defining functions with dragging and measuring angle. EL also suggested developments that would integrate the Spreadsheet page as a means for collating results and checking conjectures. Private utilisation scheme: *To draw a simple linear function on a graphs page. To construct a line that crosses this at 90°o. To measure this angle and draw a selection of parallel lines to this drawn line. To record findings.* [CEL5(Quest2)] Results were recorded on a notes page and we brought together the many concepts related to gradients, parallelness, perpendicular and y=mx+c. [CEL5(Quest2)] Social utilisation scheme: This idea was 'mis-remembered' by the teacher from a lesson developed by another teacher and reported at the third project meeting.

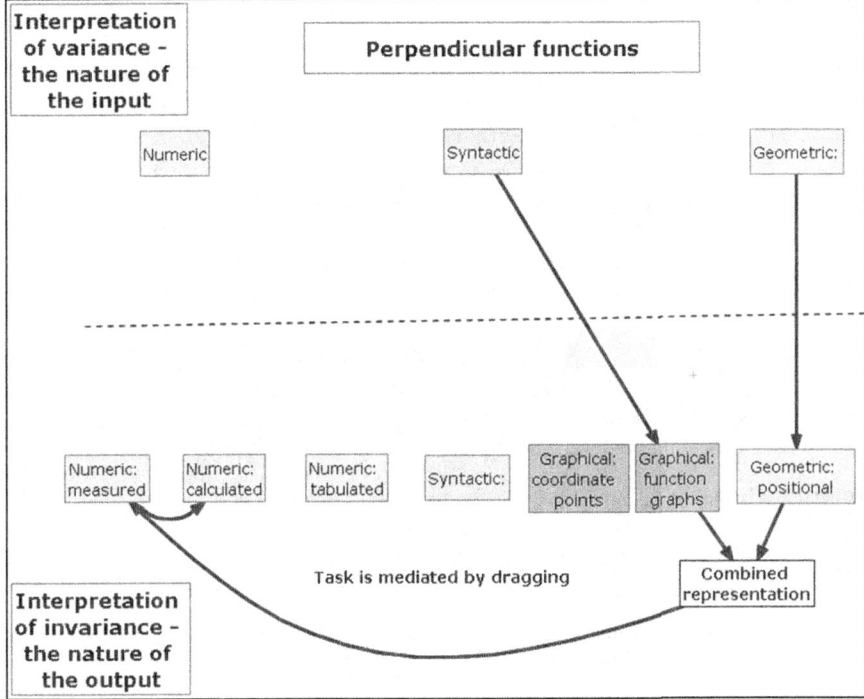

Fig. 2 The instrument utilisation scheme (IUS7) for the lesson 'Perpendicular functions'

TI-Navigator, which facilitated the real-time data collection process without interrupting the flow of the lessons. Following this, I used elements of Pierce and Stacey's (2008) pedagogical map as a tool to support the writing of a summary of each lesson from the three perspectives they describe as 'layers of pedagogical opportunities', namely the task layer, the classroom layer and the subject layer. This led to a detailed set of interpretations of the teachers' actions within the individual lessons alongside a map of their enacted instrument utilisation schemes as observed during the second phase of the study.

Hence, over time, evidence of the individual teacher's development began to emerge. The development of each teacher's instrument utilisation schemes was made visible by overlaying the individual lesson analyses from the Phase One and the Phase Two of the study (Figs. 3 and 4).

It was immediately apparent that Eleanor's activities incorporated a greater diversity of representations and each activity had its own sequential flow. This was sufficient evidence to conclude *that* Eleanor's practice had developed but it gave little indication of *how* this development had evolved.

Whilst I was writing the detailed narratives of the observed lessons, I became aware of the incidents within the lessons where the teachers experienced perturbations, triggered by the use of the technology, which seemed to illuminate discontinuities

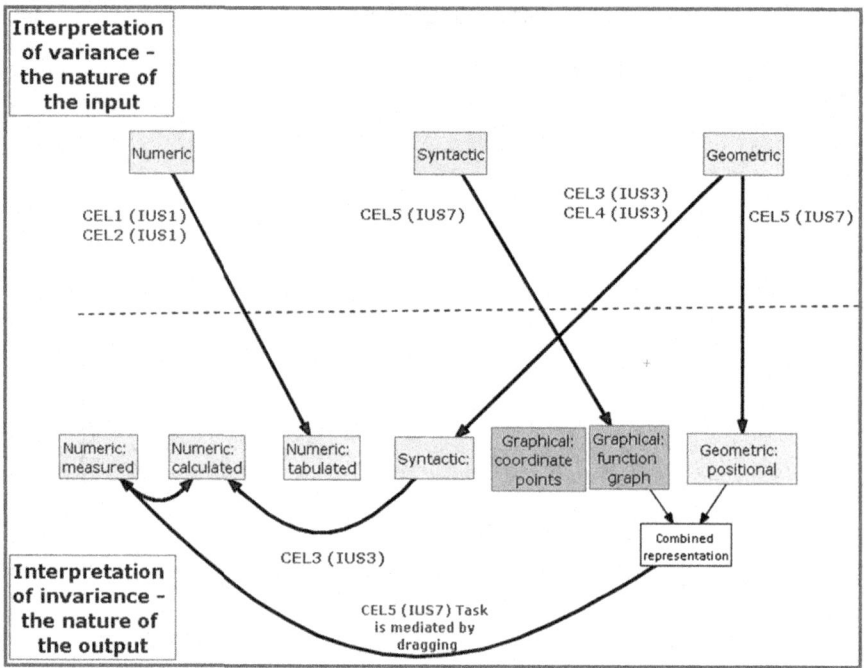

Fig. 3 The summary of Eleanor's Instrument Utilisation Schemes produced from the analysis of her Phase One lesson data (5 lessons, coded CEL1 to CEL5) (The codes that begin with IUS refer to the different categories of instrument utilisation scheme that emerged during the whole study. These are described more extensively in Clark-Wilson (2010))

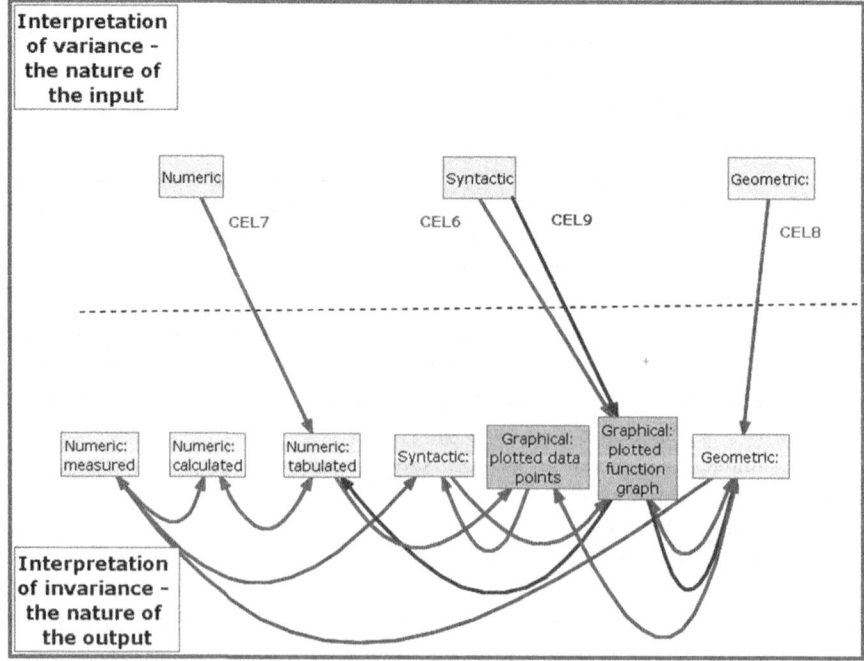

Fig. 4 The summary of Eleanor's Instrument Utilisation Schemes produced from the analysis of her Phase Two lesson data. (4 further lessons, coded CEL6 to CEL9)

in their knowledge. I defined these as the lesson *hiccups* and I viewed these hiccups as opportunities for the teachers' epistemological development within the domain of the study. They were highly observable events as they often caused the teacher to hesitate or pause, before responding in some way. Occasionally the teachers looked across to me in the classroom in surprise and, particularly in the case of hiccups relating to what they considered to be unhelpful technological outputs, they sometimes expressed their dissatisfaction verbally. Consequently, I also started to code each activity for hiccups within NVivo.

Identifying, Coding and Categorising Hiccups

In order to make sense of what follows, it is necessary to include a detailed description of a lesson activity. For this purpose I have selected an early activity that was designed and taught by Eleanor during the second phase of the study, which I called *Transformations of functions*. This activity took place during a single one hour lesson with a group of 29 higher achieving girls aged 14–15 years working from the English and Welsh General Certificate of Secondary Education (GCSE) higher tier examination syllabus. Eleanor's lesson objective was for students to develop 'An understanding of standard transformations of graphs' and she expanded on this by saying 'I wanted the students to explore the effects of different transformations of linear and quadratic functions to enable them to make generalisations for themselves'. In the lesson the students were given a worksheet devised by Eleanor that included six sets of linear, quadratic and cubic functions laid out as three pairs. Each pair was intended to encourage students to compare particular transformations, for example the first set compared the effects of $y = f(x) \pm a$ with $y = f(x \pm a)$. There were thirty-nine different functions in total and the activity sheet did not label the sets of functions in any way (Fig. 5).

The students were asked to enter the functions syntactically into a Graphing application on their handhelds and to describe the transformations they observed within each set of functions. Eleanor questioned the students about different types of transformations (reflection, translation, rotation and enlargement) and encouraged them to use these words when describing their observations. They were not instructed as to how they should communicate their observations, however, it seemed to be an established classroom practice that they would discuss their outcomes with their neighbours. The Smart Notebook file that Eleanor developed to present the activity to the students included the suggestion that the students should 'use 2 graphs per page'. A typical student's response to the first stage of the activity is shown in Fig. 6.

During the lesson Eleanor moved around the classroom and responded to questions initiated by the students. These were mainly related to instrumentation issues concerning graphing the functions such as, "where is the squared key?" and "how do I insert a new page?". Ten minutes prior to the end of the lesson, Eleanor instigated one episode of whole class discourse in which she asked the students to open "your page where you've explored this set" whilst gesturing to the set of functions shown in Fig. 7.

Fig. 5 The student task sheet for the activity 'Transforming functions'

y = f(x)	y = f(x) ± a	y = f(x)	y = f(x ± a)
y = x²	y = x² + 2	y = x²	y = (x + 2)²
y = x²	y = x² - 2		y = (x + 2)²
y = x³	y = x³ + 1	y = x³	y = (x + 1)³
y = x³	y = x³ - 1		y = (x + 1)³
y = 1/x	y = 1/x + 2	y = x	y = x + 4
y = 1/x	y = 1/x - 2		y = x - 4

y = f(x)	y = -f(x)	y = f(x)	y = f(-x)
y = x	y = -x	y = x²	y = -x²
y = x²	y = -x²	y = x² + 3	y = (-x)² + 3
y = x² - 1	y = -(x² - 1)	y = x³ - 1	y = (-x)³ - 1
y = x³ + 1	y = -(x³ + 1)	y = 3x + 4	y = -3x + 4

y = f(x)	y = kf(x)	y = f(x)	y = f(kx)
y = x²	y = 3x²	y = x²	y = (3x)²
y = x² - 3	y = 2(x² - 3)	y = x² - 3	y = (2x)² - 3
y = x³	y = 2x³	y = x³	y = (2x)³

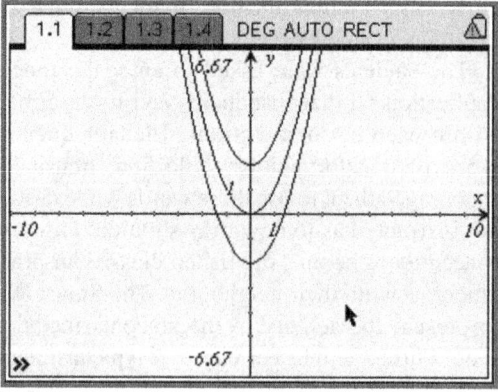

Fig. 6 A student's TI-Nspire screen in response to the task 'Transformations of functions'

The resulting screen capture view (see Fig. 8) was on public display in the classroom. Eleanor attempted to use Mason's idea of *funnelling* (Mason 2010) in order to elicit from the students the key generalisation for this transformation, i.e. that it resulted in a 'sideways shift' of ±a. No other mathematical representations were used during this discussion to justify or explore why this was true.

Fig. 7 Function set selected for whole class display

y = f(x)	y = f(x ± a)
y = x²	y = (x + 2)²
	y = (x + 2)²
y = x³	y = (x + 1)³
	y = (x + 1)³
y = x	y = x + 4

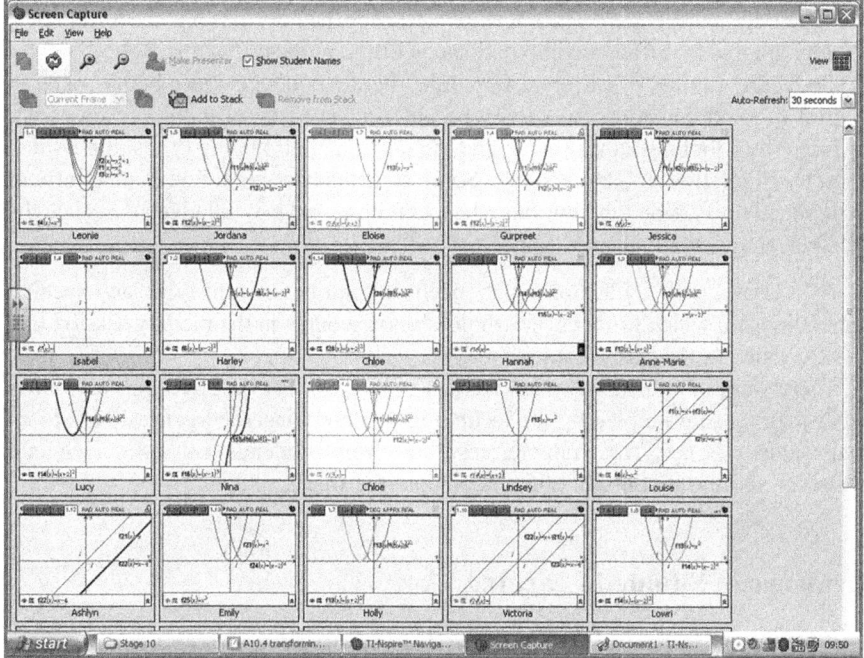

Fig. 8 The students' handheld screens on public display during the class plenary

Hiccups Identified from the Lesson Data:

During this lesson a total of nine hiccups were observed and they were grouped into the six broad categories as shown in Fig. 9.

The omission of any labelling of the sets of functions as they were laid out on the worksheet (or related teacher explanation) seemed to trigger the following hiccups during the lesson:

- Difficulties experienced by the students in making global sense of the activity and noticing the invariant properties as Eleanor had intended through her activity design.

Name
EL6 Hiccup01 - Students' reluctance to focus on the outcomes related to their inputs
EL6 Hiccup02 - Students' struggling to see 'sets' of transfornations
EL6 Hiccup03 - Instrumentation (S) - 'How do you draw them'
EL6 Hiccup04 - Instrumentation (S) - Entering x^3
EL6 Hiccup05 - Instrumentation(S) - all pages change the same
EL6 Hiccup06 - Diverse student responses make generalisations difficult

Fig. 9 The observed hiccups and their raw codes for the activity 'Transformations of functions' as captured within Nvivo8

- Whilst the students were competent with entering the functions into the technology, they did this in different combinations on different pages.
- The large number of different functions that the students were being asked to plot focused the students' activity on entering as many of them as they could, rather than looking closely at any individual set and discussing or making written notes in relation to the outcomes. Some students had worked very diligently to input all thirty-nine functions into the technology, but had failed to appreciate the 'sets' as Eleanor had envisaged.

As a consequence, Eleanor experienced difficulties in identifying any specific generalities on which to focus the whole-class discourse in the plenary session that she convened as the lesson came to a close.

There were of course many other types of hiccups that occurred during lessons other than those prompted by the technology. These concerned general classroom management issues, for example, resulting from students' off-task behaviour. However, these were outside of the domain of the study.

Evidence of Situated Learning

In response to the identified hiccups, there was evidence for the teachers' *situated learning* (as defined by Lave and Wenger, 1991) in the form of the list of seven actions taken by Eleanor during the lesson, which are summarised in Fig. 10.

Although the actions were observed during or shortly after the lessons, it was only through our discussions in the subsequent interview that the evidence for the situated learning was clarified.

Eleanor was confident in her responses to the students' instrumentation difficulties, giving quick tips such as 'control escape to undo' and 'press escape' and loading the teacher edition software to demonstrate how to input functions. However, the hiccups experienced by Eleanor in this lesson led her to reflect on aspects that she felt she would change, which she articulated during our post-lesson discussion. Reflecting on her activity design, Eleanor commented,

> I did not need all of the students to work through many similar problems – it was actually much more memorable to look at screens that appeared different, but, because of

A Methodological Approach to Researching the Development...

Name
EL6 Action01 - Loaded TE to show how to input functions
EL6 Action02 - Led discussion about types of transformation
EL6 Action03 - Noticed students' expression of the generality
EL6 Action04 - Appreciated that the comparisons the students could make related to their existing knowledge of y=
EL6 Action05 - Attempts to be specific - but a lack of common 'labelling' led to issues....
EL6 Action06 - Realised that the students don't all need to do so much -
EL6 Action07 - Suggested revisions to the activity design wrt her own questioning strategies

Fig. 10 Evidence of the teacher's actions in response to the hiccups ('wrt' is an abbreviation of 'with respect to' and 'TE' is an acronym for 'Teacher Edition', the TI-Nspire software that the teachers used for whole class display)

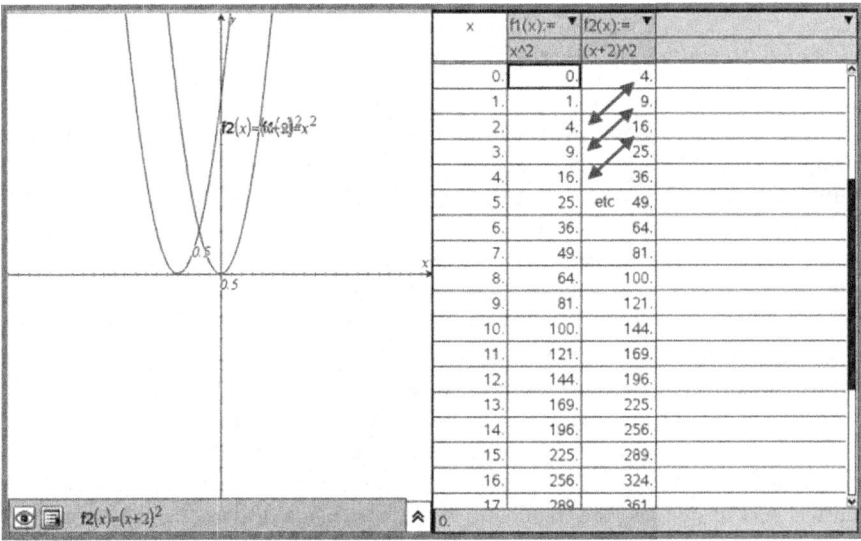

Fig. 11 Using the multi-representational technology to explore the function table

an underlying mathematical concept had something similar about them. This meant that I could have let the students choose their own functions to transform in particular ways – something that I will try next time. [Interview transcript]

Eleanor and I agreed that the underlying approach for the lesson was sound. However we discussed a redesigned format for the lesson, which responded to Eleanor's comment that she could allow the students to explore their own functions. We also incorporated an element of the lesson that I had felt was a constituent part in developing the students' understanding of the outcomes of each of the transformations. To exemplify this, when the function $y=f(x)$ is compared with $y=f(x\pm a)$, the visible horizontal shift in the graph is linked with the apparent shift in the corresponding values of x within the table of values for the functions when viewed side-by-side. This is shown for the function $y=x^2$ and $y=(x+2)^2$ in Fig. 11.

In our discussion, when I showed this to Eleanor, she commented that she had never thought about this connection before, partly because she had learned the various transformations herself by rote. As she considered how she would approach this topic next time, Eleanor suggested that she might ask pairs of students to focus on particular transformation types with a view to them being able to summarise and justify the outcomes of their explorations to other members of the class. Eleanor's epistemological development concerned: her reconceptualisation of the nature of the variant and invariant properties within her chosen example space; the use of the technology to represent an appropriate set of functions; and the way in which she could coordinate the whole-class discourse to support the students to notice the chosen generality.

Global Categories of Hiccups

By repeating the process described previously for each of the lessons observed during the second phase, the cross-case analysis, supported by the functionality within Nvivo8, led to a conclusion that all of the hiccups could be attributed to one of seven considerations (Table 3).

This set of classifications has implications for the ways in which we consider both the formal and informal support for teachers as they begin to use multi-representational technology in the classroom. For example, the emphasis within most professional development support and training, when introducing new mathematical technologies to teachers, concern the technical steps to achieve the desired functionality or 'key pressing' with a view to avoiding the occurrence of students' instrumentation issues (Hiccup type 7). However, often far less time is spent considering the mathematical and pedagogical implications of the activities that teachers design and the implications of their design decisions on the possible student outcomes.

The implications for these findings concern the nature of in-class support for teachers in addition to the global design of professional development initiatives concerning new technologies.

Conclusion

In conclusion, the study provided deep insight into teachers' technology-mediated epistemological development over a 24 month period as they began to integrate a complex new technology within their classroom practices. Their mathematical, pedagogical and technical knowledge developed through a multifaceted journey, which was centralised on their classroom-based experiences and the professional exchanges that we had before and after their lessons. The longitudinal nature of the research enabled the fragments of this epistemological development to be pieced

Table 3 The emergent types of hiccups experienced by secondary mathematics teachers learning to use a multi-representational technology.

Hiccup type	Exemplification
1. Aspects of the initial activity design:	Choice of initial examples
	Sequencing of examples
	Identifying and discussing objects displayed by the technology
	Unfamiliar pedagogical approach for the students
2. Interpreting the mathematical generality under scrutiny:	Relating specific cases to the wider generality
	Appreciating the permissible range of responses that satisfy the generality
	The students fail to notice the generality
3. Unanticipated student responses as a result of using the technology:	The students' prior understanding is above or below the teacher's expectation
	The students' interpretations of the activity's objectives differ from the teachers
	The students develop their own instrument utilisation schemes for the activity that differ from the teacher's planned scheme
4. Perturbations experienced by students as a result of the representational outputs of the technology:	Resulting from a syntactic output
	Resulting from a geometric output
	Doubting the 'authority' of the syntactic output
5. Instrumentation issues experienced by students when making inputs to the technology and whilst actively engaging with it:	Entering numeric and syntactic data
	Plotting free coordinate points
	Grabbing and dragging dynamic objects
	Organising on-screen objects
	Navigating between application windows
	Enquiring about a new instrumentation
	Deleting objects accidentally
6. Instrumentation issue experienced by one teacher whilst actively engaging with the technology:	Displaying the function table
7. Unavoidable technical issues:	Transferring files to students' handhelds
The teachers were using prototype classroom network technology that did result in some equipment failures during some lessons	Displaying teacher's software or handheld screen to the class

together to show how their actions changed over time as they re-encountered *known hiccups* but had developed appropriate *response repertoires*.

Moreover, the adaptation of Verillon and Rabardel's framework provided a useful construct for the research as it focused the research lens onto teachers' classroom practices and demanded a robust set of methodological tools to evidence the different interactions. However, the key purpose of this chapter was to provide insight into one researcher's approach to the study of teachers' epistemological development through a detailed description of the methodology that led to the conclusion that it was the *contingent moments* or *hiccups* that the teachers experienced when integrating the multi-representational technology into their classroom practices that provided both rich contexts for their situated learning and fruitful foci for professional discourse.

Acknowledgements The data collection carried out during Phase One of the study (and part of the data collection in Phase Two) was funded by Texas Instruments within two evaluation research projects, which have been published in Clark-Wilson (2008a) and (2009).

References

Ahmed, A., & Williams, H. (1997). Numeracy project: A catalyst for teacher development and teachers researching. *Teacher Development, 1*, 357–384.

Arzarello, F., & Robutti, O. (2010). Multimodality in multi-representational environments. *ZDM – The International Journal on Mathematics Education, 42*, 715–731.

Artigue, M. (2001). *Learning mathematics in a CAS environment: The genesis of a reflection about instrumentation and the dialectics between technical and conceptual work.* Computer Algebra in Mathematics Education Symposium. Netherlands: Utrecht University

Bednarz, N., Kieran, C., & Lee, L. (Eds.). (1996). *Approaches to algebra: Perspectives for research and teaching.* Dordrecht: Kluwer.

Clark-Wilson, A. (2008a). *Research Report. Evaluating TI-Nspire in secondary mathematics classrooms.* Chichester: University of Chichester.

Clark-Wilson, A. (2008b). *Teachers researching their own practice: Evidencing student learning using TI-Nspire.* Day conference of the British Society for the Learning of Mathematics, 28 (2). University of Southampton.

Clark-Wilson, A. (2009). *Research Report. Connecting mathematics in the connected classroom: TI-Nspire Navigator.* Chichester: University of Chichester.

Clark-Wilson, A. (2010). *How does a multi-representational mathematical ICT tool mediate teachers' mathematical and pedagogical knowledge concerning variance and invariance?* Ph.D. thesis, Institute of Education, University of London.

Gueudet, G., & Trouche, L. (2009). Towards new documentation systems for mathematics teachers? *Educational Studies in Mathematics, 71*(3), 199–218.

Guin, D., & Trouche, L. (1999). The complex process of converting tools into mathematical instruments: The case of calculators. *International Journal of Computers for Mathematical Learning, 3*, 195–227.

Haspekian, M. (2005). *Intégration d'outils informatiques dans l'enseignement des mathématiques, étude du cas des tableurs.* Ph.D., Doctoral thesis, University Paris 7.

Haspekian, M. (2006). *Evolution des usages du tableur.* Rapport intermédiaire de l'ACI-EF Genèses d'usages professionnels des technologies chez les enseignants [Online]. Available: http://gupten.free.fr/ftp/GUPTEn-RapportIntermediaire.pdf

Jaworski, B. (1994). *Investigating mathematics teaching.* London: Falmer.

Kaput, J. (1986). Information technology and mathematics: Opening new representational windows. *Journal of Mathematical Behavior, 5*, 187–207.

Kaput, J. (1998). Transforming algebra from an engine of inequity to an engine of mathematical power by 'algebrafying' the K-12 curriculum. In National Council of Teachers of Mathematics & Mathematical Sciences Education Board (Eds.), *Proceedings of a national symposium.* Washington, DC: National Research Council, National Academy Press.

Kieran, C., & Wagner, S. (Eds.). (1989). *Research issues in the learning and teaching of Algebra.* Reston: Lawrence Erlbaum/NCTM.

Lave, J., & Wenger, E. (1991). *Situated learning: Legitimate peripheral participation.* New York: Cambridge University Press.

Mason, J. (2002). *Researching your own practice: The discipline of noticing*. London: Routledge-Falmer.
Mason, J. (2010). Asking mathematical questions mathematically. *International Journal of Mathematical Education in Science & Technology, 31*, 97–111.
Moreno-Armella, L., Hegedus, S. J., & Kaput, J. J. (2008). From static to dynamic mathematics: Historical and representational perspectives. *Educational Studies in Mathematics, 68*, 99–111.
Pierce, R., & Stacey, K. (2008). Using pedagogical maps to show the opportunities afforded by CAS for improving the teaching of mathematics. *Australian Senior Mathematics Journal, 22*(1), 6–12.
Polanyi, M. (1962). *Personal knowledge: Towards a post-critical philosophy*. Chicago: The University of Chicago Press.
Polanyi, M. (1966). *The tacit dimension*. Chicago: The Universty of Chicago Press.
QSR International. (2008). *Nvivo8*. Melbourne: QSR International.
Rowland, T., Huckstep, P., & Thwaites, A. (2005). Elementary teachers' mathematics subject knowledge: The knowledge quartet and the case of Naomi. *Journal of Mathematics Teacher Education, 8*, 255–281.
Ruthven, K. (2002). Instrumenting mathematical activity: Reflections on key studies of the educational use of computer algebra systems. *International Journal of Computers for Mathematical Learning, 7*, 275–291.
Schön, D. (1984). *The reflective practitioner: How professionals think in action*. New York: Basic Books.
Shulman, L. (1986). Those who understand: Knowledge growth in teaching. *Educational Researcher, 15*, 4–14.
Stacey, K. (2008). *Pedagogical maps for describing teaching with technology*. Paper presented at Sharing Inspiration Conference, Berlin. 16–18 May 2008.
Sutherland, R., & Mason, J. (1995). *Exploiting mental imagery with computers in mathematics education*. Berlin: Springer.
Thompson, A. (1992). Teachers' beliefs and conceptions; a synthesis of the research. In D. Grouws (Ed.), *Handbook of research on mathematics teaching and learning*. New York: Macmillan.
Verillon, P., & Rabardel, P. (1995). Cognition and artefacts: A contribution to the study of thought in relation to instrumented activity. *European Journal of Psychology of Education, 10*, 77–102.
Zodik, I., & Zaslavsky, O. (2008). Characteristics of teachers' choice of examples in and for the mathematics classroom. *Educational Studies in Mathematics, 69*, 165–182.

Teachers and Technologies: Shared Constraints, Common Responses

Maha Abboud-Blanchard

Abstract This chapter presents a synthesis of a set of studies focusing on teachers' technology-based activity at the classroom level. Each of the studies is contextualised, singular and deals with individual teachers. Cross-analysing the findings of these separate situations aims to identify common characteristics in terms of common responses to shared constraints (in the French context) related to the use of technology by *ordinary* mathematics teachers. The synthesis is developed with the aim of analysing regularities in the practices of ordinary teachers integrating technologies into their teaching. These regularities are structured along three issues: How to simultaneously teach mathematics and use technology in class? (cognitive axis); How to teach mathematics in new teaching environments? (pragmatic axis); How to manage the time of teaching and learning when using technology? (temporal axis).

Keywords Technology integration • Teachers' practices • Mathematics teaching • Teaching environments • Didactical approach • Professional constraints

Introduction

In recent years, an increasing interest has been paid by educational research to teachers' practices in technology environments. Constraints and difficulties encountered by mathematics teachers' integration of technologies has also been an ongoing issue. Researchers have investigated different aspects of teachers' practices in technology-rich classrooms by using or developing different theoretical frames. Kendal and Stacey (2002) studied the discrepancies and variability in the ways

---

M. Abboud-Blanchard (✉)
LDAR, University of Paris Diderot, Paris, France
e-mail: maha.abboud-blanchard@univ-paris-diderot.fr

teachers use technology in their mathematics classrooms. Ruthven and Hennessy (2002) investigated teachers' ideas about their own experience surrounding lessons incorporating the use of digital technologies and developed a model that included different levels of teachers' expectations and ideals. In order to understand the key factors of teachers' activities and roles through a holistic approach, Monaghan (2004) used Saxe's cultural model centred on emergent goals under the influence of four parameters. Drijvers et al. (2010) investigated, within the general frame of the instrumental approach (Vérillon and Rabardel 1995), the types of orchestrations that teachers develop when using technology. More generally, in the latest ICMI study (Hoyles and Lagrange 2010), research studies addressing the theme of teachers and technology revealed that integrating technology is not an easy task for teachers who have to cope with an increasing complexity in preparing lessons and managing the classroom while taking into account several features going beyond familiar formats and routines in a paper and pencil environment.

The aim of this paper is not to present the results of a single research study related to these same concerns, but rather to offer a synthesis of a set of studies that I have conducted over the past decade and that have yielded outcomes focusing on the teacher's activity at the classroom level. Each of these studies is contextualised, singular and deals with individual teachers. Through the study of these singular situations, I aim to identify common characteristics related to the integration of technology by *ordinary*[1] mathematics teachers, to analyse certain regularities in teaching practices and to investigate the factors that determine them. Of course, the professional group 'secondary mathematics teachers who use technology' is not homogeneous. My goal is to try to identify, beyond this heterogeneity, some homogeneity in responses to shared constraints and various institutional incentives (in the French context) to integrate digital technologies in mathematics teaching.

Three Research Studies

Background

In the early nineties I was engaged in a project where researchers worked with a group of teachers, who were experts in digital technology, to identify the potential offered by Computer Algebra Systems for teaching and learning (Artigue 1997). One of the results of this work was that technology-expert teachers have a poor sensitivity to the changes that technology integration implies, due essentially to their technology expertise. We highlighted a complex balance between achieving learning goals and working with technology that is unfamiliar to students, where the

---

[1] 'Ordinary teachers' means in this chapter, teachers who are not technology-experts and who are not involved in experimental projects.

role of the 'expert teacher' is essential in maintaining the mathematics activity and management of students in a satisfactory manner.

A few years later, I was a member of a group of French researchers leading a review of research literature that looked at more than 600 international publications (published before 2000) dealing with Information and Communication Technology (ICT) in the teaching and learning of mathematics (Lagrange et al. 2003). My own contribution within this group was to examine publications that focused on the 'teacher dimension'. The major finding was the relative paucity of systematic studies investigating mathematics teachers' integration of ICT into their classroom practices. Most of the existing studies aimed at studying beliefs and knowledge of teachers regarding technology integration or at examining innovative technology-based activity of teachers working in experimental situations.

The findings of these two research studies and the questions they raised led me to focus on investigating ordinary teachers' use of digital technology in their lessons with an emphasis on their classroom practices. By ordinary teachers, I mean teachers who are neither technology-experts nor participating in experimental research projects but whose daily professional contexts reflect real school conditions. I then participated in several studies about teaching practices in technology environments that involved experienced teachers using either dynamic geometry or online exercises and trainee teachers who experimented with several technological tools during their first year of teaching. The three studies presented in this paper are qualitative in nature and are based on direct observation of classroom practices or on traces of classroom practice, as reported by teachers.

First Study: An Experienced Teacher's Practice

The first study involved an ordinary teacher using dynamic geometry (Abboud-Blanchard 2009). The teacher observed was not engaged in any innovation or research project and had an episodic, as opposed to significant use of technology with her students. The lesson was on spatial geometry with a grade 9 class (fourth year of the lower secondary level, aged 14/15 years) and it took place in the computer room. The students used dynamic geometry software in assigned groups of two or three working with one computer. The lesson observation was videotaped. The topic concerned the cutting of a pyramid by a plane parallel to the base, and the teacher used an activity that was pre-designed by the software developers.

This case study sought to investigate the approaches that an experienced teacher develops when using dynamic geometry system in an ordinary classroom context in order to characterise the teacher's activity and its impact on students' learning with technology. The analysis provided findings that related to: the tasks proposed for the students' learning; the management of the students' groups; and the teacher's discourse and the interaction with students. These findings were contrasted with the results of a similar analysis of a non-technology-based lesson with a class of the same level, on the same aspect of problem solving in order to highlight the characteristics of a technology-based lesson (Abboud-Blanchard and Paries 2008).

Second Study: Experienced Teachers' Practices

The second study involved five secondary mathematics teachers using online Electronic-Exercise-Bases (EEB) with grade 10 students (first year of the upper secondary level, aged 15/16 years). These specific technological tools are software applications that mainly consist of classified practice tasks within a tutoring environment that can include guidance, corrections, explanations and sometimes reminders of mathematics courses.[2] The research questions were addressed within the context of a regional French project focused on encouraging mathematics teachers to use the EEB (Artigue and al. 2008). The aim of the project was pragmatic in that it involved observing the potential of such tools in ordinary classes, with an emphasis on helping the weaker students (Abboud-Blanchard et al. 2007).

The general issues related to the investigation of teachers' practices within the project were: Why and how do teachers use EEB?; What effect does this use have on their teaching activity? To answer these questions, we observed volunteer teachers using EEB over a period of 3 years. Most of the teachers were familiar with classroom use of technology at the beginning of the project (Abboud-Blanchard et al. 2009). The data analysis was qualitative and it related to: lessons preparations; class observations and answers to questionnaires and interviews. All of the observed lessons were EEB lessons on the topic of algebra and took place in the computer room. The students worked on a common on-line worksheet that had been prepared by the teacher before the session.

Third Study: Beginner Teachers' Practices

The aims of this third study were to investigate the initial professional uses of technology by pre-service mathematics teachers in order to understand the conditions in which these uses take place. In France, pre-service mathematics teachers benefit from a one year professional course in order to obtain their master's degree in secondary education. They teach mathematics, advised by a tutor, in one or two classes throughout the year. Over the last few years training teachers to the use of technologies has become more and more significant and a set of competences in the area of ICT for teaching have to be fulfilled by the trainees at the end of the training year. However, obstacles to technology use still persist even if, during their pre-service training, the

---

[2] See for example: http://mathenpoche.sesamath.net/.

trainees benefit from conditions that might help them develop professional uses of technology (Abboud-Blanchard and Lagrange 2006).

The study focused on five pre-service mathematics teachers as case studies (Abboud-Blanchard et al. 2008). The data was of two types: professional dissertations about using technology in the classroom and interviews carried out with them at the end of the training year. During this period, pre-service teachers have to write a professional dissertation about their teaching practices as part of their final assessment and they are free to choose the topic. Some trainees, as in these five cases, choose to deal with the use of technology in the classroom. In this study, these professional writings are considered as *traces* of genuine practices (Van Der Maren 2003), as a way to approach what the pre-service mathematics teachers consider as significant practices and also their reflections on these practices. The interviews provide complementary information on how they deal with technology-related potential and possible restraints within technology-based lessons.

The five mathematics pre-service teachers considered in the research had various profiles with regard to technology (profiles drawn throughout the analysis of the interviews). The training in the use of technology had various effects depending on the one hand on these profiles, and on the other hand on differences relating to their original didactical concerns.

The analysis of the lessons reported in the dissertations enabled an exploration of the uses of technology by trainees in two phases of the teacher's work, which are preparation work and classroom work. Thus, the research provided a closer look at what pre-service mathematics teachers' technology-based activity developed throughout the year of training and how they reflect on these first teaching experiments.

Theoretical and Methodological Considerations

The studies presented above used the same theoretical frame as a route to better understand the complexity of teachers' technology-based practices within the general frame of the *double approach*, developed in France by Robert and Rogalski (2005). It is this frame that is presented in the first sub-section. My aim is to build on these studies by synthesising their results in order to emphasise the regularities in the way that teachers integrate technology into their classroom practices. This gives rise to a new theoretical construct introduced in the second sub-section.

The Double Approach Framework Used in the Three Studies

The general framework used is the *double approach*, which combines both a didactical and an ergonomic perspective in analysing the teacher's activity in classrooms, as well as the factors that determine it. Rogalski (2008) argues that the frame of reference for the *double approach* is that of *activity theory*, which was initiated by Leontiev (1978), enriched by Vygotsky (1986), and then exploited and developed within the context of ergonomic psychology (Leplat 1997; Rogalski 2004) before being articulated within the context of teaching mathematics.

The *double approach* was introduced and developed by Robert and Rogalski (2002, 2005) to incorporate, on the one hand, a didactical perspective, which views the teachers' activities that involve task choices and classroom management as a key to accessing students' activities, and on the other hand an ergonomic perspective, which considers that in order to study their activity, teachers must be seen as professionals having craft knowledge, beliefs and previous experience whilst working in given institutional and social conditions. On a methodological level, Robert and Rogalski distinguish five components that can be observed or questioned, and whose reconstitution provides access to the teacher's practices. The *didactical perspective* takes into consideration the fact that there are two main types of channels used by the teacher for and during classroom activity; the organisation of tasks prescribed to students (cognitive component) and the direct interactions through verbal communication (mediative component). The *ergonomic perspective* of analysis is associated with the teaching profession. It considers the teacher as performing a given professional activity. His/her performance depends on a multiplicity of factors, the main ones being: professional history, knowledge and beliefs (personal component); institutional constraints and rules (institutional component); and social interactions in the work environment (social component). The five components of the double approach are thus:

1. The *cognitive component* is linked to the mathematical intentions and goals of the teacher. The analysis relating to this component focuses mainly on the scenario the teacher sets for students in terms of mathematical tasks. These scenarios include the time allocated for the students to work on tasks, the form of this work and the tools to be used, such as paper-and-pencil, technological tools and blackboards.
2. The *mediative component* is related to all of the interactions, verbal or not, observed as the lesson progresses, such as the interactions between teacher and students (explaining the tasks or giving aid), and interactions between students. The data analysis focuses on how the teacher engages and maintains the students' activities and on the type of help he/she provides to enable the students to achieve the tasks. Robert (2008) distinguishes two types of help, depending on whether they modify the activities scheduled or promote directly mathematical knowledge. The first type, *procedural help*, deals with the prescribed tasks by modifying activities with regard to those planned from the presentation of the task. It corresponds to indications that the teacher supplies to students before or during their work. The second, *constructive help,* adds something between the strict activity of the student and the expected construction of the mathematical knowledge that could result from this activity.
3. The *personal component* deals with the teacher's conception of mathematical knowledge, of teaching processes and of the way students learn mathematics, as well as his/her own professional history. In the case of using technology, more

specific features could be added to the former, such as familiarity with technology or beliefs related to the impact of technology on mathematics learning.
4. The *social component* is about how teacher adapts to the conditions of the work environment in a given school, to the habits of the class, to the colleagues as individuals and also as a community. For instance, if it is not a rule of action in the school, the teacher might not let the students work in small groups although he/she is convinced of the usefulness of this type of class management.
5. The *institutional component* mainly concerns the influence of institution, for example, via the curriculum, institutional guidelines, hierarchy requirements, and so on. It might also concern compulsory textbooks or assessment forms. In the case of pre-service mathematics teachers, it could also depend on what is highly recommended by teacher educators or training programmes. These factors are often considered by the teacher as constraints to deal with while practicing the teaching profession.

An analysis using the double approach aims to locate the characteristics of each component within the activity of the teacher in situ. The recombination of these components provides access to a teacher's practices. The double approach postulates that these practices are both complex and stable, that is, a teacher's activity in classroom has its own logic and consistency, and practices do not change easily. For pre-service teachers. It is less clear whether their practices have stabilised but we assume that the coherence of their practices is already established. Indeed, Lenfant (2002) shows that the practices of pre-service mathematics teachers develop and organise into a coherent system in the early months of teaching career and stabilise quickly during the first year. Stability does not, however, mean invariance as practices evolve over time, especially depending on external constraints, but in a coherent manner specific to each individual teacher.

In my work, I consider the complex articulation between the stability of practices and the evolution of the activity in the classroom due to the use of technology. My study of technology-based lessons focuses on the analysis of tasks and scenarios (cognitive component) and on the development of the lesson in the classroom (mediative component). This analysis makes it possible to understand what occurs in classroom when integrating technologies. The interpretation of the regularities and discrepancies of the findings relates to the three other components that reflect the personal determinants of the teaching practices and those related to the teaching situation. These components are accessed indirectly since they are mainly deduced from what the teachers declare (through interviews or questionnaires) about their activity and work conditions.

Synthesising the Results: An Emerging 3-Axis Structure (CPT)

The practices analysed in the three studies are certainly shaped by the socio-educational and institutional conditions in which each teacher accomplishes his/her job as well as by the personal trajectory. Even though the research questions and contexts were diverse, a close examination of the results discloses some regularities that go beyond this factual diversity. These regularities seem to be directly related to the common constraints and difficulties that teachers face when using technology and the way that they handle them. Variety does certainly exist as it could be related in the first place to personal history and professional experience (personal component) but also to belonging to a professional group (institutional and social component), such as for the pre-service mathematics teacher group.

This view of the outcomes as a whole aims to provide a means to analyse both the constraints felt by the teachers in their work and the responses they give in their technology-based practices that are consistent with the usual paper and pencil practices. These responses reveal what seems possible with regard to the stability and coherence of practices. In other words, these are choices (though certainly related to the personal component) that reflect how teachers invest the few options left, given the institutional and social constraints.

The cross-analysis of the findings of the studies shows regularities that crystallise around three major issues:

- How to simultaneously teach mathematics and use technology in the classroom?
- How to teach mathematics in new teaching environments?
- How to manage the time for teaching and learning when using technology?

In other words, the search for regularities in the results lead to a structure along three axes that relate to: the mathematical content taught with technology; what the teacher does and says when implementing a class situation using technology; and different aspects of time management of this situation. The synthesis is therefore organised in accordance to this structure: *Cognitive axis, Pragmatic axis* and *Temporal axis*.

The results referring to the first two axes are derived from the analysis of the cognitive and mediative components of practices. Although the first axis is naturally named cognitive, the second one's name (pragmatic) reflects that it is first based on the effective observation of teacher classroom activity, i.e. what really happened and not what might have, enabling subsequent access to its interpretation. Examining the results with respect to this axis certainly incorporates elements of the mediative component (articulated with the other four components). Nevertheless, the study of practices in technological environments shows ubiquity of transversal aspects in the lesson management that go beyond the single achievement of tasks, which is the primary objective of analysis within the mediative component.

As to the third axis, class observation analyses and teacher interviews reveal the complexity of teaching in technology environments with respect to time. This complexity concerns several aspects: the length of time needed for the organisation of teacher's work (preparing lessons, planning lessons, evaluating the outcomes of lessons); the dynamic time of the class; and the didactical time of learning. Of course, the question of time is recurrent in education research and it is present either as an explicit object of study or as an implicit element in the analysis. In our work, the issue of time was a study parameter that was taken into account during the analysis. Cross-analysing the results brings me to highlight the crucial role that time plays when it comes to technology-based lessons, and which sometimes allows a better understanding of the choices and actions that relate to the other two axes.

This synthetic structure is therefore a means to describe, in a global way, the results obtained from the analysis of practices by successively following each of the three axes. Moreover, these three axes are intertwined and some interpretations relating to one of them could relate to the other. In the next section, I will define each axis and corresponding results more precisely.

Result Synthesis According to the Three Axes

How does an ordinary teacher cope with the increased complexity that arises from the implementation of technology? The outcomes of the three studies are synthesised in terms of individual or collective responses to conditions and constraints related to technology integration. What regularities emerge from studies in various contexts in dealing with different technology tools? What are the possible determinants of these regularities? What remains variable among teachers and why? The synthesis that follows will emphasise these considerations within the descriptions of the three structuring axes.

Cognitive Axis: How to Simultaneously Teach Mathematics and Use Technology in Class?

The institution has various means to encourage teachers to use technology, such as the curricula, assessment recommendations, training and institutional resources. This kind of incentive determines some of teachers' choices when preparing student tasks and the way that teachers address the role of technology in learning activities. The results would then reflect the balance that teachers achieve, consciously or not, between institutional incentives to use technology, interpretations they make of curricula, and their own routines of teaching mathematical topics (or even their own experience as learners in the case of pre-service mathematics teachers).

In our studies, and despite the diversity of tools and contexts, all of the observations showed that the tasks in technology environments are essentially identical to those in paper and pencil environments. These findings concur with other research findings that have addressed similar issues. They are close to what Kendal and Stacey (2002) underline about CAS, which is that mathematical knowledge and skills stay globally within the range of those expected in non-technological environments. Moreover, research dealing with experienced teachers shows that they view the use of technology firstly through the lens of their usual practices (Ruthven and Hennessy 2002) and tend to integrate it *a minima* in their classroom sessions (Lagrange and Erdogan 2009). Of course, many studies have recently shown teachers implementing challenging related-technology situations into their teaching (Hoyles and Lagrange 2010). Still, the teachers investigated in almost all of these studies were involved in collaborations with researchers or educators, and therefore could not be considered as 'ordinary' teachers as stated previously.

In the case of teachers using EEB, student tasks are usually the same as those proposed in paper and pencil environments, although facilitated in the EEB environment which contributes by improving the graphic and geometric dimension and providing fewer repetitive exercises on the same mathematics topic. Moreover, only knowledge in development or previously acquired knowledge is the topic of such tasks. It seems unlikely to make students work on wholly new knowledge with EEB because such tools are essentially designed only for skills practice activities. However, we observed similar phenomena with more open software that embodied different principles of design and architecture. Teachers who use dynamic geometry set tasks where the contribution of the software is limited to improving spatial awareness through the dynamic manipulation of more familiar paper and pencil figures, so as to support the proof process. Laborde (2001) examining the tasks that teachers made with Cabri, noted also that they started by using the software mainly as an amplifier for visualising properties, but not really as the source of the tasks that they gave to students.

In the case of trainee teachers, analyses show that they almost all choose dynamic geometry environments to carry out their first technology-based lessons, which might be because dynamic geometry has been emphasised in their training programmes. We note on the one hand, referring to the discourse of teacher-educators, the potential abundance of these environments for student activity, especially for the visualisation of mathematical phenomena. But on the other hand, this declarative intention does not necessarily translate into actual uses in classrooms. Indeed, when a pre-service mathematics teacher uses dynamic geometry software with the intention of allowing the student to make the right conjectures by himself/herself through experimentation, observations show that this supposed experimental activity of the student is often reduced to him/her following a well-guided worksheet, with manipulation instructions, thus considerably lessening the potential of the software in the student activity.

More generally, we observe that in order to take full advantage of technology tools, teachers prepare mathematical tasks that are globally more complex since they require many adaptations, such as the construction of stages in geometric reasoning with dynamic geometry software or the articulation of algebraic and graphical frames with EEB. However, analyses of classroom observations reveal that the teacher's interventions almost always lead to a division of the tasks into simple sub-tasks, thereby reducing the opportunities for students to achieve enriched mathematical tasks with technology tools. This last observation can also be attributed to the difficulties related to classroom and time management (pragmatic and temporal axes). This issue is discussed in the next sections.

However, long-term studies of EEB and pre-service mathematics teachers (see also Laborde 2001) show that changes will take place and seem to affect mainly the cognitive component of teachers' practices. These trends emerge from a perceived need to find a better articulation between technology-based sessions and paper and pencil sessions in order to reduce the student's perception of the former as an unusual session and to take advantage of the potential of technologies to improve learning in the latter.

For example, pre-service teachers do not feel this need early in their training year, but highlight it at the end of the year, as was the case for this pre-service mathematics teacher interviewed at the end of the training year:

> at the beginning, I did not see usefulness of technology, in the sense that, for me, it was doing the same exercise using computer instead of using paper-and-pencil. To me, it was nothing more than a change of tool without any other change. Now I see that I can do something else with technology and thus complete what I do with paper-and-pencil.

All of the observations emphasise the fact that teachers promote quickly the use of paper records within the students' activity involving technology. For instance, teachers using EEB insist that students use a sheet to keep notes and some of them promote the use of a specific notebook devoted to technology sessions. This use of paper evidences an aspect of the articulation of technology activity with the ordinary activities. The written forms enable work which has been completed with technology to remain accessible within the whole learning process. The integration of technology activities in the ordinary sessions can also influence the assessment phase, i.e. most teachers who develop significant uses of EEB also incorporate similar EEB exercises within their traditional tests.

Pragmatic Axis: How to Teach Mathematics in New Teaching Environments?

Technology-based lessons often involve changes in the working environment, particularly when technology facilities are not available in classrooms (Ruthven 2007). The observations on which this synthesis relies all took place in a computer room with generally two students to a computer. In addition, the use of technological tools is, by itself, a source of difficulty, especially when teachers are not familiar with its handling. What influence does this specific environment have on the lesson in progress? One can assume that the management of the lesson will combine both the difficulties of organising work in small groups and those of technical work with the computer, the implications of which will now be examined more closely.

The Teacher's Role

In general, computer environments seem motivating for students and the teacher's interventions may be much less frequent than those observed in paper and pencil sessions. Nevertheless, we note that the teacher's presence is essential for the students to get started in their work even with software designed to be used autonomously (EEB, for example). Indeed, many students could not progress without the teacher's assistance, but also because they have difficulty interpreting the feedback of the software, which sometimes does not correspond with their expectations. Thus the teacher is kept very busy interacting with students, often in response to their difficulties, throughout the session. Indeed, when the software itself incorporates guidance to solve exercises (e.g. EEB), one might expect to see teachers acting more as observers of students' work and being less interactive with them. The observations show this is not the case. It is the same when it comes to more open software such as dynamic geometry, where the teacher is constantly asked to help interpret the phenomena observed on the screen in terms of geometric conjectures, although it was planned that students would discover the conjectures for themselves.

However, even if this heightened interactivity with students seems to be prevalent amongst the experienced teachers, we note that trainee teachers prepare highly structured worksheets, allowing them to lessen their interventions while students interact directly with the software without their mediation. This role of beginner teachers might be due to several factors including the fact that (these teachers) have not yet developed classroom management routines enabling them to incorporate a new environment. That is to say that the mediative component of practices is in progress but not yet stabilised. Another factor is the low degree of familiarity with the use of the technology tool, which does not allow teachers to have confidence in their ability to know how to manage learning using software that they have not yet fully mastered. Indeed, didactic research in the field of technology has shown that supporting the instrumental genesis of students is a complex task for the teacher (Trouche 2004). A teacher's degree of familiarity in the use of the tool is one of the factors inherent in this complexity. Our research has shown that teachers who are unfamiliar with the software organise the students' tasks in a fairly guided way. The low level of students' instrumentation reinforces this trend. This is particularly observable amongst trainee teachers. The student tasks are often specified in a written worksheet distributed to students at the beginning of the session. This document typically includes a large number of technical tips for the handling of the software as well as questions related to mathematical issues to guide the individual student's work. In the case of EEB, because of the apparent simplicity of these tools, teachers tend not to consider the instrumentation question as a central obstacle to their uses. However, during the first uses of EEB by experienced teachers, we observe similar phenomena as above, that is, when using EEB for the first time the teachers propose guided worksheets for their students.

Finally, when some experienced teachers planned a marginal role relative to students' interaction with the software, this was due to an expressed desire to give more autonomy to students. It is in fact consistent with a feature of the personal component of these teachers who perceive that the teacher's role is to help students be more autonomous in their learning. Let us examine, for example, the case of a teacher working with EEB. She tries to make the students commit themselves to solving a mathematical task by using the software as a privileged partner that controls and validates answers. She considers that her primary role is to help the students to use the software correctly in order to perform mathematics tasks. Her intervention within mathematical tasks consists of providing only constructive help to students. This finding relates to her desire to help the students be more autonomous. Indeed, when she started to work with EEB, she stressed that her main goal was to enable students to work by themselves without any external intervention, in order to acquire 'good solving processes'.

The Teacher's Interventions

Analyses of observations show few collective interventions and a majority of individual interventions to assist the students' work. The teacher focuses on providing local mathematical help without decontextualising the students' work, that is, his/her assistance consists almost exclusively in procedural help aiming at simplifying the students' activities. They are of various kinds: controlling the solutions to problems and associated calculations; validating an answer or helping to find the error (often at the request of students); and structuring the solution or asking students to do it. They sometimes reduce the efficiency of a student's activity, for instance, when the teacher indicates the theorem to be used or questions the student about the mathematical rule referred to by the exercise.

In some cases, breaking tasks into simple sub-tasks is so evident that sometimes the teacher has practically dictated the work that the students needed to do. Often, when the teacher is interacting with a group, students only follow his/her instructions or even finish a sentence that he/she begins. This type of support is partly motivated by the teacher's concern about the progress of the students' work, in order to ensure that all the tasks prepared for the session are completed. This echoes a strong trend in teaching practices in the computer room highlighted by several researchers (see, for example, Monaghan 2004). It is worth noting that the teacher stays with every group for a very short time and thus his/her assistance must allow the students to pursue their work on their own. This last issue is also related to time constraints which are discussed in the next section (the temporal axis).

Some interventions are rather technical and related to the use of the software. They consist primarily of explaining how to resolve a technical problem such as how, in EEB, to switch from one exercise to another. They are usually brief, local, and allow the student to continue towards a solution. Other interventions consist of helping the student in the meticulous execution of a set of software commands (which are sometimes even provided in the worksheet) in order to perform a mathematical task. The latter could not be qualified as procedural help, since there is no modification/simplification of the planned student activity. It is not characterised in the typology defined within the frame of the double approach. This leads me to define a new type of help (add to the existing procedural and constructive types): *handling help* that consists of supporting the student to use the software in order to achieve the planned mathematical task without modification. This type of help is directly dependent on the use of tools. It is present in technology-based lessons (but it also can be observed in a non-technology environment when a tool is used for the first time), especially when the students cannot all handle the software with ease. The frequency of this type of help, which is not common in a mathematics course, disrupts the usual class management and adds to the previous difficulties that teachers have encountered in technology environments.

Furthermore, to provide effective help the teacher must be familiar with the tasks proposed in both their dimensions related to the technical use of the software and to doing mathematics with the software. Indeed, to understand the difficulty encountered by the student, it is often not enough to look at the computer screen, notably when there are few traces providing information on the progress of the student prior to the arrival/intervention of the teacher.

Finally, the use of support materials or teaching aids is less frequent. These aids provide an opportunity to support the students to accomplish the tasks on which they are working and, at the same time, to retain knowledge that goes beyond what is directly mobilised to solve the problem. The analysis of the aims of this kind of help often shows that the need for these aids is motivated by the fact that the sole didactic interaction implemented within the software is insufficient for the students to achieve the learning objectives set by the teacher. It also shows that these aids are all the more difficult to predict by the teacher as they should be adjusted to the particular path of each pair of students working with machine. Moreover, within computer based sessions, the generalisation of constructive help to the whole class, as it is often the case within a paper and pencil environment, seems very difficult for teachers to achieve, as explained in the following paragraph.

The Class Split into 'Mini-Classes' and the Disappearance of Collective Phases

Working in a computer room generally entails students working together in groups of two or three per machine. After an initial collective phase (where the teacher explains the work to be done), which is frequently very brief, we observe that the class splits into several 'mini-classes' (one, two or three students per computer) with whom the teacher interacts separately from the remainder of the class. For each of the mini-classes, the teacher adapts to whatever the students are doing and to their current reasoning, whereas in paper and pencil lessons, it is more often the students who have to adjust themselves to the teacher's path (Abboud-Blanchard and Paries 2008). This appears to be an important characteristic of the class management of a technology-based lesson which differentiates it from a non-technology one. Monaghan (2004) also pointed out this difference by specifying that the teacher's talk is generally directed to groups of students around a computer.

Moreover, the analysis of the teacher's discourse shows similarities in the successions of his/her interventions among the mini-classes that could be described as follows:

- The teacher arrives at a mini-class;
- The teacher finds out how far the students have progressed;
- The teacher tutors the students in their problem solving activity by structuring the reasoning and introducing sub-tasks;
- When students start to execute these sub-tasks correctly, the teacher moves on to another mini-class.

Indeed, the time the teacher spends with each mini-class is actually limited (see below temporal axis), which might explain this systematic division of tasks in order to enable students to have clear work to be completed even in the teacher's absence. This mode of student management seems to be a feature of computer room sessions, which can be tiring and uneconomical for the teacher. We observe teachers repeating the same comment several times, making the same suggestion, giving the same help.

It is also to be noted that working in the computer room implies a special pattern of how the teacher moves around the classroom and manages students' work. Drijvers (2011) identifies this type of teaching practice as the *work-and-walk-by* practice, that is the students work individually, or in pairs, and the teacher walks around the room and monitors the students' progress. Of course, this pattern is not specific to ICT environments. However, we agree with Drijvers when he stresses that within an ICT environment, this practice puts high demands on the diagnostic skills of the teacher. Indeed, a look at the computer's screen is not always enough to understand what the student has already done and to determine the most appropriate form of help.

Consequently, there is a quasi-disappearance of collective phases when technology based-lessons take place in the computer room. The students work at different paces and the teacher cannot, in certain cases, generalise the support that is given only to some mini-classes whereas they could be useful to many others. Artigue and al. (2008) encountered the same feature, notably that individual interactions substitute for collective interactions and that institutionalisation phases are non-existent because of the different trajectories of students. Furthermore, the final stages of the sessions do not give rise to any institutionalisation of knowledge. However, this regularity has a relative significance given that the sessions we observed did not aim to introduce new knowledge and were designed as revision sessions by the teachers.

Looking through the lens of evolutions of practice we note that teachers, after only a few sessions, move towards an awareness of the absence of these phases. Indeed they tried to compensate for this void when it seemed necessary, by returning in the following session to collective phases in order to unify the students' knowledge that was involved in the previous technology-based lesson.

Temporal Axis: How to Manage the Time of Teaching and Learning When Using Technology?

The issues concerning time management need to be taken into account when analysing the teacher's activity, whether within one lesson or several lessons organised over time. It concerns not only what happens in the classroom but also includes the time outside the class, i.e. preparing lessons, searching for resources, collaboration with other teachers, and so on. The notion of time requires a distinction between two types of time, didactic time and physical (clock) time. Didactic time is the time that regulates the learning process and involves knowledge construction, which can be one of two kinds, *meso time* and *micro time* (Chevallard and Mercier 1987). Meso time is somewhat linear and relates to the scheduling by the teacher of the learning objectives in a sequential way whereas micro time takes into account the dynamics of practices in the context of the classroom (Chopin 2005). Our analyses of classroom observations lead us to consider the micro didactic time in relation to the physical time and in our analyses of the evolution of practices we also consider the meso didactic time.

First, we observe that preparing technology-based lessons with new software, or software not yet enough explored could be costly because it requires a time of appropriation, to determine its potential for learning and to anticipate the aid to provide to students both at mathematical and technical levels. For example, teachers using EEB declare that they had to test all messages and feedback displayed by the software for nearly every task. Thus, even teachers who became familiar with this type of technology stated, during interviews, that the preparation and updating of work plans is very time consuming.

Secondly, on investigating the time management during the sessions, we observe in all cases a difference between the time expected by the teacher and the actual time taken. In addition to the technical problems that can sometimes interfere within the session, disparities in the students' pace when performing the tasks are magnified in technology based lessons in particular, as shown above, because of the mini-classes and the multiplicity of individual paces. In the French context, classes are usually mixed ability classes. Thus, teachers generally plan long lists of tasks in order to keep fast learners fully occupied until the end of the session. It is slow learners who are responsible for the low pace of the class. This slowness may be due to less able students who experience difficulty in performing mathematical tasks, which often leads the teacher to support them, to help them and sometimes to even execute the task with them so they can reach what he/she considers to be the minimum objective of the lesson. This slowness can also be the result of meticulous students who are interested in detailed tasks not planned by the teacher. For example, students try to draw precise geometric figures although the objective of the teacher is rather to explore properties of this figure, regardless of its conformity to precise measures. Often, when the teacher realises the gap between the planned and actual time, he/she reminds the students that they have to speed up or asks them to skip some tasks and move on to others. This observation of class management seems characteristic to technology-based lessons where the different paces of students determine the pace of the whole class. This contrasts with the management of paper and pencil tasks where the pace of the lesson prescribed by the teacher impacts on the pace of individual students. (Abboud-Blanchard and Paries 2008).

Let us examine the case of the teacher from the dynamic geometry study. She had prepared simple technology based tasks in the form of a guided worksheet in order to help the students to move on quickly to mathematical tasks. The time devoted to the former was intended to be limited to 5–10 min. Perceiving that these tasks were taking more time than expected, she tried to accelerate their completion by doing the work herself or by coaching students step by step in the execution. We note, however, that the teacher failed to reach the goal of students doing all the mathematical tasks within the allotted time. Indeed, some of them were still trying to accomplish the first tasks at 10 min from the end of the session. This is also due to the division of the class into groups and the fact that she could not stay with each group for more than a few minutes at a time.

As to trainee teachers, the time issue turned out to be very important in their reflection on their first steps within the teaching profession. They quickly realised that the preparation and implementation of a technology-based lesson are time-consuming, especially in terms of scheduling mathematics lessons over the year. It seems difficult for them to reach an acceptable balance between two kinds of institutional incentives, namely integrating technology-based sessions on a regular basis and fulfilling all of the curriculum recommendations over the course of the year:

> "*ICT lessons, while remaining useful and interesting, are difficult to implement and costly in both time and energy*". "*On one side I am told to advance in my learning program, and on the other to do ICT. Where could I find time to do all this?*" (Interview of pre-service mathematics teachers at the end of the training year).

At the same time, exploring the potential of software leads pre-service mathematics teachers to perceive a time economy of didactical time over the long term:

> "*If I wanted to do exactly the same thing with paper-and-pencil, it would have taken a much longer time*". "*I realised all the time I can gain by using dynamic 3D geometry*" (Interview of pre-service mathematics teachers at the end of the training year).

Finally, changes observed in practices are consistent with a search for a cost-benefit balance between time gain in terms of learning when the potential of technology is well exploited and time loss in the preparation and management of sessions (see also Ruthven 2007). The impact of the latter, however, tends to decrease with an improved appropriation of technologies. Evolutions of technology uses considered relative to the issue of time are also present at an economic or an institutional level, that is to say teachers invest in the development of technology-based lessons only when they estimate the existence of real benefit for learning or when they are strongly encouraged or prompted by the institution.

Discussion

The challenge in doing this synthesis was, and still is, to understand better what characterises ordinary teachers' technology-based-practices, what is shared, what is different, what may evolve and under what conditions? Aiming to investigate teachers' practices in a qualitative way gives rise to local and contextualised research. This is true for the studies presented in this text and of others quoted throughout the synthesis, which could limit the generalisation of results to other teachers working in other contexts and using other technologies. Despite such limitations, I believe that the similarities between the findings of all of these studies is a good argument for such a generalisation.

Furthermore, trying to synthesise the results of a set of research, beyond the issues, contexts and theoretical frameworks that produced them, seems legitimate at the current time. It is supported on the one hand by the fact that the integration of technology in mathematics teaching is still weak and problematic, and on the other hand by the existence of a body of research on teachers' practices which brings insightful analyses and outcomes which can help understand the barriers to a wider integration. In addition, identifying the collective dimension in teachers' responses to constraints that do exist in professional *ordinary contexts* and pointing out common features and routines which take place, could have a direct impact on teacher education by enriching the body of knowledge available to teacher educators (Abboud-Blanchard 2011).

In my development of the 3-axis synthetic structure (CPT) there are some aspects that bear a similarity to the *double approach*, but with an emphasis on the technology-based practices. From the consideration of a wide literature base, Ruthven (2009, 2010, chapter 14 of this volume) has developed a conceptual framework that identifies five key structuring features shaping patterns of technology integration into classroom practice: *working environment; resource system; activity format; curriculum script*; and *time economy*. How may these structuring features relate to my original theoretical frame (the double approach) and to the structuring of my resulting synthesis? The first two key features (working environment and resource system) are not explicitly present in the double approach but could be related in particular to institutional and social components of practices. Indeed, the physical environment in which the lesson takes place and the resources used affect directly the cognitive and mediative components. Work environment and technology resources are, however, generally dependent[3] on the school equipment and the institutional decisions and on collective decisions of the group of mathematics teachers of the school. In the studies presented in this text, the nature of the technology environment was considered in the sense of an environment in which teachers and students act, and how that impacts on the activity of each of them. In the synthetic stucture (CPT), the new teaching environment indicates an environment in which teacher develops his practices and it surrounds the issues raised in the three structuring axes. It therefore may refer to both the work environment and resources system defined by Ruthven. My understanding of the second two key features (activity format and curriculum script) lead me to relate them to a central idea in the double approach, which is the stability of practices. According to Ruthven, experienced teachers repeat general models for action in the classroom and their lessons are constructed and conducted around these familiar patterns. This observation refers, within the double approach, to the stability of mediative and cognitive components. In addition, Ruthven considers the curriculum script feature from a cognitive perspective, globally similar to some aspects of the cognitive component defined within the double approach. However, Ruthven's construct relates more specifically to aspects of technology. The application of the double approach in the three studies to technology environments, examined not only the scenarios related to a given mathematical topic and the nature of the tasks prescribed to students, but also the considerations specific to technological aspects. In the synthetic structure, the cognitive and pragmatic

---

[3] At least in the French context.

axes, both specified to technology environments, might be seen as a sort of meshing of these two features from Ruthven's frame and the cognitive and mediative components of the double approach. Finally, the time economy feature of Ruthven's frame within the double approach refers mainly to the study of mediative component, but it could also relate to the study of the institutional component. In the synthesis above, I was also sensitive to time issues in teachers' practices that related to preparing the lesson and to carrying it out in the class, and also to programming lessons over the year, which led me to define a temporal axis.

More generally, Ruthven's frame is structured around five key features, each illustrating the professional adaptation on which technology integration into classroom practice depends. Starting from the hypothesis that it is not sufficient to study technology integration through the lens of learning objectives and technology affordances, my synthetic stucture (CPT), organised along three interrelated axes, aims also to shed light on these adaptations. The fact that these two approaches for the synthesis of research on the practices of teachers who are integrating technology lead to a convergence of views, regardless of the difference of cultural and theoretical contexts, is encouraging considering their common aim to provide a 'meta-view' of teachers' practices. However, I join Ruthven when he points out that this type of conceptualisation, which describes developments that are closer to the teachers' experiences, may be of limited theoretical scope. It aspires, rather, to fulfil a mediating role helping to translate insights from more decontextualised theories into practical ideas and action (ibid.).

Other researchers have also tried to overcome the local contextualised view of their own research with the ambition of creating a coherent lens for looking at teachers' technology-based practices. Lagrange and Monaghan (2009) associated, in a useful way, the double approach and Saxe's four parameter model in an attempt to understand the difficulties which the teachers they were researching experienced. Drijvers et al. (2010) found some of the concepts within the double approach helpful to underpin their findings about orchestrations types. These kinds of initiatives could be a fruitful way to gain greater insight into the complexity of technology-based practices.

The last point I would like to make is with regard to teachers' evolutions in practice. This synthetic presentation not only indicated homogeneity in the answers brought by the teachers to some shared professional constraints, but also served to stress common evolutions of practice. Do these answers and evolutions occur at the same time in a trajectory of technology integration in practices? Or are they rather milestones in this trajectory which do not correspond to a temporal order common to all the teachers? Indeed, the specificity of technology integration sometimes requires long-term studies to make it possible to identify regularities in the evolutions of practices, to interpret them and to find their corresponding determinants. To investigate these evolutions, we are currently developing a new frame based on the concept of 'geneses of technology uses' (Abboud-Blanchard and Vandebrouck 2012; Abboud-Blanchard et al. 2012). This perspective assumes that the teachers' uses of technologies develop via a dynamic path linked to both a personal and professional appropriation of these technologies and to a growing awareness of their potential and limitations.

References

Abboud-Blanchard, M. (2009). How mathematics teachers handle lessons in technology environments. In W. Carl (Ed.), *Nordic research on mathematics education* (pp. 237–244). Denmark: Sense Publishers.

Abboud-Blanchard, M. (2011). Mathematics and technology: Exploring teacher educators' professional development. *Proceedings of the 10th international conference on technology in mathematics teaching* (pp. 48–56). UK: University of Portsmouth & University of Chichester.

Abboud-Blanchard, M., & Lagrange, J. B. (2006). Uses of ICT by pre-service teachers: Towards a professional instrumentation? *International Journal for Technology in Mathematics Education, 13.4*, 183–191.

Abboud-Blanchard, M., & Paries, M. (2008). Etude de l'activité de l'enseignant dans une séance de géométrie dynamique au collège. In F. Vandebrouck (Ed.), *La classe de mathématiques: activités des élèves et pratiques des enseignants* (pp. 261–292). Toulouse: Eds Octarès.

Abboud-Blanchard, M., & Vandebrouck, F. (2012). Analysing teachers' practices in technology environments from an activity theoretical approach. *International Journal for Technology in Mathematics Education, 19.4*, 159–184.

Abboud-Blanchard, M., Cazes, C., & Vandebrouck, F. (2007). Teachers' activity in exercices-based lessons. Some case studies. In D. Pitta-Pantazi & G. Philippou (Eds.), *Proceedings of the fifth congress of the European society for research in mathematics education* (pp. 1827–1836). Larnaca.

Abboud-Blanchard, M., Fallot, J.-P., Lenfant, A., & Parzysz, B. (2008). Comment les enseignants en formation initiale utilisent les technologies informatiques dans leurs classes. *Teaching Mathematics and Computer Science Journal, 6.1*, 187–208.

Abboud-Blanchard, M., Cazes, C., & Vandebrouck, F. (2009). Activités d'enseignants de mathématiques intégrant des bases d'exercices en ligne. *Quadrante, 18*(1/2), 147–160.

Abboud-Blanchard, M., Cazes, C., Chenevotot-Quentin, F., Grugeon, B., Haspekian, M., Lagrange, J. B., & Vandebrouck, F. (2012). Les technologies numériques en didactique des mathématiques. In M. L. Elalouf & A. Robet (Eds.), *Les didactiques en questions. Etat des lieux et perspectives pour la recherche et la formation* (pp. 232–256). Bruxelles: Eds De Boeck.

Artigue, M. (1997). Le logiciel DERIVE comme révélateur de phénomènes didactiques liés à l'utilisation d'environnements informatiques pour l'apprentissage. *Educational Studies in Mathematics, 33*, 133–169.

Artigue & Groupe TICE IREM Paris 7. (2008). L'utilisation de ressources en ligne pour l'enseignement des mathématiques au lycée: du suivi d'une expérimentation régionale à un objet de recherche. In N. Bednarz & C. Mary (Eds.), Actes du Colloque EMF 2006, *L'enseignement des mathématiques face aux défis de l'école et des communautés* (pp. 1–11). Sherbrooke: Université de Sherbrooke.

Chevallard, Y., & Mercier, A. (1987). *Sur la formation historique du temps didactique* (Vol. 8). Marseille: Publication de l'IREM d'Aix-Marseille.

Chopin, M. P. (2005). Le Temps didactique en théorie anthropologique du didactique. Quelques remarques méthodologiques à propos des "moments de l'Étude". *Ier Congrès International sur la Théorie Anthropologique du Didactique : "Société, École et Mathématiques : Apports de la TAD"*. Baeza.

Drijvers, P. (2011). From 'work-and-walk-by' to 'sherpa-at-work'. *Mathematics Teaching, 222*, 22–26.

Drijvers, P., Doorman, M., Boon, P., Reed, H., & Gravemeijer, K. (2010). The teacher and the tool: Instrumental orchestrations in the technology-rich mathematics classroom. *Educational Studies in Mathematics, 75*, 213–234.

Hoyles, C., & Lagrange, J. B. (Eds.). (2010). *Digital technologies and math education. Rethinking the terrain. The 17th ICMI study*. New York: Springer.

Kendal, M., & Stacey, K. (2002). Teachers in transition: Moving towards CSA-supported classrooms. *ZDM, 34*(5), 196–203.

Laborde, C. (2001). The use of new technologies as a vehicle for restrucuring. In F.-L. Lin & T. J. Cooney (Eds.), *Making sense of mathematics teacher education* (pp. 87–109). Netherlands: Kluwer.

Lagrange, J. B., & Erdogan, E. (2009). Teacher's emergent goals in spreadsheet based lessons: Analysing the complexity of technology integration. *Educational Studies in Mathematics, 71*(1), 65–84.

Lagrange, J. -B., & Monaghan, J. (2009). On the adoption of a model to interpret teachers' use of technology in mathematics lessons. *Proceedings of the sixth conference of the European society for research in mathematics education* (pp. 1605–1614). France: University of Lyon, INRP.

Lagrange, J. B., Artigue, M., Laborde, C., & Trouche, L. (2003). Technology and math education: a multidimensional overview of recent research and innovation. In J. Bishop, K. Clements, C. Keitel, J. Kilpatrick, & F. Leung (Eds.), *Second international handbook of mathematics education* (pp. 237–270). Dordrecht: Kluwer. Part 1.

Lenfant, A. (2002). *De la position d'étudiant à la position d'enseignant : l'évolution du rapport à l'algèbre de professeurs stagiaires*. Thèse de doctorat. Université Paris 7.

Leontiev, A. N. (1978). *Activity, consciousness and personality*. Englewood Cliffs: Prentice Hall.

Leplat, J. (1997). *Regards sur l'activité en situation de travail*. Paris: PUF.

Monaghan, J. (2004). Teachers' activities in technology-based mathematics lessons. *International Journal of Computers for Mathematical Learning, 9*(3), 327–357.

Robert, A. (2008). La double approche didactique et ergonomique pour l'analyse des pratiques d'enseignants de mathématiques. In F. Vandebrouck (Ed.), *La classe de mathématiques : activités des élèves et pratiques des enseignants* (pp. 59–68). Toulouse: Eds Octarès.

Robert, A., & Rogalski, J. (2002). Le système complexe et coherent des pratiques des enseignants de mathématiques: une double approche. *Revue canadienne de l'enseignement des sciences, des mathématiques et des technologies, 2*(4), 505–528.

Robert, A., & Rogalski, J. (2005). A cross-analysis of the mathematics teacher's activity. An example in a French 10th grade class. *Educational Studies in Mathematics, 59*, 269–298.

Rogalski, J. (2004). La didactique professionnelle: une alternative aux approches de "cognition située" et "cognitiviste" en psychologie des acquisitions. *@ctivités, 1*(2), 103–120. http://www.activites.org/v1n2/Rogalski.pdf

Rogalski, J. (2008). Le cadre général de la théorie de l'activité. Une perspective de psychologie cognitive. In F. Vandebrouck (Ed.), *La classe de mathématiques : activités des élèves et pratiques des enseignants* (pp. 23–30). Toulouse: Eds Octarès.

Ruthven, K. (2007). Teachers, technologies and the structures of schooling. In D. Pitta-Pantazi & G. Philippou (Eds.), *Proceedings of the fifth congress of the European society for research in mathematics education* (pp. 52–67). Larnaca.

Ruthven, K. (2009). Towards a naturalistic conceptualisation of technology integration in classroom practice: The example of school mathematics. *Education et Didactique, 3*(1), 131–149.

Ruthven, K. (2010). Constituer les outils et les supports numériques en ressources pour la classe. In G. Gueud & L. Trouche (Eds.), *Ressources vives* (pp. 183–199). France: Presses Universitaires de Rennes.

Ruthven, K., & Hennessy, S. (2002). A practitioner model of the use of computer-based tools and resources to support mathematics teaching and learning. *Educational Studies in Mathematics, 49*(1), 47–88.

Trouche, L. (2004). Managing complexity of human/machine interactions in computerized learning environments: Guiding student's command process through instrumental orchestrations. *International Journal of Computers for Mathematical Learning, 9*(3), 281–307.

Van Der Maren, J. M. (2003). En quête d'une recherche pratique. *Sciences Humaines, 142*, 42–44.

Vérillon, P., & Rabardel, P. (1995). Cognition and artifacts: A contribution to the study of thought in relation to instrumented activity. *European Journal of Psychology of Education, 10*(1), 77–101.

Vygotsky, L. S. (1986). *Thought and language*. Cambridge, MA: MIT Press.

Didactic Incidents: A Way to Improve the Professional Development of Mathematics Teachers

Gilles Aldon

Abstract In this chapter the professional development of teachers is observed through the joint work of researchers and teachers. In the particular context of the European project EdUmatics, which focuses on mathematics education in a computer environment, the collaboration between researchers and teachers has helped both to build innovative situations and also to better understand the difficulties involved in the introduction of technology in classrooms. The theoretical framework of the theory of didactic situations, didactic incidents and documentational genesis allows the construction of analyses in order to better understand the students' and teacher's joint action and so to enhance teachers' professional development. We highlight both the consistency of the framework and the contributions of our findings to the professional development of teachers.

Keywords Didactics incidents • Documentational genesis • Milieu • Theory of didactic situations

Introduction

The EdUmatics project[1] was a place of multiple collaborations: collaboration between researchers; collaboration between researchers and teachers; and collaboration between teams of different European countries. At the beginning of the project these collaborations could not be taken for granted and their achievement has depended on a set of local and global conditions. One of the most important challenges was to

---

[1] 50324-UK-2009-COMENIUS-CMP; European Development for the Use of Mathematics Technology in Classrooms, http://www.edumatics.eu.

G. Aldon (✉)
IFÉ, École Normale Supérieure de Lyon, Lyon, France
e-mail: Gilles.aldon@ens-lyon.fr

take into account the professional development of teachers involved in the project. As full partners of the project, the schools play an important role in the development of the EdUmatics resources, and the teachers not only experimented in their classrooms with new and original lessons, but also participated to the global construction of an in-service on-line course for others. The particular context of technology added a complexity even if most of the teachers involved in the project were, from the beginning, highly experimental teachers.

In this chapter, I would like to emphasise the relationship between professional development and analysis of classroom situations. To this end, I will present the frameworks of didactic incidents and perturbations, which describe and help understand the dynamics of the relationship between teaching and learning in a perspective of documentary genesis of teacher and students.

Theoretical Framework to Approach the Complexity

The starting point of the EdUmatics project resided in the premise that '*recent studies in Mathematics Education show that, despite many national and institutional actions within the EU aiming to integrate ICT into mathematics classrooms, such integration in secondary schools remains weak.*' Research has shown that beyond some contextual problems (computer availability, technical difficulties…), the professional activity of teachers who integrate technology in their lessons is complex, both in terms of internal reasons (linked to the mathematical and technological knowledge, to the conceptions of mathematics as well as of teaching mathematics) and external ones (institutional, social or material constraints) (Rodd and Monaghan 2002; Lagrange and Degleodu 2009). In order to understand and describe this complexity and to facilitate the dissemination of professional skills leading to integration of technology into mathematics classes, the two theoretical approaches of the Theory of Didactic Situations (TDS) and of documentational genesis appeared to be appropriate. The first, through the concepts of milieu and of didactic incidents, makes it possible to take both the point of view of teachers and students in a given situation, from the design of the situation to its implementation in the class. The second considers the technology, not only as an artefact tending to become an instrument, but also more widely as a resource tending to become a document.

The Concept of Milieu

The Theory of Didactic Situations (Brousseau 1986, 2004) provides powerful tools to describe the dynamics of the interactions between teacher and students in the classroom. This theory develops a model of teaching and learning of mathematics through the description of a 'game' where teachers and students win when students learn, that is to say, when students modify their knowledge. The game must lead to new knowledge that replaces or completes a previous knowledge, and the game must encompass all the possibilities of the teaching situation. Obviously, speaking of game involves speaking of players, of playground and of rules; the players are both the teachers and the students, with different roles. The rules are modelled by the *didactic contract*, that is to say the part of the relationship between students and teachers concerning the knowledge and the responsibility of each of them in the construction of this knowledge. In a Piagetian perspective, knowledge is built in a process of adaptation and equilibration in response to environmental constraints. The environment, the playground in which (and against which) the players play is called the *milieu*; the *milieu* is thought of, designed, organised and observed by the teacher and when students play the game, the *milieu* responds to the students' actions. The didactic situation can initially be defined as the description of the interactions between players in a particular playground. The model gives the situation a central role and, obviously, the *milieu* is an important part of the success of the game. But the previous definition of situation is not sufficient to describe the activity of the actors and their positions within the milieu. Different paradoxes become apparent: a student can develop knowledge in some situations without knowing that this knowledge is socially shared; the role of the teacher is then to recognise and to institutionalise this knowledge (in a phase of institutionalisation) and this important and often neglected part of the situation will be revisited later in the chapter. A second paradox of teaching situations is that all didactic systems possess the project of their extinction, the built knowledge having to be used outside the interactions with the teacher in the particular institution of the school. To this end, *a-didactic* situations are, in a sense, a model in a didactic situation of the real interactions between subject and environment. In such a situation, students are face to face with the milieu and act on it in a situation of action, formulating the knowledge in a situation of formulation and building relationships between mathematical objects in a phase of validation. Defined for the first time as "*the antagonist system of the previously taught system*" (Brousseau 1986, p. 340), the milieu appears to be more complex when the positions of teachers and students are included within the model. "*But a milieu without didactic intentions is clearly insufficient to infer all of the student cultural knowledge that you want it to achieve*" (Brousseau 1986, p. 297).

A didactic situation is, by definition, not static and the dynamic has to be represented relative to the position of the players in the playground. This shows the necessity of structuring the milieu relative to these positions. The concept of milieu and its structuring is well adapted to understanding the situation from its design to its implementation in classrooms (Margolinas 2004) and makes it possible to analyse 'ordinary' classrooms. We speak of ordinary classrooms to distinguish didactic engineering where the construction of the situation is devolved to the researcher in contrast to those where the construction of the situation is devolved to the teacher. At each level, the milieu includes not only material objects but also *naturalised knowledge*, conceptions, beliefs, artefacts, numerical tools and so on. The naturalised knowledge is defined as the knowledge which is familiar enough to be used naturally, for example elementary arithmetic for students starting to learn algebra, or Euclidean geometry in the context of learning hyperbolic geometry.

A didactic situation is thus defined as the interactions between players (teacher and students), and the playground, including knowledge and other artefacts, and it responds according to the position of players. The construction of knowledge moves

Table 1 The structuring of the milieux (From Margolinas 2004)

Level	Student	Teacher	Situation	Milieux
M+3: Design	–	T+3: Noospherian	S+3: Noospherian situation	Upper-didactic levels
M+2: Project	–	T+2: Developer	S+2: design situation	
M+1: Didactic	St+1: Reflexive	T+1: Projector	S+1: Project situation	
M0: Learning	St0: Student	T0: Teacher	S0: didactic situation	
M−1: Reference	St−1: Learner	T−1: Observer	S−1: Learning situation	Lower didactic levels
M−2: Objective	St−2: Acting	–	S−2: Reference situation	
M−3:	St−3: Objective	–	S−3: Objective situation	

through a dynamic process that takes into account both students and teacher in different positions, from the teacher in the situation of designing activities to the situation of the student confronted with the *material milieu*. At almost every level, teacher and students have a role to play. Table 1 summarises the structuring of the milieu. It is a nested structure, the level n situation being the milieu of the level $n+1$ situation. Thus, for example, the didactic situation (S0) is the description of the interactions between the teacher in the position of teaching, the students and the milieu. The milieu is, in that case, the learning situation where the teacher in the position of observer (T−1) interacts with a student (St−1) in the position of learner, discovering new knowledge through the interaction with the reference situation.

It is possible to read this table from the bottom, taking the point of view of students who face a material milieu made up of objects (files, geometrical tools, calculators…), knowledge or conceptions, and which is devoid of any didactic intention. Typically, when students come into the classroom and discover the theme of the lesson, from a sheet of paper with the wording of a problem or exercises, or a file uploaded on a computer, all this is part of the material milieu. Before any interaction, this milieu has no didactic intention. The interactions with the teacher, the feedback of this material milieu make sense when the students are confronted with knowledge and are able to access a reference situation, which is the situation of experiments with material objects (computer, calculator, ruler, compass etc.) and mathematical objects (circle, equality, equation, operation, which are constitutive of the mathematical situation or problem). In the learning situation, students relate the result of the experiments with knowledge, the milieu of this situation being constituted of the relationships of the mathematical experiments, their results and the student's knowledge. The didactic situation, S0, is the situation in which the teacher's teaching intentions encounter the student's learning will. It is the place of institutionalisation where the operational knowledge becomes a social and shared knowledge in a particular institution.

Symmetrically, the situation S+3 is called the '*noospherian*' situation. The word 'noosphere' (from the Greek νόοϛ: intellect or intelligence and σφαίρα: field, social circle), originating from the theory of didactic transposition (Chevallard 1985) designates a level of institutional organisation where knowledge to be taught is defined separately from academic knowledge in a social construction. The S+3 situation, as

> Ca: [...] What are you doing?
>
> JC: I don't know, I try... you must find something... (He is calculating with letters.)
>
> Ca: ab minus a minus b over ab; a square, 2 is missing...
>
> [...]
>
> JC: b minus a equals ab, well... No, b plus a equals ab, so minus b minus a equals minus ab
>
> S: That doesn't get us anywhere!
>
> JC: Hence, um, then... (he continues the calculation and writes $a=b/(b-1)$ and $b=a/(a-1)$...)
>
> S: What are you doing?
>
> JC: I don't know.
>
> S: It's impossible to find something!

Fig. 1 Three students exploring the problem of Egyptian fractions

well as the S−3 situation is not finalised, that is to say it is not directly linked to a particular situation, but more generally refers to the teachers' conceptions both of mathematics (epistemological conceptions, mathematical knowledge) and of teaching (learning hypothesis: constructivism, situated learning, transfer of learning).

The S+2 situation or design situation is the situation in which the teacher designs an activity for generic students building on work already done in the classroom. It is the situation where the teacher makes choices (didactic variables, elements of the material milieu) using his/her set of resources (see below). The S+1 situation takes into account the actual students' interpretations of the didactic intentions alongside the mathematical knowledge of concern. In this situation, the student is conceptualised as an actor engaged in his/her own learning and this position is directly linked to the didactic contract built between the didactic intentions of the teacher and the student's desire to learn as illustrated in the abstract of Fig. 1. This description of the *milieux* provides an opportunity to conduct two kinds of analysis: one, starting from the point of view of the teacher, called the descendant analysis; and the second starting from the point of view of the student, called the ascendant analysis.

This table has to be considered in a dynamical way, each actor moving from one position to another in and outside school. The three situations S−1, S−2 and S−3 constitute what Margolinas called the *lower didactic levels* which differ from the *a-didactic* situations in the context of ordinary classrooms. The lower didactic levels of a situation may lead students to meet new knowledge but sometimes, lead students to operate with almost consolidated knowledge without encountering the new knowledge. In that case, the situation brings into play only two levels of the situation, the levels −3 and −2 in which the confrontation with the material and objective milieu involves only naturalised knowledge and a stationary or static process. Such situations are called *nil-didactic situations* and can be illustrated by the following episode in which students try to solve the following problem: Is it possible to find two different natural integers a and b such that $1/a+1/b=1$?

The three students Ca, JC and S try to calculate algebraically without success because their algebraic knowledge is not sufficient, for example it should be possible to extend the reasoning of JC:

$a = b/(b-1)$ but b and $b-1$ are relatively prime because of the Bezout's relation: $b + (-1)(b-1) = 1$ hence $b-1$ divides b if and only if $b = 2$ and $a = 2$ which cannot be kept because a and b are distinct.

The material milieu of students does not allow them to carry on calculating and the objective milieu lets them calculate without any chance of reaching an algebraic solution. They continue to calculate with no result but they are not in contradiction with the didactic contract because this kind of calculation can be considered as legitimate in the classroom. This particular phenomenon is called *didactic bifurcation* and results from a gap between the teacher's intention and what comes to the students' minds to do. When the teacher gives students a problem, he/she plans on his/her teaching intentions, that is to say his/her will to modify the system of knowledge of students. He/she builds a didactic situation by designing the milieu of the situation. In their position as objective students, students may ignore or be ignorant of the teacher's intentions but may, however, guess them as reflexive students and in turn project their own objective situation. There is bifurcation when, confronted with this material milieu, students invest a different reference situation from that specified in the teacher's intentions as illustrated in the previous analysis.

Documentational Genesis and Incidents

Resources taken in a general meaning "not limited to curriculum material, but including everything likely to intervene in teachers' documentation work: discussions between teachers, orally or on line; students' worksheets, etc." (Gueudet and Trouche 2009, p. 200) are part of the milieu either for teachers in the upper levels and for students in the lower levels. The documentational genesis is an extension of instrumental genesis (Rabardel 1995; Rabardel and Pastré 2005), which has been adapted to mathematics education (Artigue et al. 1998; Artigue 2007; Drijvers and Trouche 2008). In this model an artefact (a tool, a thing…) becomes an instrument as the result of a long process in which the artefact modifies the activity of the actor (instrumentalisation) while the actor shapes the artefact for his/her use (instrumentation).

Considering the available resources as artefacts, documentational genesis models a process where instrumentalisation conceptualises the appropriation by the subject of the resource and the instrumentation describes the influence of the resources on the subject's activity. At a given time, resources become a document when combined with schemes of utilisation. However, the process is ongoing and the document becomes a resource for the ongoing process. Combining *documentational genesis* and the concept of *milieu* provide an opportunity to follow two dynamical processes, making it possible to better understand the game of knowledge construction. Particularly, new types of calculators are artefacts tending to become instruments but also resources tending to become documents because the internal properties are more than mere properties of calculation or representation. For example, the possibility to organise and share files within the machine and with other calculators or computers adds to the calculator documentational properties. In our experiments, students worked with TI-Nspire CAS, which is a novel handheld device for several reasons:

- The handheld exists as an extension of the software available on computer;
- Files can be organised into a directory tree;
- Different representational environments (graphical, geometrical, CAS, spreadsheet) can be easily connected.

When considering a dynamic process, it is natural to focus on moments of rupture or of clashes, moments where the dynamics changes, where a new direction is followed. Such an event can be seen as an event that the actors did not foresee. Clark-Wilson (2010) has introduced the concept of 'hiccup'. "The hiccup is defined as a perturbation experienced by the teachers during lessons that is stimulated by their use of the technology and which illuminates discontinuities in their knowledge" (p. 217). In the perspective of professional development, the hiccups conceptualise the moment where a teacher becomes aware of a phenomenon. This notion appears as a methodological tool to emphasise an epistemological rupture in the development of professional skills of mathematics teachers in connection with an IT environment based on multi-representation. Either the teacher does not have an available answer and simply postpones the treatment of the hiccup or seeks to provoke a dialogue in order to overcome the difficulty, or alternatively has a *well-rehearsed* repertoire of responses that are used to deal with the problem. This repertoire is built over time and can be considered as a part of the teacher's set of resources. Sabra (2011) distinguishes between individual and community incidents and explores the relationship between the individual and community documentations of mathematics teachers. He defines and *individual documentary incident* as an event, which can be seized by the teacher, leading to a reorganisation of his/her system of resources, and *collective documentary incident* as an event bringing in a community documentation system as a resource that leads to the reorganisation of the community documentation. While building on this work, I diverge from it by considering the incident from the point of view of the teacher and the student or, more precisely, from the point of view of the interactions within the couple (teacher, student) in relationship with the didactic milieu.

Another difference builds on the fact that a didactic incident can be 'invisible' for both teachers and students and the *didactic perturbations* that follow can be a source of misunderstanding between them. The concept of *didactic incident* (Aldon 2011) has been defined as an event of the didactic system that modifies the dynamics of the situation. I have distinguish different types of didactic incidents:

- An *outside incident* corresponds to an event not directly linked to the situation but often important in the classroom, for example the presence of an observer in

the classroom, the mobile phone of a student that is ringing. This type of incident can amplify a previously caused perturbation;
- A *syntactic incident* linked to the conversion between semiotic registers of representations; in a technological environment, the incidents mainly come from feedback from the machine, or from the conversion of a register into the specific language of the software;
- A *friction incident* corresponds to the confrontation of two situations of different levels (cf. Table 1); such an incident may be caused by a change in the position of the teacher who moves from a position T-n to a position T-$n+1$ or T-$n-1$ or when, in the interactions, students' positions are different;
- A *contract incident* occurs when an event breaks or modifies significantly the didactic contract; this modification disrupts the trajectory of the dynamics and is strongly correlated with the appearance of didactic bifurcations where students invest a nil-didactic situation;
- A *mathematical incident* when a mathematical question is asked without answers.

Following the incidents, and in a perspective of joint action (Sensevy 2007), actors (students and/or teacher) may have different answers, modifying the milieu, or reorganising the development of the lesson or changing responsibilities within the situation. In the relationship between student and teacher, the kind of answer (or the absence of an answer) can deeply change the dynamics of the class and lead to a didactic bifurcation.

Analysis of a Situation

The Context and the Methodology

Methodology can be defined as the shape that is given to research to try to answer a question in a given framework. Choices have to be made and are interrelated with context and research questions. In the research presented in this Chapter, I wanted to: observe an 'ordinary' classroom in the sense that the responsibility for the teaching lies with the teacher; and focus on the uses of the technology in the class without being distracted by mathematics teaching difficulties.

I also wanted a *micro-view*, allowing me to capture events as they happened and a *macro-view*, allowing me to track changes over time. It is the reason the methods were chosen in order to address this challenge, that is to say, to catch incidents that are unpredictable and to follow their dynamics during the school year. Three different classes from two schools have been observed over a period of 3 years. Classes have been chosen from 16–18 year old students on a scientific course (last two grades of high school in France). In each school, one teacher was observed. The teachers were both experienced. In the first school, the teacher did not have much expertise of technology integration but in the second school, the teacher was an expert at teaching with technology. The timeframe for the data collection is summarised in Table 2. Two kinds of data were

Table 2 Data collection's timetable

First year (T: teacher, St: student, Obs: Observation in the classroom)

	Dec.	Jan.	Feb.	March	April	May	June
Observations	–	Obs 1		Obs2	Obs3	Obs 4	–
Handheld contents	X	x	x	x	X	x	x
Interviews	–	T	–	T	T	T	St
Questionnaires (St)	Q1	–	–	–	–	–	Q2

Second year (T: teacher, St: student Obs: Observation in the classroom)

	Dec.	Jan.	Feb.	March	April	May	June
Observations	–	–	–	–	Obs1	Obs2 Obs 3	–
Handheld contents	–	–	X	X	X	X	X
Interviews	–	–	–	T	T	T	St

Third year (T: teacher, St: student Obs: Observation in the classroom)

	Oct.	Nov.	Dec		April
Observations	Obs1	Obs 2	Obs 3	–	–
Handhelds contents	X	X	X	–	X
Interviews	T	T	T	–	St

collected, the first by direct classroom observations during the year and the second by asking the teacher and students to provide additional data that included:

- Teachers were asked to fill in a small journal and agreed to answer interviews before and after observations in class;
- Students agreed to send me the content of their handhelds at regular intervals and were interviewed at the end of the year.

The following analysis integrates the interviews (students and teacher), the content of the handheld device and classroom observations with a focus on one particular lesson that was conducted during the European project EdUmatics. In this project a university-school pairing in one country worked closely with a university-school pairing from a second country. The pair ENS de Lyon-Lycée Parc Chabrière was coupled with the pair University of Torino-Liceo Scientifico Copernico. From a perspective of international experimentation, a classroom activity was designed by the Italian team and adapted to the French context. In the text that follows, Jean, the (male) French teacher started from the original idea to build his own didactic situation, taking into account the French curriculum and his 16–17 years old students (who were all following a scientific pathway). Before and after the lesson Jean took part in an interview and the lessons were videotaped. The analysis has been constructed from these interviews and on the transcripts of the lessons.

The mathematical situation was developed around the notion of sequences, aiming to lead students to find a mathematical description of sequences of natural numbers from the following prompt, which was presented in the written scenario shown in Table 3.

Table 3 Task given to the students

Alberta (A), Bruno (B), Carla (C), Dario (D), Elena (E) and Federico (F) (pseudonyms) are exploring the set of natural numbers and each one identifies a sequence. Here are the sequences identified

A: 1, 2, 3, 4, 5, ..., ...
B: 3, 6, 9, 12, 15, ..., ...
C: 5, 8, 11, 14, 17, ..., ...
D: 1, 3, 6, 10, 15, ..., ...
E: 3, 9, 27, 81, 243, ..., ...
F: 2, 3, 5, 7, 11, ..., ...

Individually

What is, in your opinion, the sixth number that each of the six friends will insert next and how do you find it?

Do your previous answers change if we ask you to write the tenth number in each sequence? And the fortieth? Why?

By groups

Explain your answers to your friends. Discuss the different solutions and give a common answer for the group. If you can't, express your disagreement

In your opinion, will someone among A, B, C, D, E, F, eventually find the number 1275 in his/her sequence? If yes, after how many steps?

Describe the method you use

Can you answer the same questions with the number 2187?

The students were invited to discuss in groups the solution and answer some supplementary questions leading to a more formal definition of the sequences. The complete scenarios in Italian and in French are available in Appendices 1 and 2.

The Analysis

In this section, we develop the analysis of the French situation, starting from descendant and ascendant analysis (an *a priori* analysis) and continuing with the analysis of the incidents (and *a posteriori* analysis).

Descendant Analysis

In the situation level S+3 the teacher considers that each new concept requires a process of discovery on the part of students and a preliminary research problem will highlight the students' knowledge and their difficulties. This research problem and the class situation aim at supporting a future lesson by providing a point of references within the students' memories throughout the sequence of lessons. In the interviews, Jean said: "Later in the class; I just have to refer to the problem and for students it makes sense".

In addition, T+3 considers that technology has to be integrated into the everyday functioning of the classroom. All students in his classroom possess a TI-Nspire™ handheld and students have the opportunity to use them in almost every lesson as a natural and familiar tool. As Jean explains: "I use calculators very often, not necessarily for a long time, but just to verify something or to illustrate a property, or... Students are used to working with it".

In his role as T+2: Developer the teacher is in a particular position because of his work in the EdUmatics project since he agreed to adapt the activity proposed by the Italian team. T+2 organised the situation as a research problem based on the initial wording and adapted to his students' knowledge according to the French curriculum. In the French curriculum students had not had any formal lessons on sequences previously, but this kind of problem (What is the next term?) is often used in magazines for young people. The teacher therefore organised the wording as a challenge, taking into account this cultural familiarity, but with precise questions in order to lead students to a formal definition of the concept of sequence. He also sought different mathematical possibilities to answer the questions, noticing in particular that there is no unique answer. For example, the first sequence 1, 2, 3, 4, 5 can be continued with the number 7 considering the sequence of prime power p^k (p prime and k ≥ 0) and the second sequence (3, 6, 9, 12, 15) can be continued with 20 considering the sequence defined by $a(1)=3$ and $a(n)=a(n-1)+$ greatest prime factor of $a(n-1)$ and so on.

The teacher in the position of T+1, within the project, chose to conduct this lesson in a particular room where computers were available along with space for group work. Computers as well as calculators were available but with no direct instruction about how to use them. It is interesting to note that the material milieu that the teacher wanted for students included different kinds of tools but also the freedom to consider these tools as useful or not for the resolution of the problem and for the documentation. Also, the teacher in position T+1 wrote worksheets for students in order to allow and to encourage them to write their answers individually and in groups; these sheets are part of the material milieu as well as the common knowledge about sequences described above.

Ascendant Analysis

In the material milieu of students there are digital artefacts, namely the two-page description of the problem and sheets on which they would produce their report. Knowledge of students in the position St−3 on the subject of sequences is non-existent in the school context, but as already said, present in a cultural and emotional context. The students' ability regarding the technology is sufficiently high to consider the artefact as an instrument permitting calculation in a familiar context. It is also a part of the set of resources that students may use if needed.

In the reference situation, it is possible to think that St−2, playing with sequences, is going to construct criteria for a valid response and confront them with the objective situation at level S−3. The material milieu in itself is unable to validate the answer (and the validation is impossible because of the different possible answers). However, the wording asking for an answer for large numbers or questioning the presence in the sequence of large numbers (1275, 2187, …) is an element of the material milieu that generates feedback, not on the values but on the process of calculation. In particular, the use of software (spreadsheet, computer algebra system, numerical calculating) brings immediate feedback inasmuch as the correct syntax can be implemented in the machine. There is a necessary translation in $S-2$ from the semiotic register of representation of natural language into the semiotic register of representation of the software, if software is used, or of the algebra, if specifications are algebraically performed.

In the learning situation, the interactions 'against' the reference milieu permit both individual validation (it is possible to find a method of calculation to obtain the nth term of the sequence) and collective validation (it is possible to clarify and to explain this process). In the $S-0$ situation, the intentions of the teacher may meet the learning acquisition of students accomplished during the a-didactic phase. Formalising and institutionalising results can then restart the problem in a new situation that includes, within the material milieu the institutionalised knowledge that has to be assimilated in order to promote its naturalisation.

Analysis of Key Incidents in the Classroom

The two previous analyses are *a priori*, but now I turn to analyses made after observations in the classroom. This approach reveals a potential gap between the analysis and the contingency and makes it possible to analyse the cause of the bifurcations and the role of the teacher in maintaining the dynamics of the situation. The complete analysis is not reported in this chapter but concentrates on the different kinds of incidents, in an attempt to illustrate the different types and the perturbation that follows.

The teacher's introduction was short, less than 3 min. During this time, Jean gave out the first worksheet (Appendix 1) and students worked individually for 5 min. Before asking students to work in groups, Jean placed the calculator and the software in the material milieu of students, saying, "The software is installed on computers, OK, you can use either computers or your handheld".

The research observation within this lesson concerns a group of four students, two boys (B1 and B2) and two girls (G1 and G2) working as shown on Fig. 2

The observation in the group of students shows that the devolution of the problem is properly executed, even if the goal is not yet clear (Fig. 3):

The word 'people' designates here the future readers of the report, including the teacher, of course, but also other students and this refers to the established didactic contract in Jean's classroom. It is interesting to see in this short extract different positions in the structure. B2 seems to be in the $S+1$ situation, thinking about the situation given by the teacher ('we must explain', 'It helps explain!') whereas G2 and G1 focus on the objective situation ('the gap between numbers') which is characteristic of an incident of friction.

Fig. 2 The group working together

> B2: People will say, yes but how many, we must explain!
> G1: No matter!
> G2: Yes, precisely, we explain, just here, the gap between numbers.
> [...]
> G2: But, the difference between two numbers, we found it, at the beginning of the sequence.
> B2: Yes, but people don't know,... It helps to explain! [...]

Fig. 3 The devolution and the negotiation of the didactic contract

> B1: You do the first gap, it's equal to three minus two, and after, little by little, you add.
> G2: The next one is easy because we just have to multiply by three.

Fig. 4 An incident of contract

> B2: I don't find anything, at least we can say one, one, two, two, four four...
> G2: I conclude like that, but... perhaps is it one, two two, four, six six. Perhaps there is only one four as there is only one one.

Fig. 5 A discussion as a consequence of the incident

The first incident of contract occurs very quickly when B1 gives an answer for the sequence D (Fig. 4):

In this episode, B1 is seeking a closed formula whereas G2 is seeking a recursive definition. They cannot understand each other and this might be due to a discrepancy between their individual understandings of the aim of the problem. But the perturbation offers possibilities of discussions and lasts a long time until they realise that the two definitions are possible (Fig. 5):

The consequence of these incidents is a discussion in the group about the problem itself, which helps the students to make the goal of the problem clearer and contributes to the devolution of the situation. The proposed milieu is sufficiently adapted to support changes in the position of students but also strong enough to interact with students and to facilitate discussions. In this case, incidents encountered in the lower didactic levels were the driving force behind the dynamic making it possible for the students to engage thoughtfully on the problem.

Fig. 6 The mime with fingers to indicate the recursion

G2: It is not billion, million, perhaps? [...] What comes after billion?

B2: There's a trillion?

Fig. 7 A digression as a nil-didactic situation

Another interesting and important set of incidents comes in the phase of action when students try to answer the question: "is 2187 present in the different sequences?" B1 and B2 try to use the calculator whereas G1 and G2 work with paper and pencil. The difficulty for B1 and B2 is to translate the recursive definition of sequences given in the register of natural language into an algebraic register and finally into the register of the calculator's syntax. Figure 1 shows the gesture that goes with the trial of translation (Arzarello & Robutti 2010) (Fig. 6).

B1 takes his calculator: I'm sure, there are sequences in it...
B2: Yes sure!
B1: But where?

The calculator is part of the material milieu and the syntactic incident leans towards a nil-didactic situation: B1 and B2 use their calculator to evaluate 3^{40} and digress by talking about the huge number they obtain and reading the number aloud (Fig. 7):

The consequences of the incident diverts the students from the aim of the problem and the difficulties of translation between registers of representation lead students back to the objective situation.

In the lower didactic levels, didactic incidents play two different roles depending on whether the milieu reacts. In the first example, the feedback of the milieu constitutes a guide and the incidents present an amplification of the dynamic whereas in the second example, the technical incidents bring the students back to the objective situation. The calculator's syntax is not sufficiently naturalised to become the place of experiments and remains an obstacle to reaching the learning situation.

> St1: Thirteen!
>
> St2: Fourteen!
>
> Teacher (joking): Thirteen, fourteen, well, good prices!
>
> St3: Fifteen!
>
> Teacher: Another answer? "What do you say St"?
>
> St4: We can't know...
>
> Teacher: We can't know? Well, what does it mean, we can't know? (Hubbub) Wait, wait, one after the other!
>
> St1: We don't have enough information.
>
> Teacher: Why do you have enough information for the others?
>
> St1: They were linear.
>
> Teacher: You say, they were...?
>
> St1: Linear.
>
> Teacher: Linear?
>
> St3: Yes, you know, at the beginning there's two, then...
>
> St2: It's always the same thing...
>
> St5: Constant.
>
> Teacher: It is always the same thing. It is constant, ... yes?
>
> St6: For the others, there was a logical sequence
>
> Teacher: And now, why are you sure it is not a logical sequence?
>
> St: We are not sure. We have not enough information.

Fig. 8 The debate about the prime numbers' sequence

The second part of the observation concerns a common phase where Jean wants to institutionalise, firstly the two possible definitions of a sequence (recursive or using a closed formula) and, secondly, the possibility of having several different and correct answers for a problem. Despite all his efforts, Jean does not succeed in the second aim with the first five sequences. From the point of view of students, there is only a unique possibility:

A: 1, 2, 3, 4, 5, **6**
B: 3, 6, 9, 12, 15, **18**
C: 5, 8, 11, 14, 17, **20**
D: 1, 3, 6, 10, 15, **21**
E: 3, 9, 27, 81, 243, **729**

On the other hand, the discussion is strong when the sixth sequence comes up for discussion; students have never formally studied the prime numbers even if they know the definition. The result is that different responses emerge around the classroom (Fig. 8).

The debate is about the place of different didactic incidents, which are all visible and allow Jean to institutionalise the second point even if he does not take into account the vagueness of vocabulary. Linearity is seen as regularity or a logical sequence as a result of a known formula. The incident of contract occurs because of the distance of the students from the mathematical thinking; in a 'typical' mathematics class each problem has a unique answer and thinking about the possibility of having different answers goes against the students' conceptions of mathematics. This conception is unsettled by the prime number sequence which is not sufficiently familiar to students to remain in the material milieu and the experiments on numbers lead to negotiation about the incident of contract to a new didactic contract.

Even if all the details of the analysis of incidents cannot be reproduced here, it is possible to draw a conclusion that is illustrated by the previous extracts. An important issue that is raised by this observation is the confirmation of the constructive dimension of didactic incidents, which in several cases have revived the students' work. Mathematical incidents, provided that they become visible for students and teacher, appear as prompts that link knowledge and experiments on mathematical objects and they facilitate the transition from the objective situation to the learning situation. Incidents of contract allow a renegotiation of the contract in the classroom and promote a step back in relation to the didactic situation. In contrast, syntactic incidents have not been able to be overcome and instead have played out, in this observation, as a brake on the dynamics of research. This conclusion points to the need to better understand the place of technology in the set of resources of both the teacher and the students.

Technology in the Set of Resources

In previous research, I concluded that:

> [...] the documentational geneses become distinct and separated processes for teacher and students. These processes are confronted with each other only in a collective domain and concern mainly the property of the creation. The communication and cognitive properties (memorization and organization of ideas) seem to remain private but are important parts of the documentational genesis. (Aldon 2010, p. 746)

Incidents created by the gaps between the private, collective and public use of calculators had been highlighted by looking at the content of calculators and the activity of students working on a task. It is quite clear that the calculator belongs to the set of resources of the teacher and in this sense, it is part of the milieu of design. At the same time, it belongs to the material milieu of the objective situation. More precisely, the calculator can become a document useful in the situation of reference and the situation of learning if, and only if, it belongs, for the teacher, to the milieu of the project and to the didactic milieu. In other words, in the perspective of the integration of the calculator in the mathematics lesson, it remains compulsory to negotiate the didactic contract, including the different properties of calculators, not

> E6: Well, for the functions, with my old calculator, I type the function, Graph and I have the curve, whereas, with this one, I don't know, you must define it...
> E5: There are many steps...
> E6: Yes, there is a lot of things to do, just for one result, whereas with my calculator, you type your calculation, you have your result, that's all!
> E5: It's faster...
> I: And do you remember the moment you said: I don't want this calculator!
> E5: Very quickly, yes, we must use menu, then this place, then click everywhere, we had a long course to do a calculation that can be done very quickly with our calculator.
> E6: Yes, it was a lesson at the beginning of the year, about functions, we spent two hours with the calculator, it really bugged me. It put me off this calculator.

Fig. 9 Interview of two students who do not use the TI-Nspire technology

T: Then you open the catalogue and type the first letter of the command, well for the moment, R and you just have to go down, OK, you see Randint, it's here. Well. (he is doing on the computer whilst speaking)	E1: We have to type a blank.
	E2: Do you think that?
	E1: It's six.
	E2: Yes, randint one six minus randint one six?
	E1: And, how do you type the absolute value?
	E1: It doesn't work.
T: Well. I have simulated the throw of a dice. The question now is: how are you going to simulate the throw of two dice and how will you obtain the value of the difference of the greater minus the smaller?	E2: (watching to the screen of E1's calculator) Missing?
	E1: and now it gives six, Ahhh!
	E2: Ahhhh!
	E1: It doesn't work!
	E2: Too many arguments!
	E1: I can't do that!

Fig. 10 Crossed dialogues of teacher and students

only as a tool becoming an instrument in specific situations, but also as a resource becoming a document available in the set of resources of students and the teacher.

In the following example, the global consequence of an incident is illustrated. E5 and E6 are two students who do not want to use the TI-Nspire and prefer their old calculator, in fact a TI-82 (E5) and a Casio Graph 35 (E6). See (Fig. 9):

It is interesting to set this dialogue against the observation which took place at the beginning of the year where the teacher is speaking to the whole class whilst students work with their calculator (Fig. 10):

The discrepancy between the talk of the teacher and the students' difficulties is clear. The syntactic incident is caused by students' incomprehension of the machine's feedback. At first, instead of typing randint(1,6), E1 typed randint 1 6. The feedback of the machine was *Missing*), but the bracket was not read by the students. E1 tried to type brackets but finally obtained fresh feedback, which he could not interpret. This kind of incident may lead to a rejection of the technology, as E5 and E6 said.

Clearly the syntactic incidents are inherent in the use of technology in the classroom. Taking into account the perturbations, consequences of incidents are essential to limit their long-term effects from the point of view of:

- teachers' professional development by increasing the *response repertoire* (Clark-Wilson 2010);
- students by increasing the registers of representation of studied mathematical objects.

The classroom management and the orchestration of a mathematical situation in a digital environment (Trouche 2004) accentuate the importance of the teachers' responsibilities with respect to the instruments and show the necessity of including the analyses of such situations' in the process of teacher development.

Teacher Development

In this section, I would like to emphasise the links between the analysis, the observation, the feedback and the professional development of teachers. Starting from observations in the classroom and interviews with Jean, a French teacher involved in the EdUmatics project, I will show how and why the collaborations introduced at the beginning of the chapter contribute to the professional development of teachers as well as to strengthening theoretical approaches.

Collaboration Between Researchers and Teachers

One of the important aspects of the EdUmatics project was to enable teachers and researchers to work together on the implementation of lessons using technology. Even though these work habits are already widely implemented in France in the network of IREM (Institut de Recherche sur l'Enseignement des Mathématiques/ Research Institute on Mathematics Education), the particular experience of the EdUmatics project provided valuable information for the professional development of teachers. The confrontation of teachers' professional skills with analyses based on theoretical frameworks helped both to increase the skills and refine the theoretical tools.

An *a priori* analysis sufficiently complete to embrace the mathematical aspects, the didactic characteristics and the pedagogical modalities give the design of a lesson a new dimension, as Jean says: "To predict, to analyze and to find solutions to all the difficulties, pedagogical as well as technical is something demanding and interesting". The contrast between this *a priori* analysis and the reality of the classroom shows that the theoretical tools are consistent and the possibility to see '*live*' the occurrences of predicted events modifies significantly the teaching approach. During the interviews with Jean, before and after the class observations, the two analyses (*a priori* and *a posteriori*) were shared and discussed with him; as shown in Table 2, the data collection tended to catch the evolution either for students or the teacher over the long term. Jean commented on the benefits of this *a priori* analysis "When I see in the classroom some attitude that the *a priori* analysis had predicted and for which a solution was already ready, it's reassuring and very satisfying in my professional practice [...] Several times, later, in my classroom, I surprise myself in remembering this moments and I modified my attitude to take into account the observations".

The work done in the project and the collaboration between researchers and teachers developed an awareness of professional gestures. The analysis using the concept of incidents illuminates different processes occurring in the classrooms and, more particularly, the place and the role of technology in the development of both teaching and learning. In addition, the observations highlight the role of didactic incidents in the students' construction of knowledge, particularly in the lower didactic levels. But in order to become shared knowledge in the classroom and, more generally, to become a potential naturalised knowledge, the knowledge that students encounter must be recognised as legitimate. The institutionalisation of knowledge in the course of acquisition is essential and this institutionalisation is typically the responsibility of the teacher who needs to recognise, to interpret, to organise and to transform the *knowledge in action* from what the teacher at level $T-1$ observed in the learning situation into what students must know and learn. Players win not only because they reach the end of the game but also because they know how and why they win. Students have to transform their *knowledge in action* into shared knowledge and teachers have to understand the key elements of the situation allowing this knowledge construction, or perhaps the key elements that prevent them reaching their initial didactic intentions.

Working in a technological environment adds to this institutionalisation knowledge, being directly linked with the technology in use. One of the main difficulties is surely to recognise the different knowledge that students act upon during the phase of action in lower level situations. The *a priori* analysis and the feedback of what happens in real classrooms shine light on the actual activity of students and the knowledge that has to be institutionalised. In the last interview, Jean said: "In fact, when you are in my classroom I see things that I didn't see usually. Sometimes, I'm not happy with my lesson, but you say that a student or a group of students work on this or that; I know then that I've not wasted my time."

Giving teachers this opportunity, at least once, is surely a fundamental aim of teacher development, but in an 'ordinary' classroom this awareness is a key element of the modification of schemes. Giving tools which make it possible to observe and analyse what happens in the class can augment the *response repertoire* of teachers. The framework of *didactic incident* may increase the awareness of teachers interpretations of students' work when they are in a position of observer (T−1) and facilitate the institutionalisation of knowledge directly linked with the actual activity of students. The design of our part of the EdUmatics course takes into account the common analysis. Future research should concern the construction of tools facilitating the incident analysis by teachers themselves.

The Documentational Genesis

A second aspect that occurs from the observations in Jean's school concerns the documentary role of the digital artefacts. The different properties of digital documents are described by Pédauque (2006):

> The two cognitive functions, mnemonics and organization of ideas, seem to be the fundamental basis for the documentary production. […] The function of creativity comprising enrichment due to the domain of interest related to the document surpasses that kind of organization just mentioned. […] The third and last constituting function of the documentary production is the transmission function. (pp. 5–6).

The technological context shows the four dimensions present in this handheld device seen as a resource, and the phenomenon of documentational genesis builds on these properties in different meditational contexts. The cognitive properties of storage and organisational ideas are built in parallel and remain in a private domain, both for teachers and for students. On the other hand, the properties of creation and communication are built in the collective domain. The method(ology) allows the researcher to follow the joint documentational genesis of teacher and students by entering into private domains, particularly regarding the contents of handheld devices. The handheld with its computational and representational properties, along with its properties of storage and communication, prefigures digital resources that may be available in coming years. The documentational genesis of such an artefact may not be understood without taking into account the domains of mediation, whether private, collective or public. The handheld appears then to be at the crossroads between the teacher's teaching intentions and the students' learning intentions, that is to say, at the core of the didactic game. Different trajectories are sources of tension and generate didactic incidents that deeply affect interactions in the classroom, interactions between teacher and students, and also interactions between teacher, students and artefact. The integration of digital resources in the mathematics classroom cannot be achieved by considering only one property but, on the contrary, by thinking globally about the integration of all properties in the learning game. In the upper didactic levels, incidents call into question the teacher's personal epistemology and contribute to professional development.

Conclusion

The exploitation in teacher education of the frameworks of the Theory of Didactical Situations (TDS), didactic incidents and documentational genesis, should make it possible to build a detailed analysis of situations in ordinary classrooms in a technological environment. The observation of interactions within the classrooms through didactic incidents and the understanding of joint documentational geneses of students and teachers are two parts of the same methodological tool aiming at better understanding the didactic game.

The descendant and ascendant analyses assist the *a priori* analysis to take into account the role and the place of both teacher and students in the didactic game, and the incident analysis refines the *a posteriori* analysis. Inter-connecting the two analyses constitutes a tool for teachers in the preparation of lessons and in the understanding of what happens in the classroom. The typology of didactic incidents can be extended and refined to allow easy and more operational identification for new teachers in a perspective of understanding the dynamics of the classroom. It can also become a tool for regulating those dynamics within the classroom. Finally, connecting local incidents to global phenomena resulting from differences in the documentational geneses of teachers and students makes it possible to better understand the place of digital artefacts in the classroom.

New hypotheses that result from this research are about documentational geneses and the possible conflicts between the point of view of students, teachers and society as a whole. Further research might involve clarifying the role and the learning potential of digital artefacts in a digital age and reorganising the importance of teacher development in their usage.

Appendix 1

N.B. In questa attività, sia nei lavori individuali, sia in quelli di gruppo, potrai utilizzare, se lo desideri, gli strumenti informatici che ritieni più opportuni. Nei lavori di gruppo, nel caso in cui opinioni discordanti dovessero rimanere tali anche dopo un confronto, riportatele sul foglio di lavoro.

Situazione

Alberta (A), Bruno (B), Carla (C), Dario (D), Elena (E) e Federico (F) stanno esplorando la successione dei numeri naturali, studiando le proprietà dei numeri che la costituiscono. Le modalità di esplorazione, pero, sembrano molto diverse fra loro, anche se tutte sono caratterizzate da una forte sistematicità. Ecco i numeri che i sei amici prendono in considerazione:

A: 1, 2, 3, 4, 5, …, …
B: 3, 6, 9, 12, 15, …, …

C: 5, 8, 11, 14, 17, ..., ...
D: 1, 3, 6, 10, 15, ..., ...
E: 3, 9, 27, 81, 243, ..., ...
F: 2, 3, 5, 7, 11, ..., ...

Proposta di lavoro

Attività 1 (individuale)

Qual è, secondo te, il sesto numero che ciascuno dei sei amici prenderà in considerazione? In caso di risposta affermativa scrivilo e cerca di spiegare come/cosa hai fatto. In caso di risposta negativa, spiega perché non riesci a individuarlo.

Le tue precedenti risposte cambierebbero se ti venisse chiesto di individuare il decimo numero? E il quarantesimo? Spiega perché.

Attività 2 (di gruppo: 3 studenti)

Parlando uno alla volta, spiegate ai vostri compagni di gruppo come avete risposto alle domande dell'attività 1. Discutete sulle eventuali differenze. Riuscite a produrre una risposta condivisa di gruppo? In caso di risposta affermativa, riportatela sul vostro foglio; in caso di risposta negativa, riportate i punti di dissenso rimasti dopo la discussione.

Attività 3 (di gruppo)

C'è qualcuno, fra A, B, C, D, E, F che, secondo voi, prima o poi, troverà, nella sua successione, il numero 1275? In caso di risposta affermativa, dopo quanti passi?

Giustificate la risposta e precisate le strategie utilizzate per rispondere. Come cambierebbero le vostre risposte se le domande fatte sul numero 1275 fossero fatte sul numero 2187?

È possibile trovare un numero naturale diverso da 0 tale che nessuno, fra A, B, C, D, E ed F, prenderà mai in considerazione? Giustificate la vostra risposta.

Esiste almeno un numero naturale che non potrà mai essere raggiunto da B, né da C, né da D, né da E, né da F? In caso di risposta positiva, trovatelo e spiegate come avete fatto. In caso di risposta negativa, spiegate perché, secondo voi, tale numero non esiste.

Appendix 2

> À suivre...

Partie 1

En travaillant sur l'ensemble des nombres naturels, Alberta (A), Bruno (B), Carla (C), Dario (D), Elena (E) et Federico (F) ont chacun créé une suite de nombres. Ils ont tous suivi un processus de construction différent mais systématique.

Voilà les cinq premiers nombres que chacun des six amis a écrit:

- A : 1, 2, 3, 4, 5, ..., ...
- B : 3, 6, 9, 12, 15, ..., ...
- C: 5, 8, 11, 14, 17, ..., ...
- D: 1, 3, 6, 10, 15, ..., ...
- E: 3, 9, 27, 81, 243, ..., ...
- F: 2, 3, 5, 7, 11, ..., ...

1. Êtes-vous capable d'écrire le sixième nombre qui selon vous a été créé par chacun des six amis ?
   Si oui, expliquez comment vous avez fait.
   Si non, expliquez les raisons qui vous empêchent de répondre.
2. Vos réponses précédentes changeraient-elles si on vous demandait d'écrire le dixième nombre ? Et le quarantième ? Pourquoi ?
3. Y a t-il quelqu'un parmi A, B, C, D, E, F qui selon vous, tôt ou tard, trouvera dans sa suite le nombre 1275 ? Si oui, lequel (ou lesquels) et après combien d'étapes ?
   Justifiez votre réponse et décrivez la méthode qui vous a permis de répondre.
   Pouvez-vous alors répondre aux mêmes questions avec le nombre 2187 ?

> À suivre...

Partie 2

4. Les méthodes que vous avez utilisées précédemment vous permettent-elles de calculer le 70ème, le 200ème, le 1000ème nombre de chaque suite ?
   Si oui, calculez ces nombres, si non essayez de modifier vos méthodes pour les obtenir.
5. Essayez, en utilisant la calculatrice, de donner une représentation graphique de ces suites.

6. Les méthodes que vous avez utilisées précédemment vous permettent-elles de demander à votre calculatrice de calculer ces nombres ? Si oui, écrivez le calcul demandé.
Sinon, dire pourquoi ces méthodes utilisées ne le permettent pas.

References

Aldon, G. (2010). Handheld calculators between instrument and document. Dans P. Drijvers & H. -G. Weigand (Éd.), *The role of handheld technology in the mathematics classroom* (Vol. 42, pp. 733–745). ZDM Mathematics Education, Karlsruhe.
Aldon, G. (2011). *Interactions didactiques dans la classe de mathématiques en environnement numérique: construction et mise à l'épreuve d'un cadre d'analyse exploitant la notion d'incident*. Doctorat, Université Lyon 1.
Artigue, M. (2007). Conference: Teaching and learning mathematics with digital technologies: The teacher perspective. In *International meeting Sharing Inspiration*. Brussel.
Artigue, M., Defouad, B., Dupérier, M., Juge, G., & Lagrange, J. -B. (1998). L'intégration de calculatrices complexes à l'enseignement des mathématiques au lycée. *Cahier DIDIREM, IREM Paris VII, Spécial no 4*.
Arzarello, F., & Robutti, O. (2010). Multimodality in multi-representational environments. Dans P. Drijvers & H. -G. Weigand (Éd.), *The role of handheld technology in the mathematics classroom*. (Vol. 42, pp. 715–731). ZDM Mathematics Education, Karlsruhe.
Brousseau, G. (1986). *Théorisation des phénomènes d'enseignement des Mathématiques*. Doctorat, Université Bordeaux 1.
Brousseau, G. (2004). *Théorie des situations didactiques*. La pensée sauvage éditions.
Chevallard Y. (1985). *La transposition didactique – Du savoir savant au savoir enseigné*, La Pensée sauvage, Grenoble (126 p.). Deuxième édition augmentée 1991.
Clark-Wilson, A. (2010). *How does a multi-representational mathematical ICT tool mediate teachers' mathematical and pedagogical knowledge concerning variance and invariance?* Ph.D., Institute of Education, University of London.
Drijvers, P., & Trouche, L. (2008). From artifacts to instruments: A theoretical framework behind the orchestra metaphor. Dans G. W. Blume & M. K. Heid (Éd.), *Research on technology and the teaching and learning of mathematics* (Vol. 2, pp. 363–392). Charlotte: IAP (Information Age Publishing).
Gueudet, G., & Trouche, L. (2009). Towards new documentation systems for mathematics teachers? *Education Studies in Mathematics, 71*, 199–218.
Lagrange, J.-B., & Degleodu, N. C. (2009). Usages de la technologie dans des conditions ordinaires. le cas de la géométrie dynamique au collège. *Recherches en Didactique des Mathématiques, 29*(2), 189–226.
Margolinas, C. (2004). *Points de vue de l'élève et du professeur Essai de développement de la théorie des situations didactiques*. Habilitation à Diriger des Recherches, Université de Provence.
Pédauque, R. T. (2006). *Le document à la lumière du numérique*. Caen: C & F éditions.
Rabardel, P. (1995). L'homme et les outils contemporains. A. Colin, Paris
Rabardel, P., & Pastré, P. (2005). *Modèles du sujet pour la conception*. Octares, Toulouse.
Rodd, M., & Monaghan, J. (2002). Graphic calculator use in Leeds schools: Fragments of practice. *Journal of Information Technology for Teacher Education, 11*(1), 93–108.

Sabra, H. (2011, en cours). *Contribution à l'étude du monde et du travail documentaire des enseignants de mathématiques: les incidents comme révélateurs des rapports entre individuel et collectif*. Université Lyon 1.

Sensevy, G. (2007). Des catégories pour décrire et comprendre l'action didactique. Dans G. Sensevy & A. Mercier (Éd.), (pp. 13–49). Presses Universitaires de Rennes.

Trouche, L. (2004). Managing complexity of human/machine interactions in computerized learning environments: Guiding students' command process through instrumental orchestrations. *International Journal of Computers for Mathematical Learning, 9*, 281–307.

Part III
Theories on Theories

Meta-Didactical Transposition: A Theoretical Model for Teacher Education Programmes

Ferdinando Arzarello, Ornella Robutti, Cristina Sabena, Annalisa Cusi, Rossella Garuti, Nicolina Malara, and Francesca Martignone

Abstract We propose a new model for framing teacher education projects that takes both the research and the institutional dimensions into account. The model, which we call *Meta-didactical Transposition*, is based on Chevallard's anthropological theory and is complemented by relevant elements that focus on the specificity of both researchers' and teachers' roles, while enabling a description of the evolution of their praxeologies over time. The model is illustrated with examples from different Italian projects, and it is discussed in light of current major research studies in mathematics teacher education.

Keywords Meta-Didactical Transposition • Communities of inquiry • Research for innovation within institutions • Teacher education practices • Meta-didactical praxeologies • Mathematics laboratory

Introduction

The education of teachers is a relevant issue in the evolution of a society and is even more significant at particular historical moments of social or political change. Since the 1960s, with the progressive diffusion of socio-constructivism as a cognitive model, social interaction in the classroom came to the fore, resulting in an increased attention to the social dynamics of learning. This progressive change of attention, from the individual to the social construction of meaning, along with an increasing

---

F. Arzarello (✉) • O. Robutti • C. Sabena
Dipartimento di Matematica, Università di Torino, Via Carlo Alberto 10, Turin 10123, Italy
e-mail: ferdinando.arzarello@unito.it

A. Cusi • R. Garuti • N. Malara • F. Martignone
Università di Modena e Reggio Emilia, Modena, Italy

use of technological artefacts, led to a corresponding interest in teacher education. Particularly in the last decade, attention to teacher education has increased (Ball and Bass 2003; Ball et al. 2008; Clark and Hollingsworth 2002; Even and Ball 2009; Wood 2008) and digital technologies have an increasing relevance in this context (Drijvers et al. 2010; Hoyles and Lagrange 2009; Lagrange et al. 2003). In Italy, we have witnessed the multiplication of teacher education programmes involving digital technology at the European,[1] national, regional, and local levels. As researchers, we are involved both at the level of teacher education programme development and management, and in studying teaching and learning processes in the classroom. This has prompted the emergence of a deeper reflection on the resulting complexity.

We began to recognise the importance that institutions play in the school context, including the national curriculum, national assessment tools and the constraints of teachers' time and space, and textbooks. Our attention was directed toward the theoretical elements that could adequately frame these, which we found in Chevallard's (1985, 1992, 1999) Anthropological Theory of Didactics (ATD), particularly with respect to his notion of *didactical transposition*.

The complexity arising from the intertwining of the processes involved during a teacher education programme has led us to introduce a descriptive and interpretative model, which considers some of the main variables in teacher education (the community of teachers, the researchers, the role of the institutions), and accounts for their mutual relationships and evolution over time. We call the overall resulting process *Meta-didactical Transposition*. We offer the model as a tool for studying the complexity of teacher education as a research problem that involves a transposition from the practice of research to that of teaching.

In the following sections, after some theoretical background on teacher education, we present the Italian context in which our research is situated. Then we present the *Meta-Didactical Transposition* model (in short, MDT). We use this model to analyse the different variables listed above and their dynamic relationship, contextualised within three Italian teacher education programmes that use digital technologies. The three programmes are used as 'generic examples' that we hope will find resonance within other international contexts. Finally, we discuss the results of our analysis, pointing to the model's potential with respect to current research in the field.

Teacher Education and the Italian Context

In 2000, the International Commission on Mathematics Instruction (ICMI) commissioned a study that was coordinated by Anna Sfard on the relationships between research and teaching practice in mathematics education. The results of this study were presented at ICME in Copenhagen, 2004. It highlights three main periods in

---

[1] One of the European projects in which we have been involved is the EU funded project *EdUmatics* (*50324-UK-2009-COMENIUS-CMP; European Development for the Use of Mathematics Technology in Classrooms*), http://www.edumatics.eu.

the evolution of issues addressed by mathematics education research: the *era of the curriculum*, mainly focused on the study of education programmes; the *era of the learner*, focused on student's learning and difficulties; and, the *era of the teacher*, focused on teachers and teacher education.

Sfard (2005) stresses that the advent of *the era of the teacher* has brought about a re-conceptualisation of the relationship between the teacher and the researcher, which constitutes "a big leap toward research that plays a genuine role in shaping and improving practice" (p. 405). She argues that in most of the international research studies, the question is not *what* is taught in classrooms, but *how* it is taught: "rather than trying to arrive at a mechanistic view of 'what works in classrooms', I focus on how things work and try to make myself aware of alternative possibilities" (p. 406). This shift of attention to teaching practices is due in part to international comparative tests (TIMSS, PISA), which often show poor results, despite the quantity of resources devoted to curricular changes.

In the last years, many publications have focused on teacher education. They have been concerned with teachers of different school levels, addressing issues such as the relationship between teachers and both curricular or methodological innovation and technology integration. In particular, research on teacher education programmes has intensified, gradually changing the focus from pre-service to in-service education, with an emphasis on the role played by specific tools and methods on the professional development of teachers. An overview on this wide-ranging research can be found in the 15th ICMI study on teacher education (Even and Ball 2009) and the four volumes of the *International Handbook of Mathematics Teacher Education* (2008).

Much of the research on teacher education has focused on identifying the knowledge that is necessary for the teaching of mathematics. Researchers generally agree that this knowledge consists of three main components, which progressively interrelate to each other: knowledge about mathematics content; general pedagogical knowledge; and the mathematical-didactical knowledge. These components can be related to those introduced by Shulman (1986), who was the first to identify the notion of *pedagogical content knowledge* (PCK) as the particular knowledge for teaching: "the particular form of content knowledge that embodies the aspects of content most germane to its teachability" (p. 9). In the case of the teaching of mathematics, PCK concerns the intertwining of mathematics and pedagogy in relation to the different conditions for and ways of teaching and learning specific content.

Taking Shulman's studies as a starting point, Ball and Bass (2003) propose a finer and more effective characterisation of what they refer to as the *mathematical knowledge for teaching* (MKT), which Bass (2005) defines as "the mathematical knowledge, skills, habits of mind, and sensibilities that are entailed by the actual work of teaching" (p. 429), that is "the daily tasks in which teachers engage, and the responsibilities they have to teach mathematics, both inside and outside the classroom". Ball et al. (2008) highlight the fundamental difference between mathematics and mathematics for teaching. While the former has the capability of compressing the information into abstract forms, the latter requires a sort of decompression, in that the main ideas pertaining to the mathematical content is made more explicit. These authors choose to characterise MKT through the analysis of the daily practice of teachers:

> Instead of starting with the curriculum, or with standards for students learning, we study the work that teaching entails. […] We seek to unearth the ways in which mathematics is involved in contending with the regular day-to-day, moment-to-moment demands of teaching. Our analyses lay the foundation for a practice-based theory of mathematical knowledge for teaching. (p. 395)

They thus analyse the typical features of mathematics that are involved in teaching and identify the main components of MKT in relation to Shulman's subject matter knowledge (SMK) and pedagogical content knowledge (PCK). They distinguish three sub-domains of PCK: (a) *knowledge of content and students*; (b) *knowledge of content and teaching;* (c) *knowledge of content and curriculum.* Referring to SMK they identify *specialised content knowledge* (SCK) as an important sub-domain of mathematical knowledge. Bass (2005) stresses that SCK

> is strictly mathematical knowledge (not about students or about pedagogy) that proficient teachers need and use, yet is not known by many other mathematically trained professionals, for example, research mathematicians. Contrary to popular belief, the purely mathematical part of MKT is not a diminutive subset of what mathematicians know. It is something distinct, and, without dedicated attention, it is not something likely to be part of the instruction in content courses for teachers situated in mathematics departments. (p. 429).

Another important element that characterises the main studies on teacher education is their involvement of teachers in the joint analysis and reflection on the main features of the didactical projects being researched. Within the research literature, this involvement is described in terms of communities of practice, communities of inquiry, adaptive systems, collective participation, sustained conversation and egalitarian dialogue. The cornerstone of these studies is the notion of critical reflection, conceived not only as a fundamental attitude to be instilled in teachers but also as a professional responsibility. Drawing on Schön's studies (1987), many researchers stress the value of critical reflection as well as the importance of sharing reflections amongst teachers and between teachers and researchers (e.g. Mason 1998, 2002; Jaworski 1998, 2003; Schoenfeld 1998). These studies suggest that teachers should share their interpretations of teaching and that observing different ways of acting can lead them to re-conceiving their ideas about their role in the classroom as well as the nature of their profession. As we will show, this philosophy permeates the practice developed by Italian mathematics education research since the 1980s.

With respect to the evolution of the research on teacher education, another essential aspect is its strict interrelation with the research on the integration of new technologies in the teaching of mathematics. The focus of this research has shifted from the study of new programming languages for the implementation of algorithms (in the 1980s and 1990s), to the exploration of didactical software expressly conceived for education (in the 1990s and later), to the more recent use of new technologies not only for the teaching of mathematics but also as tools for communication and education in general, which led to the constitution of a specific research area on *educational technology* (Guin et al. 2005).

Arzarello and Bartolini Bussi (1998) provide a synthesis of the Italian research in the 1960s–1990s, which reflects the different dynamics and the changes that occurred. The authors identify four different trends, the fourth of which represents the dominant Italian research paradigm of *research for innovation*. According to this paradigm, the main features that characterise the work carried out during teacher education programmes by the teachers and the researchers are collaboration, mutually supportive and integrative of knowledge and skills. This collaboration links theory and practice, and is fundamental for the professional development both of teachers and researchers in mathematics education. A peculiar feature of *research for innovation* is the important role played by the 'teacher-researchers', that is, teachers that are deeply involved in all phases of the research process, from planning to implementation to data analysis to dissemination (Malara and Zan 2002). Whereas only a relatively small number of teachers become teacher-researchers, a greater number of them have been involved in institutions (e.g. Ministry of Education), in research communities within pre-service and in-service teacher education programmes, or as tutors or trainers for other teachers.

The model we present is strongly culturally framed in the Italian context, from which we identify the main variables. However, we are confident that it is possible to extend this model to other contexts, because of its flexibility in describing teacher education as a complex system and in highlighting the interaction between its variables.

A New Paradigm: Meta-Didactical Transposition

The model we propose, which takes into consideration the practices of mathematics educators (researchers) and those of teachers, when both communities are engaged in teachers' education activities, is based on the Chevallard's Anthropological Theory of Didactics (Chevallard 1985, 1992, 1999; Bosch and Chevallard 1999) It adapts and extends ATD to the context of teacher education. This model, called *Meta-Didactical Transposition,* considers:

(i) the complex dynamic interplay, which develops in activities involving different communities (e.g. between the teachers and the mathematics educators);
(ii) the constraints imposed by the institutions that promote such activities (including schools and Ministry of Education) in view of some specific goals (e.g. promoting teachers' knowledge of new curricula or of new technologies);
(iii) other 'institutional' constrains, including the tradition of the school(s), the related (intended, implemented, attained) curricula and the textbooks used by the teachers.

Meta-Didactical Transposition involves five intertwined features: the *institutional aspects*, the *meta-didactical praxeologies*, the *double dialectics*, the *brokering* processes and the dynamics between *internal and external components*. We describe each aspect in the next sections. Our model thus complements the MKT model described above insofar that it focuses on these main aspects of teachers' education programmes: their dynamicity; the dialectic between the communities[2,3] of teachers and those of the researchers who coach them; and the influence of the institutional components and their relationships to the communities.

Institutional Aspects

ATD focuses on the institutional dimension of mathematical knowledge, placing mathematical activity, hence the activity of studying in mathematics, within the bulk of the human activities and of the social institutions (Chevallard 1999). In our view, it is important to consider such an institutional dimension in teacher education activities since these activities are fully situated within and constrained by the context of social institutions (research communities, schools, the Ministry of Education, the policy makers, the teachers associations, etc.). In Italy, as in many other European countries, the whole educational system (from kindergarten to university) is public and is governed by several institutions at different levels (national, regional, local). Within this context, the importance of the institutional dimension is also at play within the politics of the European Union. As lifelong education is considered a strategic element for development in Europe, community programmes are promoted for prospective or in-service teacher education. These programmes assume a clear cooperation between the research world and the institutional-political world (see http://ec.europa.eu/education/llp/official-documents-on-the-llp_en.htm).

Chevallard (1992) stresses the fact that the very nature of mathematical objects in school is dependent on the person or the institution with which it is related: "An object exists since a person, or an institution acknowledges that it exists (for it itself)" (p. 9). With respect to teacher education, our model focuses on two types of communities, which sometimes intertwine: (a) the *communities of the researchers*, who design and coach the educational programmes, generally as an official task commissioned by the responsible authorities (e.g., School administration, Ministry of Education);

---

[2] We refer to this term in tune with the following characterisation of communities of inquiry proposed by Jaworski (2008): "*In terms of Wenger's (1998) theory, that belonging to a community of practice involves engagement, imagination and alignment, we might see the normal desirable state as engaging students and teachers in forms of practice and ways of being in practice with which they align their actions and conform to expectations…In an inquiry community, we are not satisfied with the normal (desirable) state, but we approach our practice with a questioning attitude, not to change everything overnight, but to start to explore what else is possible; to wonder, to ask questions, and to seek to understand by collaborating with others in the attempt to provide answers to them. In this activity, if our questioning is systematic and we set out purposefully to inquire into our practices, we become researchers.*"

[3] It derives from the Chevallard's notion of didactical transposition (Chevallard 1985), which roughly speaking, consists in the relationships between the production, the use and the teaching of the scientific knowledge and in the ways, according to which it adapts itself in order to 'work' in different types of institutions (compare for example a theorem as expressed in the Journal where it is proved by a mathematician, what Chevallard calls "le savoir savant", with the same theorem as it is written in a textbook, "le savoir enseigné").

(b) the *communities of the teachers*, who participate within the projects, either on a voluntary basis or because of an official duty. Both of these communities are in relationship with the school: the actual schools where the teachers teach, and the School as an institution with its curricula, its teaching traditions, the textbooks used, etc.

Meta-Didactical Praxeologies

ATD proposes a general epistemological model of mathematical knowledge, conceived as a human activity developed for the purpose of addressing specific families of tasks. Its main theoretical tool is the notion of *praxeology* (or mathematical organisation), which is structured in terms of two main levels (García et al. 2006): (a) The 'know how' (*praxis*), which includes a family of similar *problems* to be studied, as well as the *techniques* available to solve them (e.g. 2nd degree equations and the formulae for their solution); (b) The 'knowledge' (*logos*), which is the 'discourses' that describe, explain and justify the techniques that are used within a more or less sophisticated frame and may even produce new techniques (e.g. the justification of the formula for 2nd degree equations through the completion of squares or even the theory of algebraic equations and how it encompasses 2nd degree equations).[4] A praxeology consists of a task, a technique, and a more or less structured argument that justifies or frames the technique for that task. Hence, it encompasses both the *know-how* and the *knowledge*, with respect to a family of tasks.

In constructing our model, we consider the *meta-didactical praxeologies*, which consist of the tasks, techniques, and justifying discourses that develop during the process of teacher education. For example, consider the teacher training course described by Sullivan (2008), in which he used the question "which is bigger, 2/3 or 201/301?" (p. 3) in order to prompt teachers for ideas that might be used as the basis of a lesson. The discussion with the teachers made evident at least three points of view, according to which one can answer the question: the mathematics knowledge, the knowledge specific for teaching and the pedagogical knowledge. According to such knowledge, specific interventions could be designed to introduce the students to the task, e.g. to think of baseball statistics: if a player passes from 200/300 to 201/301 his score increases. All of this can be considered as an example of a *meta-didactical praxeology* in that the task is stimulating the teachers' reflection, and the techniques are those that Sullivan used in the course to promote discussion. During this discussion, it is possible that the two communities of mathematics educators and teachers, respectively, shared a common theoretical framework, which would justify the techniques being discussed. For example, based on one's professional experience, the teachers might discuss why the initial question presents difficulties for many students and why the baseball example makes sense in a classroom and thus help overcome these difficulties. Moreover, the teachers may scaffold their

---

[4]The 'knowledge level' can be further decomposed in two components, i.e. *Technologies* and *Theories*. The provided description is enough for our purposes.

arguments within specific pedagogical discourses, for example stressing the necessity to foster the transition from everyday to scientific and formal concepts, according to a Vygotskian approach. The theoretical side of the *meta-didactical praxeology* also includes the reflection made by Sullivan on the possible reasons why the activity was a good illustration of the way teachers can become aware of MKT, an aspect that may have been highlighted within Sullivan's exposition.

Within *meta-didactical praxeologies*, what is under scrutiny is not the didactics in the classroom but the practices and the theoretical reflections developed in teacher education activities. Of course, they are the result of the interaction between the reflections of the community of researchers about the didactic praxeologies previously designed and developed, and the concrete practices used by the teachers in their professional activities.[5]

We now have the basic ingredients that allow us to introduce the core of our model. Looking at teacher education processes from a dynamic point of view, we initially identify two communities: that of researchers, who design and coach the activities, and that of the teachers, who are engaged in an education process. For the modeling purpose, let us distinguish two kinds of praxeologies: the *researcher praxeologies*[6] and the *teacher praxeologies*. The researchers and teachers praxeologies in some cases may be shared, but we assume that in general, when the teachers encounter the researchers for the first time at the beginning of the education process, they are not. Teacher education programme aim to develop teachers' existing praxeologies towards new ones, which consist of a blending of the two initial praxeologies. This evolution is the result of an interaction with the community of researchers and, for this reason, we call it a s*hared praxeology*. For example, from the discussion of different techniques to address a problem, new ones can be acquired by the teachers, with a suitable theoretical justification, thus replacing or integrating old techniques and so as to change the nature of the teacher's MKT. Also within this dynamic evolution are some external components, which may play a crucial role. A typical example is when the activity is developed in response to changes in the official curriculum or in external assessment expectations for students.

The community of researchers generally reflects upon the nature of, and reasons for, the changes produced by the teacher education programme and possibly shares such reflections with the community of teachers. This can result in *new researcher praxeologies*. Also the teacher praxeologies may change, and develop into *new teacher praxeologies*, a process that can repeat and further refine itself. A global illustration of this is provided in Fig. 1.

---

[5] This is true for activities with in-service teachers; in the case of prospective teachers, the second component may be missing but their beliefs are active and still constitute a powerful part of the component.

[6] Of course there may be more than one praxeology referring to researchers, as well as referring to teachers: in the text we will use either singular or plural (researchers praxeologies; teachers praxeologies). In particular the researchers have their own praxeologies as researchers, which concern the praxis and the logos of their researches; but they have also their praxeologies as teachers' educators, where the praxis and the logos concern the concrete way they coach these activities, because of their theories about teachers' educational processes.

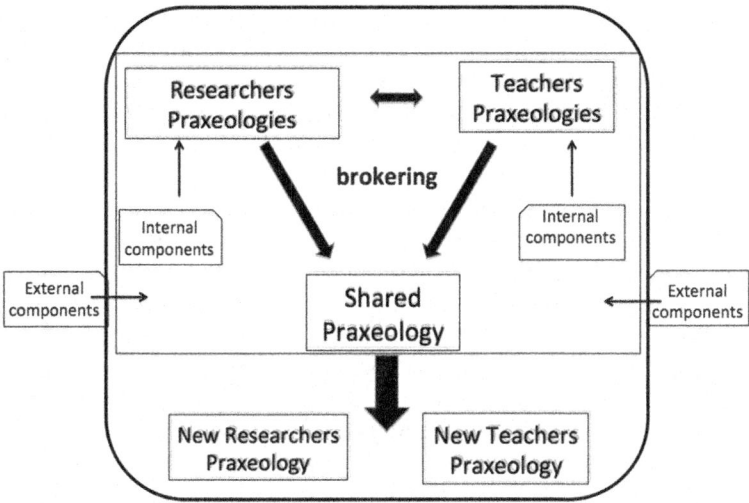

Fig. 1 The Meta-didactical Transposition model

Meta-didactical Transposition consists of a dynamic process through which, thanks to the dialectical interactions between two communities, both the didactic praxeologies of the community of researchers and of the teachers' community change within the institutional environment in which the two communities reside. This dialectical interaction leads to the development of a shared praxeology, which represents the core of our model. One of the main results of the dialectical interaction is the teachers' development of both a new awareness (on the cultural level) and new competences (on the methodological-didactical level, i.e. that of teaching practice), which lead them to activate, in their classrooms, a didactical transposition in line with recent educational trends. Therefore, the term 'meta-didactical' refers to the fact that important issues related to the didactical transposition of knowledge are faced at a meta-level.

Internal and External Components

An important feature of Meta-didactical Transposition is that some of the components of the two communities' praxeologies change their status over time. Typically they move from being *external* to becoming *internal* with respect to the community under scrutiny. To clarify this crucial point, which will be further discussed in the following sections, we give a brief example. Consider a community of teachers that starts an educational programme in which, due to some institutional situation (e.g. curriculum changes), a community of researchers introduces a specific ICT tool (e.g. a dynamic geometry software). Initially, the tool is an external component for the teachers (and possibly also for the researchers). However, at the end of the educational programme, it has become an internal component in their praxeologies, albeit possibly at different levels. Such an internalisation process, which happens via a Meta-didactical Transposition, defines a *meta-didactical trajectory*,[7] that is, the dynamic evolution of the teachers' education programme. For example, a technique (and the theory that justifies it) is initially in the hands of the researchers. Their aim is to make it shared within the community of inquiry as a technique and possibly, in addition, build an understanding of the theoretical arguments that justify its use. At the end of the process, the initial techniques (and possibly also the theoretical part) has become a new a set of shared techniques, as a result of the actions taken by the researchers and teachers.[8] As we will point out in the next section, this evolution is fostered by a dialectic interaction between these components.

The internal/external distinction is adapted from Clark & Hollingsworth (2002). They distinguish an external domain, located outside the teacher's personal world, from the internal domains, which "constitute the individual teacher's professional world of practice, encompassing the teacher's professional actions, the inferred consequences of those actions, and the knowledge and beliefs that prompted and responded to those actions" (, p. 951). Compared with their approach, our model emphasises the process of the teachers' professional evolution, according to which some of the external components become internal as a result of the process of Meta-didactical Transposition.

A Meta-didactical Transposition produces a dynamic change in the praxeologies of the community of teachers. Some components of the praxeologies of the community of researchers enter the praxeologies of the community of teachers as an outcome of the Meta-didactical Transposition. Presumably, also, the researcher praxeologies change as well, as a result of their encounters with the community of teachers. It is possible that some of these components may be external to both communities and it is the educational process that produces their transformation into internal components of the communities.

We will see below that this change is only one of the possible transformations that Meta-Didactical Transposition can produce within the praxeologies of the two interacting communities.

Brokering

The Meta-didactical Transposition model integrates the ideas of ATD with elements coming from other frameworks. The notion of brokering is an example; it is introduced because it describes the role that teachers and researchers often find

---

[7] The choice of this term to refer to teachers' education programmes is in tune with Simon definition of Learning Trajectory: "The Hypothetical learning trajectory consists of the goal for the students' learning, the mathematical tasks that will be used to promote students' learning and hypothesis about the process of the students' learning" (Simon 1995).

[8] This process has a common feature with the processes of instrumental genesis, as described by Trouche (2005). Space does not allow us to develop this issue.

themselves playing within the different communities. According to Rasmussen et al. (2009), a *broker* belongs to more than one community. Typically a teacher belongs to the community of mathematics experts, to that of her/his school teachers and to her/his classroom community:

> Brokers [...] are able to make new connections across communities of practice, enable coordination, and – if they are good brokers – open new possibilities for meaning (p. 109).

Brokers facilitate the transition of mathematical concepts from one community to the other (*boundary crossing*), which is accomplished by drawing on *boundary objects*:

> boundary objects are those objects that both inhabit several communities of practice and satisfy the informational requirements of each of them. (Bowker and Star 1999, p. 297)

Within Meta-didactical Transposition, brokering is a common habit and, frequently, researchers play a brokering role between the two communities involved. A good example of a typical boundary object is the baseball score used by Clark (cited in Sullivan 2008). Teachers can use such a boundary object to move students' thinking from the usual meaning of the score to a more mathematical comparison between two fractions (2/3 and 201/301). At the same time, used within an episode of teacher education, this example is a boundary object used by the researcher to move the teachers from the standard mathematical meaning of fractions to an everyday contextualised meaning that is useful for teaching. In this sense, the researcher makes a brokering action with respect to the teachers.

Double Dialectic

Another important element of our model is the double dialectic involved in the *Meta-Didactical Transposition*. The first dialectic is at the *didactic level* in the classroom in that it is between the personal meanings that students attach to a didactic situation, to which they are exposed in the didactic activity, and its scientific, shared sense (Vygotsky 1978). The second dialectic is at the *meta-didactic level*, which lies between, on the one hand, the interpretation that the teachers give to the first dialectic as a result of their personal meaning, which is a result of their praxeology and, on the other, the meaning that the first dialectic has according to the community of researchers, which results from researcher praxeology. The second dialectic corresponds to the *scientific shared meaning* of the first dialectic.

Typically, the second (meta-didactical) dialectic arises from a contrast between researcher praxeologies and teacher praxeologies and the first dichotomy engenders the second one as an outcome of a suitable meta-didactical trajectory, which is designed by the researchers. It is through this double dialectic that teacher praxeologies can change and align with the praxeologies of the researchers, which may cause a significant evolution of the teacher professional competences.

Examples from the Italian National and Regional Programmes

We will now discuss three different Italian teacher education programmes, which will show how the *Meta-didactical Transposition* model can help describe and analyse some important aspects of these programmes that that have not been adequately addressed in existing approaches. These aspects relate to the relationships between the Research, the Institutions, and Mathematics Education research. In particular, the three examples are meant to highlight different aspects of the model.

The first example aims to illustrate the various components of the model and the relationships between them. It concerns an ongoing Italian programme for teacher education called M@t.abel, which is based on an extensive use of ICT and, in particular, a purposeful, dedicated internet-based platform. We will show how the *Meta-Didactical Transposition* model allows the role of ICT to be adequately framed within teacher education. In particular, we will highlight the dynamics between the internal and external components and the brokering function of the tutors within the transposition.

The second example, MMLab-ER (Laboratories of Mathematical Machines for Emilia Romagna), shows how to exploit the potential of the *Meta-Didactical Transposition* in order to analyse the *development* of the project. In particular, the model allows us to identify and describe when, how and why the different components of the praxeologies changed during a teacher education programme.

The third example, which refers to the teacher education programme within the ArAl Project, is aimed at showing how, through the planning of an appropriate meta-didactical trajectory, it is possible to both highlight a first-level dialectic (didactical dialectic) and engender a second-level dialectic (meta-didactical dialectic), which enables teachers to develop a new awareness of their role in the classroom.

The three examples have been chosen being very different in scope, activities, and modalities of action. Considering the specific aspects that each example highlights can give a taste of the potential value of the MTD model in objectifying complex and different situations of teacher education projects.

A National Example of Meta-Didactical Transposition: The M@t.abel Project

The M@t.abel Project is a national teacher education programme for in-service mathematics teachers supported by the Ministry of Education. It started in 2006 and, to date, it has involved more than 10,000 secondary school teachers distributed across the whole of Italy. M@t.abel has its roots in the Italian *research for innovation* paradigm and, in particular, a previous project called '*Matematica per il cittadino*' (Mathematics for the citizen,[9] 2001–2005), which was elaborated within an

---

[9] http://www.umi-ciim.it/in_italia--28.html.

innovative curriculum in mathematics, from primary to secondary school (Anichini et al. 2004). The *Matematica per il cittadino* curriculum is based on the idea of the *mathematics laboratory*, intended as a methodology based on varied and structured activities. These activities aim toward the construction of meanings, in which the students can learn by doing, seeing, imitating and communicating with each other, under the guidance of the teacher, as in a Renaissance workshop. This methodology fosters close interaction between novices and experts, in the context of *cognitive apprenticeship*. This phrase "refers to the fact that the focus of the learning-through-guided-experiences is on cognitive and meta-cognitive, rather than on physical, skills and processes" (Collins et al. 1989, p. 458).

Although the current Italian National Curriculum mirrors in some respects the influence of the project *Matematica per il cittadino*, the school reality is quite far from being broadly influenced by the new perspectives, and the innovation is restricted to isolated cases (teachers, schools, or networks of schools) and to primary or middle, rather than secondary schools. For this reason, the M@t.abel project aims to improve school mathematics education at the secondary level, through the wide-scale dissemination of the ideas and didactic activities (i.e. the didactic praxeologies) of the *Matematica per il cittadino* curriculum. To reach this aim, a fundamental part of the M@t.abel project requires that teachers try out activities in their own classrooms that involve new didactic praxeologies (using a problem solving approach, tasks that involve discovering-conjecturing-arguing and proving, group work and discussions, and digital technologies).

In the M@t.abel Project, the institutional aspects are fundamental, because the Ministry of Education (MIUR), along with the Agency of School (Indire), is responsible for the project, which also includes researchers as members of the Scientific Committee. Researchers are called upon to plan all of the components of the teacher education programme as described above, to implement the educational meetings for the tutors and to prepare materials for the teachers.

In the project, the praxeologies of the researchers encounter teacher praxeologies by means of a two-step process. Each step can be considered as a Meta-Didactical Transposition process, where the first step concerns the tutors' education and the second one the teachers' education. The *tutors* are a small number of expert teachers who take part in research projects with University researchers. In many cases, tutors have previously participated in the *Matematica per il cittadino* project. In some cases, they may be teacher-researchers. The whole tutor community is formed at the beginning of their involvement in the project. In this first Meta-Didactical Transposition, the researchers play the role of *brokers* between the two communities (see Fig. 2a). In the second and far-reaching steps of the project, tutors themselves play the role of brokers in the Meta-Didactical Transposition that is directed toward a large number of teachers (see Fig. 2b).

Due to the limitations of space, in this Chapter we only describe this second process in more detail.

In order to develop shared praxeologies, the teachers are organised into *communities of inquiry*, composed of 15–20 teachers and supervised by tutors. Within this context, the tutors act as *brokers* between the two communities of teachers (involved

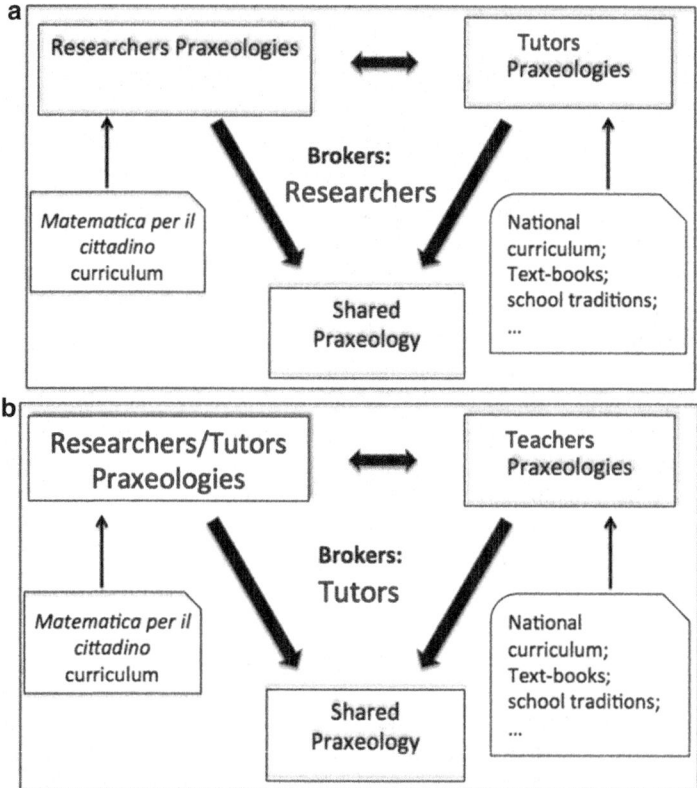

Fig. 2 (**a, b**) The two Meta-didactical Transpositions of the M@t.abel project

as learners in the educational programme) and of researchers (involved as designers of the programme). The tutors are confident with the innovative paradigms that emanate from research,[10] and they share with teachers their experience.

The communities of teachers work both remotely through an e-learning platform and during face-to-face meetings with tutors. Initially, the tutor outlines the spirit of the project, presents the activities during some meetings and asks the teachers to analyse them from a didactical point of view. Then, the tutor coordinates the groups of teachers remotely through synchronous meetings (using screen sharing) and asynchronous discussions (emails, forums). Having shared activities and methods, the teachers choose four activities and experiment with them in their own classrooms. These trials are a fundamental part of the teacher education programme and,

---

[10] According to the Italian paradigm of 'research for innovation', in this second step the tutors praxeologies may be assimilated to the researchers ones: as said above, in many cases the tutors are teachers-researchers, i.e. are experienced with research studies and methodologies, having been part of research teams in mathematics education for many years. Of course this is not always the case. For the purpose of the paper, we privilege clarity, taking the risk of over-simplification.

during the experimentations, the tutor asks the teachers to carefully observe their students' processes,[11] and to record their notes in a logbook, which is uploaded on the platform. More precisely, in the logbook the teacher is asked to:

- Make explicit the principal conceptual points of the activity;
- Describe the classroom experience and the methodology followed (worksheets, groupwork, software, ...);
- Monitor how the students participate in the activities and appreciate them;
- Signal the main student difficulties;
- Comment on the evaluation of the tasks and their conclusions.

In the researcher praxeologies, the logbook is meant to be a tool that helps the teachers plan, monitor, and control their own work and, in particular, organise the observation of what happens in the classroom, focusing their attention on processes rather than on products. In this sense, it can enable the teachers to orient or re-orient their didactic practice, and contribute to improving their teaching practice by means of self-reflection. Furthermore, logbooks can be a valuable means of exchange between teachers working around the same mathematics topic, or at the same school level, and may provide information tools for external observers. As an institutional constraint, the project enables the participating teachers to gain certification that is useful within their career progression; also, the completion of the logbook is a required element of this accreditation.

Over their years of practice, each teacher has developed her/his own individual praxeologies comprising tasks, techniques and theoretical discourse. Depending on the individual teachers, the initial praxeologies of the researchers and the teachers can be far apart. For instance, some of the teachers involved in the programme often used quite traditional tasks and techniques, consisting of lectures, exercises and applications. Their (often implicit) theoretical discourse that justified these choices was based on traditional textbooks and an old curriculum. During the educational programme, the praxeologies could evolve and change through meta-didactical trajectories, and develop toward shared praxeologies (e.g. laboratory practices and use of ICT).

Figure 3 contains a portion of a teacher's logbook, in which the teacher presents a synthesis documenting her consciousness about the changes in her praxeologies related to the teaching of geometry. The choice of the tabular format for the logbook is made to this particular teacher. In the third column we can get some evidence of a shared praxeology. We can notice technical terms originating from the researcher/tutor praxeology, which are also expressed in the *Matematica per il cittadino* curriculum, such as 'guiding students in discovering properties' (see the reference to cognitive apprenticeship above), 'discussing with them about descriptions, definitions, properties', 'institutionalising knowledge', and so on.

Unfortunately, many teachers involved in the project did not or could not annotate daily their logbook (Rapporto PON M@t.abel 2009–10). These teachers only wrote up their logbook at the end of their experience in the project, as a sort of compulsory homework that was required by the system (Institutional constraint).

---

[11] Contrasted with their products. e.g. students' reasoning, arguments, difficulties, and so on.

	What I thought before the teaching experiment on quadrilaterals	What I think now
Working modalities with students	Work in pairs with concrete materials	Work in pairs in laboratory with GeoGebra software
Teacher's role	Teaching, explaining, exemplifying	Guiding students in discovering properties of quadrilaterals Discussing with them about description, definition, properties, Coordinating discussions giving stimuli, ordering conjectures Institutionalising knowledge
Tools and their functions	Concrete materials (paper and pencil) as a model where constructing quadrilaterals according to their symmetry properties	Software GeoGebra for constructing the same model of concrete materials (Fig 4) Paper sheets forevery activity Instruction for constructing quadrilaterals in GeoGebra

Fig. 3 Excerpt from a teacher's logbook

For some, time pressures seem to have prevented thoughtful writings. They describe concisely the experiment as a finished product. Consequently, there is no important information in the log books about the real processes occurring during the development of the teaching experiment and there are only few reflections about the difficulties encountered and the planned changes. For these teachers, the interaction with the tutor was less dialogic and limited to the use of forums, email, and online resources in the platform.

As mentioned above, the Meta-didactical Transposition in the case of the M@t.abel Project has its strength in the use of a *platform* for synchronous and asynchronous activities among teachers. The platform is the environment that gives new techniques to teachers, influencing and supporting them in changing their praxeologies. In particular, if a teacher has worked for many years in a traditional and isolated way, now she is forced to discuss new methodological issues through ICT. For instance, the GeoGebra software (and the figure constructed with the software, like the one in Fig. 3) can be used as a boundary object between the community of tutors and that of teachers.

The platform, together with the brokering function of the tutors, aims to build a community of teachers with *shared praxeologies*. Besides being a *communication infrastructure*, allowing synchronous and asynchronous interactions, for sharing ideas, materials and methods, the platform works also as a *representational infrastructure* (Hegedus and Moreno-Armella 2009), fostering the use of a shared

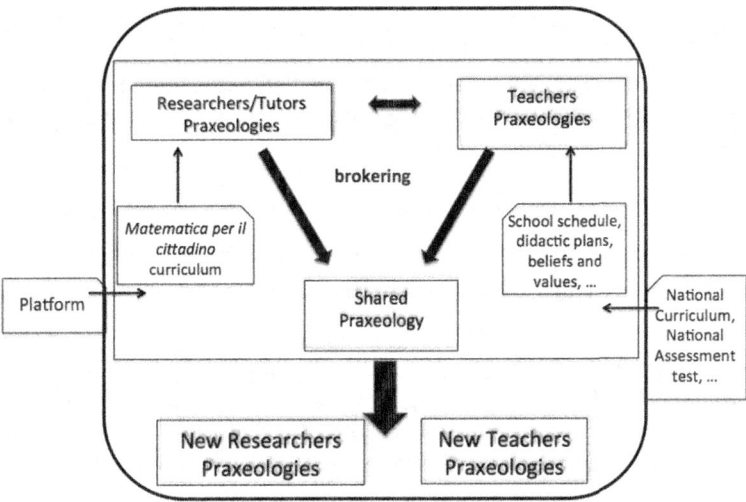

Fig. 4 Internal and external components in the Meta-didactical Transposition of M@t.abel

desktop where the teachers can work together on-line on the same topic or mathematical object. In the Meta-didactical Transposition model, the platform constitutes an example of an *external component* (for both researchers/tutors' and teachers' initial Praxeologies) that becomes an *internal* one over the course of the Project. Finally, the model includes two important effects of Meta-Didactical Transposition, which are the changes brought about by the project both in the researcher praxeologies and in teacher praxeologies. The change in the researcher praxeologies occurs through the researchers' ongoing reflection, which is prompted by considering the evolution of the system over time and the analysis of the internal/external components. Figure 4 provides an overall picture of the Meta-didactical Transposition within the M@t.abel project (the second step):

As a picture of a dynamic process, Fig. 4 cannot capture temporal evolution. If we imagine the model evolving over time, we can focus our attention on *occurred* and *not-yet-occurred evolutions*. Concerning the *occurred evolutions* in M@t.abel, we find two external components that become internal ones: the platform (as described above), and the *Matematica per il Cittadino* curriculum (which was internal for researchers/tutors' praxeologies but not for those of the teachers). The first component is a technical component and the second is a part of the theoretical discourse, which justifies certain tasks and techniques.

As mentioned above, unfortunately the logbook often constitutes a case of *not-yet-occurred* evolution on a large scale. For many teachers in fact the logbook did not function as a helpful day-to day observation tool, which means that it did not become an internal component in their praxeology. Instead, for a small number of them, it was a component that became internal thanks to their participation in the project.

In general, by looking at the evolutions in terms of the internal and external components, researchers can identify those features of the teacher education process that are in need of further reflection and work. For instance, the logbook tool is currently under investigation by the researchers, in order to understand why it did not work as expected and to set up suitable changes. This kind of consideration is part of a *new researchers' praxeology*. Further considerations of this aspect of the model will be presented in the next examples.

The Evolution of Researcher Praxeologies Over Time: The Example of the MMLab-ER Project

In this second example of a teacher education programme, we show the potential of the *Meta-didactical Transposition* model for studying the *evolution of praxeologies over time*. Specifically, we use the model to identify and describe when, how and why some different components of praxeologies changed during the MMLab-ER teacher education programme. We also use the model to plan and control the variation in praxeologies.

MMLab-ER (http://www.mmlab.unimore.it/site/home/progetto-regionale-emiliaromagna.html, Martignone 2010) is a regional project that responds to national and international standards on IBSE (Inquiry Based Science Education; see Rocard et al. 2007). It aims to construct a network of competent practising teachers using the *mathematics laboratory* method (in the sense introduced in the previous section). In this project, old and new tools are involved (for example, reconstructions of historical mathematical tools and technologies). The compass is a well-known mathematical tool but, in the history of mathematics, several other mathematical machines (e.g. pantographs for geometric transformations, curve creators, perspectographs) have been designed and used for theoretical and practical purposes (Bartolini Bussi et al. 2010). In the MMLab-ER project, teachers and students work both with real mathematical machines and with virtual reconstructions of them, created by means of dynamic geometry software.

The laboratory sessions with mathematical machines, guided by a specific methodology and particular tasks (Bartolini Bussi et al. 2011; Martignone 2011), are suitable environments for the development of fundamental mathematical activities such as problem solving, production of conjectures, argumentation processes and generation of proofs within Euclidean geometry. This is one of the underlying assumptions of the MMLab-ER project. To date, approximately 200 teachers have participated in the first cycle (mainly grades 4–8) and the second cycle (grades 9–12, in high schools and vocational schools). The researchers worked as teacher educators with small groups of teachers (15–25 teachers from each of the eight Italian provinces involved) in both face-to-face sessions (28 h) and through an e-learning platform and email. The project began in 2008 with a 2-year period of regional financial support and then, at the beginning of 2012, as a result of new financial support, it recommenced with other teachers and schools.

This development over 3 years, which includes a 1-year break during which time the researchers analysed the project results, enables study of the evolution of praxeologies in the MMLab-ER project. By means of the *Meta-Didactical Transposition* model, we analysed how the different components of the *researcher praxeologies* have developed from the beginning of the project until now. The MMLab research group analysed the experiences carried out both during the teacher education programmes and the teaching experiments, and have modified some of their praxeologies. The *researcher praxeologies* arose from a dynamic evolution of the relation between research and teaching that characterises the *Italian Research for Innovation* (Arzarello and Bartolini Bussi 1998). At the beginning of the Meta-Didactical Transposition process, the researchers had their own praxeologies linked to their studies of students' activities in the classroom and their experiences with and studies of teacher education. In the former, the *task* is to study the educational potential of the laboratory activities with mathematical machines and the *techniques* involve the design and analysis of activities for primary and secondary school students. In the latter, the *task* is to design activities that shift teachers' attentions to the processes of exploration, the resulting conjectures and the constructions of proof by means of laboratory sessions involving the mathematical machines. The *techniques* concern the development of tasks for teachers that include, for example, the selection of suitable educational paths to be discussed and the analysis of different teaching experiments. The *theoretical discourse* that describes, explains and justifies the techniques of these praxeologies is based on studies of mathematics teaching and learning by means of laboratory activities with mathematical machines (within the theoretical framework of *semiotic mediation* (Bartolini Bussi and Mariotti 2008). The *Meta-didactical Transposition* model is useful in describing and analysing the evolution over time of the different components of these researcher praxeologies. We can identify the aspects that do not change and describe how the levels of praxis and logos are modified. Concerning the level of praxis, some *techniques* were improved, such as: how new activities were introduced and elaborated with the teachers; the modification of some tasks for teachers; modification of the classroom tasks, taking into account what was discussed during the teacher education programme (*shared praxeologies*); and refining the tools for analysing teachers' and students' worksheets, logbooks, video, etc. In addition, the *theoretical discourse* was improved by refining some theoretical tools. After the first year, in order to study the exploration of the mathematical machines carried out by teachers and students in more depth, cognitive studies concerning mathematical machines were developed that identified and analysed the argumentation processes involved (Antonini and Martignone 2011). After 2 years of the project, the researchers analysed all of the documentation, which included the videos, worksheets about the laboratory activities carried out by teachers and students, teachers' reflections collected in the logbooks, and the final reports of the teaching experiments. New research was carried out in order to identify, study and characterise the main features of the MMLab-ER teacher education programme (Garuti and Martignone 2010). These studies showed that the project's main results were not only the dissemination of innovative teaching methods, but also the design and testing of activities that seemed to develop teachers' skills in analysing the cultural aspects involved in the laboratory activities with mathematical machines. In order to interpret the kind of *teacher knowledge* the

MMLab-ER programme had developed, the researchers referred to the aforementioned studies about *Mathematical Knowledge for Teaching* (MKT) (described earlier). This construct was used to identify specialised teacher knowledge related to the non-pedagogical content. In particular, we found that the teachers' Specialised Content Knowledge (SCK), which is linked to the *cultural analysis of contents* (Boero and Guala 2008), improved through the development of new praxeologies involving reflection on teaching and learning activities. A teacher has (or can acquire) SCK linked to the cultural analysis of contents if s/he come to appreciate the potential of, and can analyse the cultural aspects (e.g. attention and analysis of historical-epistemological and cognitive aspects) related to some specific mathematical content (Garuti and Martignone 2010). The identification of this specialised content knowledge modified the *logos* level of *researcher praxeologies,* enriching them with a new theoretical construct. Today, these new praxeologies, with the praxeologies of the new teachers involved, are the starting point of the *Meta-didactical Transposition processes* that are going on in the second part of the MMLab-ER teacher education programme.

The Double-Level Dialectic: The Teacher Education Programme Within the ArAl Project

As we stated before, this third section is aimed at exemplifying how the *Meta-didactical Transposition* model highlights a typical aspect of teacher education programmes, which is the activation, through appropriate *meta-didactical trajectories*, of a *first-level dialectic* and, at the same time, the engendering of a *second-level dialectic* that enables teachers to acquire a new awareness of their role in the classroom.

The teacher education programme we present here is the *ArAl Project*, whose main objective is to foster a linguistic and constructive approach to early algebra (Malara and Navarra 2003) within an integrated teacher education programme (Cusi et al. 2010). The model for teacher education developed within the ArAl project resonates particularly with research carried out by Mason (1998, 2002) and Jaworski (1998, 2003, 2006). This programme is based on the hypothesis that observation and critical-reflective study of class processes, activated both individually and among communities of inquiry, is a necessary condition to foster teachers' development of awareness (Mason 2008) about the 'subtle sensitivities' that could guide their future choices and determine their effective action in the classroom. Another fundamental hypothesis is that giving teachers the possibility to analyse and interpret the activities they conduct in their classrooms, referring to specific theoretical lenses for the observation of the role they play, can foster a shift of attention for teachers as they reflect on their own practice, enabling them to focus not only on students' difficulties, but also on the interrelation between the attitudes and behaviours of teachers and students.

For this reason, the chosen methodology for our work with the teachers, which reflects the chosen *meta-didactical trajectory*, is characterised by these main three aspects: (1) in order to foster the development of a *community of inquiry*, the teachers involved in the same teaching experiment are associated with a mentor-researcher with whom the teachers engage in face-to-face work as well as email exchanges, which becomes the starting point of a *dialectic interaction between the two communities*; (2) teachers are involved in activities of theoretical study, aimed at providing them with theoretical and methodological tools useful for interpreting, through new lenses, their own actions in the classroom (in this way, teachers and researchers start to refer to a *shared praxeology*); (3) teachers are involved in a complex activity of critical analysis of the transcripts of audio-recordings concerning classroom processes and associated reflections, which is carried out by developing what we call *Multi-commented transcripts* (MT). The experimenters-teachers send the transcripts, together with their own comments and reflections, to mentors-researchers, who make their own comments and send them back to the authors, to other teachers involved in similar activities, and sometimes to other researchers (at this level, the dialectic interaction between the two communities is particularly intense). Often, the authors make further interventions in this cycle, commenting on comments or inserting new ones (see Malara 2008; Cusi et al. 2010; Malara and Navarra 2011).

The MT methodology, which helps teachers reflect on the activities they carried out in their classrooms, reveals their attitudes and behaviours as well as the effects of their interventions on their students. Thus, they highlight: the contrast/interaction between the personal sense their students attribute to the activities and the institutional meaning of the same activities (the *first-level dialectic*); and the role they play in fostering (or not) students' development of a personal sense, which is in tune with the institutional meaning of the activities.

At the same time, the researchers' analyses of the reflections proposed by teachers in the MT and the identification of a possible contraposition between teachers' and researchers' comments, enable a *second-level dialectic* to be highlighted in relation to both: (a) the possible different interpretations of the dynamics realised during class activities and (b) the possible different uses of the same theoretical lenses made by teachers and researchers in their analysis of classroom processes. Moreover, through the *a posteriori* analysis of the different comments proposed on the MT, the teachers can also become aware of this second-level dialectic.

The reflection carried out with the teachers involved in the project provides an opportunity to notice how the tension developed as a result of this *double-level dialectic* produces an evolution in the interrelations between the different components of the praxeologies involved within the process of *Meta-didactical Transposition*. In particular: it fosters the development of *new teachers' praxeologies*, related both to roles they should activate in their classrooms and to ways of pursuing their professional development; it enables researchers to hypothesise a possible refinement of the theoretical lenses for the observation of class processes and the possibility for further evolutions of the methodology to be adopted in the work with teachers, therefore fostering an enhancement of the chosen *meta-didactical trajectory*.

Discussion

In this chapter, we introduced the *Meta-didactical Transposition* model as a theoretical tool to objectify and describe the complex dialectic between research and institutional dimensions of teacher education programmes. This model is based on Chevallard's ATD, but was complemented by additional constructs in order to account for some dynamic features occurring when teachers and researchers are engaged in teacher education activities. We outlined and analysed three examples from the Italian context, in order to illustrate how this model can be productively used to analyse diverse types of teacher education programmes.

The first example discussed is the M@t.abel Project. By means of the Meta-didactical Transposition model, we were able to observe the dynamics between external and internal components (Fig. 2), and to identify strong and weak points of the project that we had not noticed as clearly before. We found two relevant *occurred evolutions* from external to internal components in the Meta-didactical Transposition: the *Matematica per il cittadino* curriculum and the platform. The curriculum was at first internal to the researcher praxeologies, and external to the teacher ones. The passage from external to internal was fostered by the brokering actions of the tutors. The platform is a technological device, which is initially external to both teacher and researcher praxeologies. This platform not only enabled the communication of ideas, feelings and didactic plans between teachers and tutors, but it also opened up a concrete space for the development of didactic activities that involved the use of software. The logbook constitutes a more delicate element. It too was initially an external component to the teacher praxeologies. Throughout the project, for some teachers, the logbook became an internal component of the shared praxeology as they used it to organise their classroom observation and plan their work better. This dynamic transformation from an external to an internal component did not occur for all teachers. Many of them, in fact, wrote their logbook at the end of the whole project, despite the constant prompts of their tutors, and for these teachers the logbook remained an external component that did not alter their praxeologies.

In the second example, we highlighted how the Meta-didactical Transposition model offers a framework that enabled us to analyse the evolution of praxeologies over time in the MMLab-ER project. In particular, we focused on the researcher praxeologies related to the changing of logos and praxis. The Meta-didactical Transposition model was useful in analysing not only how the praxeologies changed, but also why they were modified in relation to the teacher education programme development. Furthermore, the model allowed us to objectify, through the identification of the researcher and teacher praxeologies, their evolutions over time, while also maintaining a systemic view. At the end of the Meta-didactical Transposition process, both researchers and teachers developed new praxeologies, changing some of their techniques as well as their ways of explaining and justifying these techniques. This could become the new starting point of a fresh Meta-didactical Transposition process.

The third example revealed the essential role played by an appropriate 'meta-didactical trajectory' in: helping teachers become aware of the first-level dialectic related to the contrast/interaction between the personal sense their students attribute to the activities and the institutional meaning of the same activities; and in enabling researchers and teachers highlight the second-level dialectic which is related to the contrast/interaction between the different interpretations of the dynamics realised in the classrooms, given by teachers and researchers, in relation to specific theoretical lenses. Moreover, this example showed that the tension developed out of this to the double-level dialectic could foster the evolution of both researcher and teacher praxeologies. In particular, it highlighted the strict interrelation between this evolution and the chosen methodology of work with teachers. Involving teachers in the critical-reflective study of class processes, in fact, means, as Jaworski (2003) states, making them shift from a context of perpetuation of existing practices (the communities of teachers within a school) to a new context, typical of communities of inquiry, characterised by "the importance attached to meta-knowing through reflecting on what is being or has been constructed and on the tools and practices involved in the process" (p. 256).

Globally, the three examples must be conceived as pieces of a puzzle, which shed light on specific aspects of the model. They illustrated how the Meta-didactical Transposition model responds to the challenge of studying "how different approaches to teacher development have different effects on particular aspects of teachers' pedagogical content knowledge" (Ball et al. 2008, p. 405). We have already noticed that a similar construct was used by Clark & Hollingsworth (2002) to underline that teacher education programmes can produce changes in teachers' teaching strategies, "that represented in themselves new pedagogical knowledge" (*ibid.*, p. 953) for those teachers and that "were subsequently put into practice" (*ibid.*, p. 954). In other words, teacher education programmes can produce changes in teacher praxeologies. In fact, our model is similar but not identical to that of Clark & Hollingsworth, since ours underscores the interdependence of such changes with the institutions (according to the ATD approach), and focuses on the Meta-didactical components of the processes, which remain more implicit in Clark & Hollingsworth's approach.

The Meta-didactical Transposition model is deeply related also to the MKT construct. Both models focus on the intertwining of the theoretical knowledge and the common practices needed by teachers in their work, but each stresses different aspects of this intertwining. The MKT focuses on the structure of the mathematical knowledge for teaching while the Meta-didactical Transposition stresses more the dynamic evolution of its components. In particular, as illustrated in the examples above, it shows the relevance of the double-level dialectic and of the evolution from external to internal components in promoting and supporting the processes of teacher education.

As the MKT model refines Shulman's PCK model (Pedagogical Content Knowledge, 1986), so the Meta-didactical Transposition model enriches the MKT one. In fact, it essentially adds dynamicity to its description, allowing a transition from the snapshot illustrated by the fixed categories of MKT to the film of the Meta-didactical Transposition model as shown in Fig. 1. More precisely, our model introduces the temporal dimension, the double level dialectic and the internal-external dynamics, which are all elements that allow us to focus on the dynamic evolution of teachers' educational programmes, which eventually produce the specificity of the different "domains of mathematical knowledge for teaching", to use Ball's terminology (Ball et al. 2008, p. 403). The lens of the meta-didactical praxeology allows the dynamicity of the process to be made evident. In fact, the model helps reveal the evolution of MKT, which can be hard to see because it is so imbedded in its particular institutional context. In the researchers' and teachers' hands, the MDT model can become a conscious tool in order to plan, develop and accomplish teacher educational programmes taking into account the complex interplay and dynamics between their components.

The study of this potential efficacy also introduces a fresh and promising strand of future investigation, which could produce further results concerning the nature of the domains of mathematical knowledge for teaching and the underlying processes of teachers' education.

References

Anichini, G., Arzarello, F., Ciarrapico, L., & Robutti, O. (Eds.). (2004). *New mathematical standards for the school from 5 through 18 years, The curriculum of mathematics from 6 to 19 years, on behalf of UMI-CIIM, MIUR (edition for ICME 10)*. Bologna: UMI.

Antonini, S., & Martignone, F. (2011). Argumentation in exploring mathematical machines: A study on pantographs. In *Proceedings of the 35th conference of the International Group for the Psychology of Mathematics Education* (Vol. 2, pp. 41–48). Ankara.

Arzarello, F., & Bartolini Bussi, M. G. (1998). Italian trends in research in mathematics education: A national case study in the international perspective. In J. Kilpatrick & A. Sierpinska (Eds.), *Mathematics education as a research domain: A search for identity* (pp. 197–212). Dordrecht: Kluwer.

Ball, D. L., & Bass, H. (2003). Toward a practice-based theory of mathematical knowledge for teaching. In B. Davis & E. Simmt (Eds.), *Proceedings of the 2002 annual meeting of the Canadian Mathematics Education Study Group* (pp. 3–14). Edmonton: CMESG/GDEDM.

Ball, D. L., Thames, M. H., & Phelps, G. (2008). Content knowledge for teaching: What makes it special? *Journal of Teacher Education, 59*(5), 389–407.

Bartolini Bussi, M. G., Taimina, D., & Isoda, M. (2010). Concrete models and dynamic instruments as early technology tools in classrooms at the dawn of ICMI: From Felix Klein to present applications in mathematics classrooms in different parts of the world. *ZDM Mathematics Education, 42*, 19–31. Springer.

Bartolini Bussi, M. G., Garuti, R., Martignone, F., & Maschietto, M. (2011). Tasks for teachers in the MMLab-ER Project. In *Contribution to 35th conference of the International Group for the Psychology of Mathematics Education Research Forum* (Vol.1, pp. 127–130). Ankara.

Bartolini Bussi, M. G., & Mariotti, M. A. (2008). Semiotic mediation in the mathematics classroom: Artefacts and signs after a Vygotskian perspective. In L. English, M. Bartolini, G. Jones, R. Lesh, B. Sriraman, D. Tirosh (Eds.), *Handbook of International research in Mathematics education* (pp. 746–783). New York: Routledge Taylor & Francis Group.

Bass, H. (2005). Mathematics, mathematicians, and mathematics education. *Bulletin of the American Mathematical Society, 42*(4), 417–430.

Boero, P., & Guala, E. (2008). Development of mathematical knowledge and beliefs of teachers. In Sullivan, P. & Wood, T. (Eds.), *The international handbook of mathematics teacher education* (Vol. 1, pp. 223–244). Purdue University/Sense Publishers: West Lafayette.

Bosch, M.,& Chevallard, Y. (1999). La sensibilité de l'activité mathématique aux ostensifs. Objet d'étude et problematique. *Recherches en Didactique des Mathématiques, 19*(1), 77–124.

Bowker, G. C., & Star, S. L. (1999). *Sorting things out: Classification and its consequences*. Cambridge, MA: MIT Press.

Chevallard, Y. (1985). *Transposition Didactique du Savoir Savant au Savoir Enseigné*. Grenoble: La Pensée Sauvage Éditions.
Chevallard, Y. (1992). Concepts fondamentaux de la didactique: perspectives apportées par une approche anthropologique. *Recherches en Didactique des Mathématiques, 12*(1), 73–112.
Chevallard, Y. (1999). L'analyse des pratiques enseignantes en théorie anthropologique du didactique. *Recherches en Didactique des Mathématiques, 19*(2), 221–266.
Clark, D., & Hollingsworth, H. (2002). Elaborating a model of teacher professional growth. *Teaching and Teacher Education, 18*, 947–967.
Collins, A., Brown, J. S., & Newman, S. E. (1989). Cognitive apprenticeship: Teaching the craft of reading, writing, and mathematics. In L. B. Resnik (Ed.), *Knowing, learning and instruction* (pp. 453–494). Hillsdale, NJ: Lawrence Erlbaum.
Cusi, A., Malara, N. A., & Navarra, G. (2010). Early algebra: Theoretical issues and educational strategies for bringing the teachers to promote a linguistic and metacognitive approach to it. In J. Cai & E. Knuth (Eds.), *Early algebrailization: Cognitive, curricular, and instructional perspectives* (pp. 483–510).
Drijvers, P., Kieran, C., & Mariotti, M. A. (2010). Integrating technology into mathematics education: Theoretical perspectives. *Mathematics education and technology-rethinking the terrain. New ICMI study series* (Vol. 3, Part 2, pp. 89–132).
Even, R., & Ball, D. L. (2009). *The professional education and development of teachers of mathematics*. Dordrecht: Springer.
García, F. J., Gascón, J., Ruiz Higueras, L., & Bosch, M. (2006). Mathematical modelling as a tool for the connection of school mathematics. *ZDM, 38*(3), 226–246.
Garuti, R. & Martignone, F. (2010). La formazione insegnanti nel progetto MMLab-ER, in Scienze e tecnologie in Emilia Romagna: Un nuovo approccio per lo sviluppo della cultura scientifica e tecnologica nella Regione Emilia-Romagna, Azione 1 a cura di Martignone, 73–97, Tecnodid.
Guin, D., Ruthven, K., & Trouche, L. (Eds.). (2005). *The didactical challenge of symbolic calculators: Turning a computational device into a mathematical instrument*. New York: Springer.
Hegedus, S. J., & Moreno-Armella, L. (2009). Intersecting representation and communication infrastructures. *ZDM: The International Journal on Mathematics Education., 41*(4), 399–412.
Hoyles, C., & Lagrange, J.-B. (Eds.). (2009). *Mathematical education and digital technologies: Rethinking the terrain* (pp. 439–462). New York: Springer.
Jaworski, B. (1998). Mathematics teacher research: Process, practice and the development of teaching. *Journal of Mathematics Teacher Education, 1*, 3–31.
Jaworski, B. (2003). Research practice into/influencing mathematics teaching and learning development: Towards a theoretical framework based on co-learning partnerships. *Educational Studies in Mathematics, 54*, 249–282.
Jaworski, B. (2006). Theory and practice in mathematics teaching development: Critical inquiry as a mode of learning in teaching. *Journal of Mathematics Teacher, 9*, 187–211.
Jaworski, B. (2008). Building and sustaining inquiry communities in mathematics teaching development. In K. Krainer & T. Wood (Eds.), *Participants in mathematics teacher education* (pp. 309–330).
Lagrange, J.-B., Artigue, M., Laborde, C., & Trouche, L. (2003). Technology and mathematics education: A multidimensional study of the evolution of research and innovation. In A. J. Bishop, M. A. Clements, C. Keitel, J. Kilpatrick, & F. K. S. Leung (Eds.), *Second international handbook of mathematics education* (pp. 239–271). Dordrecht: Kluwer.
Malara, N. A. (2008). Methods and tools to promote a socio-constructive approach to mathematics teaching in teachers. In B. Czarnocha (Ed.), *Handbook of mathematics teaching research* (pp. 89–102). Rzeszòw: University Press.
Malara, N. A., & Navarra, G. (2003). *ArAl project: Arithmetic pathways towards pre-algebraic thinking*. Bologna: Pitagora.
Malara, N. A, & Navarra, G. (2011). Multicommented transcripts methodology as an educational tool for teachers involved in early algebra. In M. Pytlak, E. Swoboda (Eds.), *Proceedings of CERME 7* (pp. 2737–2745). Poland: University of Rzezsow.

Malara, N. A., & Zan, R. (2002). The problematic relationship between theory and practice. In L. English (Ed.), *Handbook of international research in mathematics education* (pp. 553–580). Mahwah: LEA.

Martignone, F. (Ed.) (2010). MMLab-ER: Laboratori delle macchine matematiche per l'Emilia Romagna, Az. 1. In USR E-R & Regione Emilia-Romagna, Scienze e Tecnologie in Emilia-Romagna (pp. 16–210). Napoli: Tecnodid, [Adobe Digital Editions version]. Retrieved from: http://www.mmlab.unimore.it/online/Home/ProgettoRegionaleEmiliaRomagna/Risultatidel Progetto/LibroProgettoregionale/documento10016366.html

Martignone, F. (2011). Tasks for teachers in mathematics laboratory activities: A case study. In *Proceedings of the 35th Conference of the International Group for the Psychology of Mathematics Education* (Vol. 3, pp. 193–200). Ankara: PME.

Mason, J. (1998). Enabling teachers to be real teachers: Necessary levels of awareness and structure of attention. *Journal of Mathematics Teacher Education, 1*, 243–267.

Mason, J. (2002). *Researching your own practice: The discipline of noticing.* London: The Falmer Press.

Mason, J. (2008). Being mathematical with and in front of learners. In B. Jaworski & T. Wood (Eds.), *The mathematics teacher educator as a developing professional* (pp. 31–55). Rotterdam: Sense Publishers.

Rapporto PON M@t.abel 2009–10: Rapporto sui risultati preliminari sugli effetti del Programma PON M@t.abel 2009/2010 http://www.invalsi.it/invalsi/ri/matabel/Documenti/Report_Diari_di_bordo.pdf

Rasmussen, C., Zandieh, M., & Wawro, M. (2009). How do you know which way the arrows go? The emergence and brokering of a classroom mathematics practice. In W.-M. Roth (Ed.), *Mathematical representations at the interface of the body and culture* (pp. 171–218). Charlotte: Information Age Publishing.

Rocard, M., Csermely, P., Jorde, D., Lenzen, D., Walberg-Henriksson, H., & Hemmo, V. (2007). *Science education now: A renewed pedagogy for the future of Europe.* European Commission. http://ec.europa.eu/research/science-society/document_library/pdf_06/report-rocard-on-science-education_en.pdf

Schoenfeld, A. (1998). Toward a theory of teaching in context. *Issues in Education, 4*(1), 1–94.

Schön, D. A. (1987). *Educating the reflective practicioner: Toward a new design for teaching and learning in the professions.* San Francisco: Jossey-Bass.

Sfard, A. (2005). What could be more practical than good research? On mutual relations between research and practice of mathematics education. *Educational Studies in Mathematics, 58*(3), 393–413.

Shulman, L. S. (1986). Those who understand: Knowledge growth in teaching. *Educational Researcher, 15*(2), 4–14.

Simon, M. (1995). Reconstructing mathematics pedagogy from a constructivist perspective. *Journal for Research in Mathematics Education, 26*, 114–145.

Sullivan, P. (2008). Knowledge for teaching mathematics: An introduction. In P. Sullivan & T. Wood (Eds.), *The international handbook of mathematics teacher education, Vol. 1, Knowledge and beliefs in mathematics teaching and teaching development* (pp. 1–12). Rotterdam: Sense Publisher.

Trouche, L. (2005). Instrumental genesis, individual and social aspects. In D. Guin, K. Ruthven, & L. Trouche (Eds.), *The didactical challenge of symbolic calculators.* Berlin: Springer. Cap. 6.

Vygotsky, L. S. (1978). *Mind in society. The development of higher psychological processes.* Cambridge, MA/London: Harvard University Press. Edited by M. Cole, V. John-Steiner, S. Scribner & E. Souberman.

Wenger, E. (1998). *Communities of practice: Learning, meaning, and identity.* Cambridge: Cambridge University Press.

Wood, T. (Ed.). (2008). *The international handbook of mathematics teacher education* (Vol. 1–4). West Lafayette: Purdue University/Sense Publishers.

Frameworks for Analysing the Expertise That Underpins Successful Integration of Digital Technologies into Everyday Teaching Practice

Kenneth Ruthven

Abstract This chapter examines contemporary frameworks for analysing teacher expertise which are relevant to the integration of digital technologies into everyday teaching practice. It outlines three such frameworks, offering a critical appreciation of each, and then explores some commonalities, complementarities and contrasts between them: the Technological, Pedagogical and Content Knowledge (TPACK) framework (Koehler & Mishra, *Contemporary Issues in Technology and Teacher Education, 9*(1), 2009); the Instrumental Orchestration framework (Trouche, L. (2005). Instrumental genesis, individual and social aspects. In D. Guin, K. Ruthven, & L. Trouche (Eds.), *The didactical challenge of symbolic calculators: Turning a computational device into a mathematical instrument* (pp. 197–230). New York: Springer.); and the Structuring Features of Classroom Practice framework (Ruthven, *Education & Didactique, 3*(1), 2009). To concretise the discussion, the use of digital technologies for algebraic graphing, a now well established form of technology use in secondary school mathematics, serves as an exemplary reference situation: each of the frameworks is illustrated through its application in a study of teacher expertise relating to this topic (respectively Richardson, *Contemporary Issues in Technology and Teacher Education, 9*(2), 2009; Drijvers, Doorman, Boon, Reed, & Gravemeijer, *Educational Studies in Mathematics, 75*(2), 213–234, 2010; Ruthven, Deaney, & Hennessy, *Educational Studies in Mathematics, 71*(3), 279–297, 2009).

Keywords Instrumental orchestration • TPACK • Structuring features of classroom practice

K. Ruthven (✉)
University of Cambridge, Cambridge, UK
e-mail: kr18@cam.ac.uk

Introduction

Although the uptake of digital technologies in mathematics teaching continues to be inhibited by factors such as poor resourcing of schools, limited recognition in curricula, and lack of acceptance in examinations, such barriers are slowly diminishing. This brings to the fore what is perhaps the most crucial influence on the successful integration of digital technologies into everyday teaching practice: relevant expertise on the part of the teacher. This paper will examine three contemporary frameworks for analysing such expertise, and explore commonalities, complementarities and contrasts between them: the Technological, Pedagogical and Content Knowledge (TPACK) framework (Koehler and Mishra 2009); the Instrumental Orchestration framework (Trouche 2005); and the Structuring Features of Classroom Practice framework (Ruthven 2009). To concretise the discussion, the use of digital technologies for algebraic graphing, a now well established form of technology use in secondary school mathematics, will serve as an exemplary reference situation. Each of the frameworks will be illustrated through its application in a study of teacher expertise relating to this topic.

The Technological, Pedagogical and Content Knowledge (TPACK) Framework

Core Ideas

The first of these frameworks, originally Technological Pedagogical Content Knowledge [TPCK] (Mishra and Koehler 2006), now Technology, Pedagogy and Content Knowledge [TPACK] (Koehler and Mishra 2009), represents an extension of the now classic conceptualisation of the types of knowledge and reasoning that underpin successful subject teaching (Wilson et al. 1987). The core argument is that teachers develop a special type of 'pedagogical content' knowledge (PCK) which is more than a simple combination of subject content knowledge and generic pedagogical knowledge. Typically this knowledge is developed through solving distinctive problems that arise in the course of teaching a particular topic. These problems raise considerations both of content and pedagogy, and solutions to them are typically not reducible to the logic of either knowledge domain alone. Moreover, while solutions to such teaching problems may become crystallised as stable professional knowledge, they may equally be subject to continuing adaptation and refinement, and they will vary between teachers and across teaching settings. Finally, for reasons both of ecological adaptation and cognitive economy, such knowledge is typically organised around prototypical teaching situations. For these reasons, the subsequent development of this line of work has been criticised for an unproductive focus on a logical demarcation of types of teacher knowledge rather than on its functional organisation (Ruthven 2011a).

Fig. 1 Venn-diagram metaphor for the TPACK model as shown at http://tpack.org/

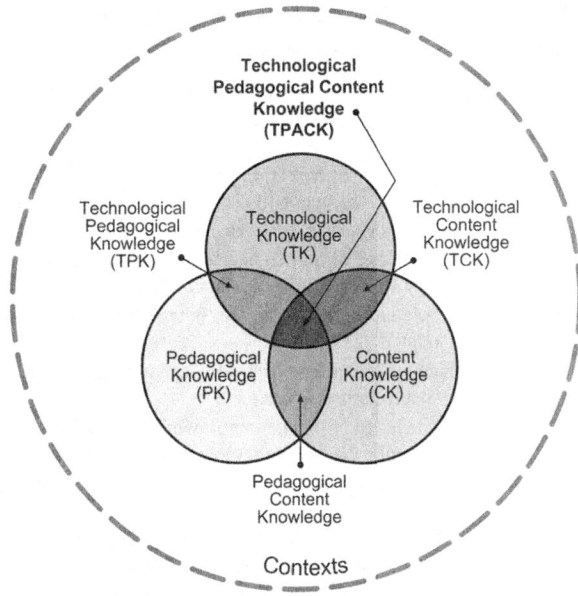

The idea of TPCK was introduced to draw attention to the way in which new technological resources reshape pedagogical knowledge, content knowledge and pedagogical content knowledge. Of course, there are already traditional forms of technology associated with established knowledge of these types, although the ways in which these technologies, such as those of written recording, routine computation and even didactic organisation, mediate thinking and action tend to be invisible to us because we take them so much for granted. It is not surprising, then, that this technological dimension is not recognised in the original PCK framework. However, the contemporary expansion in the technological media through which thinking, learning and teaching take place calls for corresponding evolution, even if still tentative, of teachers' knowledge of content, pedagogy and their interaction. The idea of TPACK seeks to make the need for such evolution visible by highlighting the existence of 'intersections', according to the Venn-diagram metaphor (shown in Fig. 1), between knowledge of technology and knowledge of pedagogy and/or content.

There are, however, some ambiguities in the way in which TPACK is – and has been – used. First, the acronym is sometimes employed to focus attention on the whole system of two- and three-way interactions between these components (as when the standard figure is referred to as 'the TPACK image'); at other times, the term is used to pick out the three-way intersection at the core (as is done within the version of the image shown in Fig. 1) that might otherwise be referred to as TPCK (following the labelling pattern for the other intersections). Second, the character of the 'intersections' or 'interactions' between knowledge domains remains underanalysed, mirroring the differing strengths of definition found in current usages of pedagogical content knowledge (PCK): from a weak definition requiring no more than some simple combination of common knowledge of content

Table 1 Elaboration of TPACK components by Mishra and Koehler (2006)

Component	Elaborated characterisation
TK	Knowledge about standard technologies, such as books, chalk and blackboard, and more advanced technologies, such as the Internet and digital video. Includes:
	The skills required to operate particular technologies
	Knowledge of operating systems and computer hardware
	Ability to use standard sets of software tools such as word processors, spreadsheets, browsers, and e-mail
	Knowledge of how to install and remove peripheral devices, install and remove software programs, and create and archive documents
TCK	Knowledge about the manner in which technology and content are reciprocally related. Includes:
	Knowledge of how technologies afford particular representations and flexibility in navigating across them
	Knowledge of the manner in which the subject matter can be changed by the application of technology
TPK	Knowledge of the existence, components, and capabilities of various technologies as they are used in teaching and learning settings, and conversely, knowing how teaching might change as the result of using particular technologies. Includes:
	Understanding that a range of tools exists for a particular task
	Ability to choose a tool based on its fitness and strategies for using the tool's affordances
	Ability to apply pedagogical strategies for use of technologies
TPCK	Emergent form of knowledge that goes beyond all three components (content, pedagogy, and technology). Includes:
	Understanding of the representation of concepts using technologies
	Pedagogical techniques that use technologies in constructive ways to teach content
	Knowledge of what makes concepts difficult or easy to learn and how technology can help redress some of the problems that students face
	Knowledge of how technologies can be used to build on existing knowledge and to develop new epistemologies or strengthen old ones

with generic pedagogical knowledge, to a stronger definition that insists that PCK be underlain by some distinctive content-specific pedagogical reasoning. Third, there is a hierarchy implicit in the labelling rules under which content is more fundamental than pedagogy, and both of these than technology. In particular, under the strong definition, an unexamined amalgamation takes place of what might have been termed PTCK – pedagogical knowledge relating specifically to the development (by students) of particular forms of technological content knowledge – with what might have been termed TPCK – technological knowledge relating specifically to particular aspects of pedagogical content knowledge. Finally, there is ambiguity about the level at which the pedagogical and the technological are conceived: between a more concrete level at which knowledge is taken as relating to some particular pedagogy or technology, and a more reflexive meta-level at which these terms are reserved for knowledge about pedagogical or technological alternatives.

Perhaps recognising some of these ambiguities, Koehler and Mishra have proposed more elaborated characterisations of those components of the model relating to technology (as shown in Table 1) which could serve to operationalise them more effectively (Mishra and Koehler 2006; Koehler and Mishra 2009). Nevertheless, some ambiguities remain. First, where technologies are content specific, such as dynamic algebra or geometry software, it can be particularly difficult to differentiate between TK and TCK. While knowledge of features and techniques that are generic to much software (such as the basic use of menus and pointers) clearly should be classed as TK, it can be hard to decide when knowledge becomes so content specific (such as the individual operations listed on menus and the particular functions for which the pointer is used) that it should be assigned to TCK. Likewise, given that understanding of certain types of representation forms part of CK, it is problematic to assign 'understanding of the representation of concepts using technologies' in general to TPCK rather than TCK. There may be a risk of confusion here with the more specific usage of 'representation' found in Shulman's original characterisation of pedagogical content knowledge, based on the idea that there are specifically 'pedagogical' forms of representation, or specifically 'didactical' organisations of representations, that go beyond those canonical forms of representation that form part of subject content knowledge. Indeed, pursuing the logic of Shulman's original argument, the constructs of CK, TK and TCK should be free of any specifically pedagogical aspect and applicable as much to the knowledge of students as that of teachers.

Turning more specifically now to mathematics, the US Association of Mathematics Teacher Educators (AMTE) has developed a Mathematics TPACK Framework (AMTE 2009), organised around four major themes: designing and developing technology-enhanced learning experiences; facilitating technology-integrated instruction; evaluating technology-intensive environments; and continuing to develop professional capacity in mathematics TPACK. Just as the way in which T is interpreted in TPACK reflects a preoccupation with new digital technologies, the way in which P is interpreted here reflects a broadly neoprogressive orientation to pedagogy, a longstanding type of association (Cuban 1989). By way of example, the second theme includes:

- Incorporat[ing] knowledge of learner characteristics, orientation, and thinking to foster learning of mathematics with technology;
- Facilitat[ing] technology-enriched, mathematical experiences that foster creativity, develop conceptual understanding, and cultivate higher order thinking skills;
- Promot[ing] mathematical discourse between and among instructors and learners in a technology-enriched learning community;
- Us[ing] technology to support learner-centered strategies that address the diverse needs of all learners of mathematics; and
- Encourag[ing] learners to "become responsible for and reflect upon their own technology-enriched mathematics learning." (AMTE 2009)

It seems, then, that both the 'technological' and the 'pedagogical' components of TPACK are open to narrower and broader interpretations: as highlighting, even valorising, a specific form of pedagogy or technology, or as acknowledging the existence of a range of pedagogies and technologies.

Let us turn now to an example of TPACK in use.

An Example

Amongst a number of recent studies employing the TPACK framework to analyse the professional learning of teachers of algebra, I have chosen the one which makes use of the full system of TPACK categories. In this study of middle-school teachers participating in a professional development programme (Richardson 2009), observational records of interactions and discussions between participants and entries extracted from their professional journals were classified as relating variously to TPK, TCK, PCK or TPCK. The study reports that it did not prove straightforward to demarcate these categories and indicates that they tended to acquire narrower operationalisations specifically related to the particular guiding rationale for the professional development programme. It appears, then, that TPACK may have been more valuable as a holistic construct inspiring the professional development course than as a research tool for analysing the process or product of knowledge construction.

Within the programme, the novel technology (graphing calculator) was viewed as supporting greater emphasis on a particular representational medium (graphic figure). Accordingly, the guiding hypothesis for the professional development was that this technology provides an effective means of supporting deeper pedagogical engagement with the content ("To make meaning of certain problem situations, it is imperative that students model these situations graphically and use graphing to find solutions to these problems"). Inasmuch as this idea invokes interaction between considerations of technology, pedagogy and content, it could reasonably be classed as technological pedagogical content knowledge. The most developed algebraic example provided in the study report arose from a project session in which project teachers were asked to solve the inequality $2(x-4) \geq {}^3/_2 (2x+1)$ using only symbolic, then only graphic, methods. This led to some teachers broadening what could be classed as their content knowledge of algebra (taken as transcending use of any particular tool system) beyond familiar symbolic methods ("to solve the inequality in algebraic form") to include unfamiliar graphic methods ("to solve the same inequality in graphic form"). Teachers also displayed what could be classed as technological content knowledge (taken as technology-specific content knowledge) relating to graphing with the two tool systems in play ("to graph… inequalities… by hand and with a graphing calculator").

Drawing on transcripts of discussion between participating teachers, the study seeks to identify what types of knowledge are under exchange and/or development, interleaving the resulting classification of specific contributions in terms of the TPACK framework:

Teacher B: We already have the graphs. We need to figure out the answer.
Teacher A: No…we already know the solution to the inequality. We found that using basic algebra. This is different. How can we verify it using only the graph? What strategy would you use to explain this to your students?
[This is an example of the teacher's PCK. She explores ways to make this notion comprehensible to her students.]
Teacher C: Let's start over. Graph the inequality on the Nspire. Well … I don't know how to graph it with the inequality.… But we can graph the two sides separately but on the same page.

	[This is an example of the teacher's TCK. She explores how to graph an inequality using a graphing device.]
Teacher A:	I'm not sure if that will help but at least we will be able to actually see the lines and move them to make one bigger than the other. [This is an example of the teacher's TCK. She understands technological content.] (Richardson 2009)

The two suggestions embedded in this extract about where the exchange and/or development of technological content knowledge has been displayed by participating teachers are rather more persuasive than the one relating to pedagogical content knowledge. Teacher A's utterance ("What strategy would you use to explain this to your students?") certainly could be framing the emergent problem as being one of pedagogical content rather than plain content, but there is no clear indication of this framing being sustained; although, by taking 'we' to serve as a projection onto 'they' a later contribution ("I'm not sure if that will help but at least we will be able to actually see...") could be interpreted in such terms.

Likewise in the extract below, classing Teacher B's concluding contribution as technological pedagogical content knowledge involves a high level of inference from an anticipation of PCK ("So how would I explain this to my students?"), followed by a more reflexive expression of TCK ("The solution could be obtained quicker from the [calculator] graph than when we solved the inequality by hand in the beginning") that might be taken as appealing implicitly to some pedagogical notion of didactical time, before returning to what might represent crystallised PCK rather than just CK ("It makes so much sense. "Greater than" means..."). However, it is not clear why this utterance from Teacher B is taken as indicative of TPCK whereas that from Teacher A, which alludes specifically to content ("no matter how I move the lines, this part of this one is always on top of this one") is classed as TPK. It may be that these classifications draw on evidence beyond the transcript, as suggested by what appears to be categorisation of Teacher C's contribution as TCK on the basis of supporting observation rather than words spoken ("She understands how to use the graphing device to explore the effect altering either graph has on changing x values").

Teacher A:	Showing students this with computer software would be great. OK, so look … no matter how I move the lines, this part of this one is always on top of this one. [This is an example of the teacher's TPK. She understands that more than one technology tool exists to help students make connections between effects of manipulating graphs and solving inequalities.]
Teacher C:	Right. Yes. You are right. Well, that's what we need to know. Right? Look – values on this line are bigger than that line anytime x is at least… [This is an example of the teacher's TCK. She understands how to use the graphing device to explore the effect altering either graph has on changing x values.]
Teacher B:	… Negative 9 and a half. So how would I explain this to my students? The solution could be obtained quicker from the graph than when we solved the inequality by hand in the beginning. It makes so much sense. "Greater than" means "When is the left bigger than the right?" [This is an example of the teacher's TPCK. She reflects on how a teacher can show students how to perform the technological procedures and relate solving inequalities in a coherent way during her teaching.] (Richardson 2009)

The example provided by this study suggests that trying to use the detailed TPACK framework to analyse naturally occurring teacher discourse is likely to founder because such utterances often provide insufficient evidence to draw inferences with confidence and to make clear discriminations about the character of the underlying knowledge in play or in the course of development. The framework might, however, prove more effective if it were employed to design a focused interview protocol and analyse the discourse arising from in-depth pursuit of specific aspects of teachers' knowledge.

Another recent study employed the TPACK framework to identify the developmental needs of a school-based lesson-study group. Over the course of two planning cycles, the researchers examined the group's evolving lesson plans for teaching the topic of systems of equations through making use of graphic calculators (Groth et al. 2009). Analysis of this evidence led to the researchers identifying various lines of development needed in the TPACK of the lesson-study group:

- How to use the graphing calculator as a means for efficiently comparing multiple representations and solution strategies;
- How to avoid portraying graphing calculators as black boxes;
- How to pose problems that expose the limitations of the graphing calculator.

In this study, it is notable that TPACK serves simply as a basic heuristic to raise questions about the interaction between technology, pedagogy and content in mathematics teaching, with the detailed framework of component intersections not used at all.

All in all, then, it seems that the idea of TPACK is used to signal the need to consider technological, pedagogical and epistemological aspects of the knowledge underpinning subject teaching and their interaction in general terms. Beyond that, the more detailed framework of TPACK components provides a rather coarse-grained tool for conceptualising and analysing teacher knowledge; one that generally needs to be supplemented by other systems of ideas to accomplish analysis to the depth required for effective professional development and improvement.

The Instrumental Orchestration Framework

Core Ideas

A further system of ideas that has attracted considerable interest as a means of analysing technology-mediated teaching and learning in mathematics is the 'instrumental approach' (Artigue 2002; Guin et al. 2005). This approach was developed in cognitive ergonomics to study the typically non-propositional and action-oriented knowledge involved in making use of tools (Rabardel 2002). The focus of the approach is on the process of 'instrumental genesis' in which tool and person co-evolve so that what starts as a crude 'artefact' becomes a functional 'instrument' and the person who starts as a naive operator becomes a proficient user. It was taken up in mathematics education as a means of analysing developmental processes underpinning the introduction of digital technologies into teaching and learning. For the student learner, in particular, development of technological and mathematical proficiency are intertwined in the process of instrumental genesis. Although some aspects of its conceptual apparatus are rather convoluted, the broad thrust of the instrumental approach has proved valuable in highlighting these processes of co-evolution and so challenging the dissociation of conceptual and technical development characteristic of much neoprogressive thinking about mathematics teaching (Ruthven 2002).

The extension of the instrumental approach through development of the idea of 'instrumental orchestration' (Guin and Trouche 2002; Trouche 2004, 2005) seeks to address a central issue of technology integration in classroom teaching and learning: the management by the teacher of what could potentially be very disparate instrumental geneses on the part of individual students so as to ensure that technico-mathematical development within a class follows a more collective path by means of which emergent knowledge is socialised into a shared form aligned with wider conventions and practices. This calls for the teacher to 'orchestrate' activity across the class with this collective development in mind. The idea of 'instrumental orchestration', then, served for Trouche as a construct covering a range of mechanisms directed towards such collective knowledge-building. Each mechanism was characterised in terms of a particular 'didactical configuration' – some disposition of tools within the classroom and allocation of user roles to participants – and the varied 'exploitation modes' – the patterns of tool use and user interaction – that could be associated with it.

The fullest account of the original construct of instrumental orchestration (Trouche 2005) incorporates four examples (Table 2). It seems that it is the didactical configuration which represents the core feature of an orchestration with the exploitation modes indicating a range, even system, of didactical variables that underpin versatile use of the configuration in ways that can be tailored to a specific stage of a planned collective instrumental genesis. In particular, the system of exploitation modes may include options to not use the configuration (as in the first mode for 'Sherpa student'), or to use it only in some limited way. Nevertheless, the first of these examples ('Customised calculator') appears somewhat different in character from the other three. As Trouche points out, this first example involves adaptation of the tool itself, whereas the other three all attend to the organisation of activity and assignment of roles associated with use of the tool. Equally, the first example depends much more explicitly on analysis of what might be described as a specific 'instrumental trajectory' of the class towards intended technico-mathematical learning outcomes, whereas this dimension is more implicit in the latter three examples. The most widely presented example has been that of the 'Sherpa student' (Guin and Trouche 2002; Trouche 2004; Trouche 2005), and so that has tended to become the prototype of an instrumental orchestration taken up by other researchers.

Table 2 Examples of instrumental orchestration from Trouche (2005)

Orchestration example	Didactical configuration	Exploitation modes
Customised calculator	Classroom calculators are 'fitted out' with a guide affording three levels of study of the limit concept These are designed to support the shift from a kinetic concept of limit to an approximative concept	Guide can be available always or only during a specific teaching phase Students can use guide freely when available, or be constrained to follow the order of the levels Components can be fixed, or updated in response to classroom lessons Recording of steps of instrumented work, can be required, or not
Sherpa student	A *sherpa* student operates a calculator projected to the whole class under the guidance of, and subject to checking and questioning by, the teacher, intended to provide a common reference in addressing the collective instrumental genesis of the class	Calculators and projector off: work with pencil and paper only Calculators and projector on: work strictly guided by the sherpa-student under the supervision of the teacher, with other students supposed to replicate the projected display on their own calculator Calculators and projector on: students work freely but are able to view the work of sherpa-student Calculators on and projector off: students work without being able to view work of sherpa-student
Paired practicals	Each student is equipped with calculator and pencil and paper. Students work in pairs to solve an assigned problem Each pair then has to explain and justify their reasoning and results, noting observations and dead-ends in a written research report	Students can be free, or not, to form pairs Students can be free, or not, to choose which one will write the research report The teacher can offer help to students during the practical, or only at the end of it, or a week after Written research reports can be handed in at the end of practical, or a week later After reading students' research reports, the teacher can give a problem solution, or only give pointers to new strategies for students to pursue
Mirror observations	Students work in pairs While one pair tackles a mathematical task, another pair, guided by an observation protocol, notes the actions carried out for later discussion and reflection	May be used only exceptionally, or be a regular tool for regulation of students' tool-using activity May fix, or not, the role of each student in the working pair (e.g. one can be in charge of the calculator, the other in charge of the report) Protocol can be modified according to the type of mathematical problem set

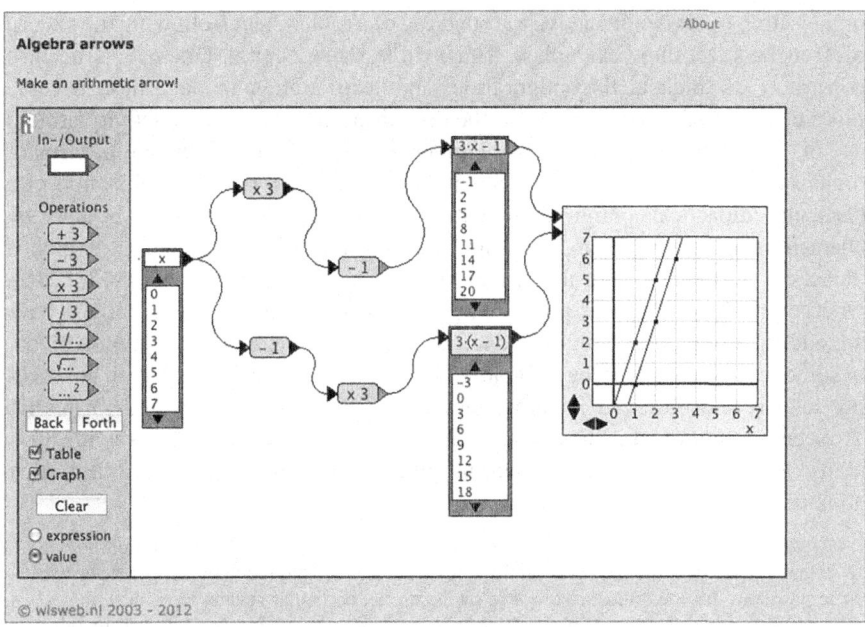

Fig. 2 A screen display showing the Algebra Arrows tool in use

An Example

A recent study has adapted the notion of instrumental orchestration to develop a typology of forms of organisation of classroom activity around use of a tool system (Drijvers et al. 2010). The context for this study was one of trialling a teaching sequence at early-secondary school level on the concept of mathematical function. The researchers write that the sequence "aimed at the development of a rich function concept, whereby functions are conceptualised as input–output assignments, as dynamic processes of co-variation and as mathematical objects with different representations" (p. 216). Their design of a Java applet called Algebra Arrows (Fig. 2) matches this agenda.

It is possible to implicitly discern the tool-adaptive form of instrumental orchestration (by analogy with Trouche's first example of the Customised calculator) in the didactical configuration of the applet to provide options to display or not display the Table and Graph components, affording the possibility of constraining lesson tasks so as to focus attention on particular types of representation and the relations between them. The applet is embedded in a Digital Mathematics Environment (DME) through which the tasks forming the teaching sequence are made available, and which allows students to access their work from any location, and the teacher to access this work in order to monitor progress and track development.

However, in Drijver's study, the notion of 'instrumental orchestration' is explicitly employed in the second activity-structuring sense to designate some particular organisation of classroom activity around use of a tool system (following the pattern of Trouche's last three examples). Thus, while Drijvers et al. take over Trouche's constructs of 'didactical configuration' and 'exploitation mode', these become more closely tied to concerns with the organisation of classroom activity around use of a tool. In particular, because Drijvers et al. wish to differentiate patterns of organisation, they take an 'instrumental orchestration' to be the combination of a particular 'didactical configuration' with a specific 'exploitation mode'. Equally, by characterising 'exploitation mode' as "the way the teacher decides to exploit a didactical configuration for the benefit of his or her didactical intentions" (p. 215), Drijvers et al. give greater prominence to such intentions. Consequently, I have added 'didactical intention' to 'didactical configuration' and 'exploitation mode' in summarising their typology (Table 3). Likewise, because Drijvers et al. are seeking to describe observed patterns, they report that they felt obliged to modify Trouche's definition of instrumental orchestration to acknowledge the way in which plans are elaborated and adapted in performance, through adding a further component:

> A *didactical performance* involves the ad hoc decisions taken while teaching on how to actually perform in the chosen didactic configuration and exploitation mode: what question to pose now, how to do justice to (or to set aside) any particular student input, how to deal with an unexpected aspect of the mathematical task or the technological tool, or other emerging goals. (Drijvers et al. 2010)

The development of the typology was influenced both by prior examples of instrumental orchestration that the developers included in the guidance materials for teachers (notably 'Sherpa-at-work' and 'Link-screen-board') and by templates identified through subsequent observation of teachers at work. While, in principle, it seems possible that there could be clashes within the typology – for example, if an episode revolved around the thinking displayed in a piece of work selected by the teacher ('Spot-and-show'), with that student nominated to act as the sherpa student ('Sherpa-student'), in effect a particular version of a more generic form ('Explain-the-screen'), in practice Drijvers et al. report that inter-rater reliability of the codings was good, although gradual shifts in classroom activity could create some difficulties of demarcation. Likewise, the researchers acknowledge that the range of orchestration types that emerged from the study might well have been conditioned by factors particular to the trialling situation

Nevertheless, developing this typology helped to identify overall patterns in classroom activity, and to pinpoint differences between the profiles of teachers, and between one teacher's enactments of the same sequence with different classes. While 'Technical-demo' was a common orchestration (in the sense both of being used by all teachers and frequently so), there were differences in the degree to which teachers made use of the more student-centred 'Discuss-the-screen', 'Spot-and-show' and 'Sherpa-at-work' orchestrations as opposed to the more teacher-led 'Explain-the-screen' and 'Link-screen-board'. This suggests that the way in which such expertise develops is also shaped by broader personal orientation to teaching mathematics.

Finally, this typology makes visible an important dimension of the professional knowledge that teachers participating in trialling had employed or developed in order to incorporate use of these digital technologies into their practice. In effect, these six orchestration types also represent the core of a collective system of

Table 3 Typology of whole-class instrumental orchestration from Drijvers et al. (2010)

Orchestration type	Didactical intention	Didactical configuration	Exploitation modes
Technical-demo	Demonstration by the teacher of techniques for using the tool	Provision to project DME. Classroom arrangement allowing students to view the projected screen	Teacher employs new situation or their own solution or earlier student work as a point of departure
Explain-the-screen	Explanation by the teacher going beyond technique, involving mathematical content	Provision to project DME. Classroom arrangement allowing students to view the projected screen	Teacher employs new situation or their own solution or earlier student work as a point of departure
Link-screen-board	Instruction by the teacher relating the representations of mathematics in different media	Provision to project DME. Classroom arrangement allowing students to view both the projected screen and the board	Teacher employs new situation or their own solution or earlier student work as a point of departure
Discuss-the-screen	Discussion between teacher and students about what is happening on the screen	Provision to project DME and preferably to access student work. Classroom arrangement allowing students to view the projected screen and favouring discussion	Teacher employs new situation or their own solution or earlier student work as a point of departure
Spot-and-show	Discussion between teacher and students in which student reasoning is brought to the fore through deliberate use of carefully chosen student work	Access to student work in the DME during lesson preparation. Provision to project DME. Classroom arrangement allowing students to view the projected screen	Teacher chooses earlier student work in advance of the lesson as a point of departure for the student to explain their reasoning, or for other students to give reactions, or for the teacher to provide feedback
Sherpa-at-work	Activity in which a sherpa-student uses the technology to present his or her work, or to carry out actions that the teacher requests	Provision to project DME. Classroom arrangement enabling sherpa to use the projected tool and other students to view the projected screen and follow contributions of sherpa and teacher	Teacher has work presented or explained by the sherpa-student, or poses questions to the sherpa-student and asks them to carry out specific actions in the technological environment

professional expertise, making the typology of particular interest to teacher educators seeking to help teachers develop practical strategies for the organisation of classroom lessons using such digital technologies.

The Structuring Features of Classroom Practice Framework

Core Ideas

The Drijvers et al. study emphasises how integration of new technologies depends on teachers adapting and developing appropriate craft knowledge to underpin their classroom work. A third framework has been explicitly designed to support the identification and analysis of this type of teaching expertise. The Structuring Features of Classroom Practice framework (Ruthven 2009) was devised by bringing a range of concepts from earlier studies of classroom organisation and interaction and of teacher craft knowledge and thinking to bear on this specific issue of technology integration. Thus this framework synthesises and extends concepts that have already proved valuable in analysing classroom practice (Table 4).

The framework identifies five structuring features of classroom practice which shape the ways in which teachers integrate (or fall short of integrating) new technologies: working environment, resource system, activity structure, curriculum script, and time economy. The introduction of new technologies often involves changes in the *working environment* of lessons in terms of room location, physical layout, and class organisation, requiring modification of the classroom routines which enable lessons to flow smoothly. Equally, while new technologies broaden the range of tools and materials available to support school mathematics, they present the challenge of building a coherent *resource system* of compatible elements that function in a complementary manner and which participants are capable of using effectively. Likewise, innovation may call for adaptation of the established repertoire of activity formats that frame the action and interaction of participants during particular types of classroom episode, and combine to create prototypical *activity structures* or cycles for particular styles of lesson. Moreover, incorporating new tools and resources into lessons requires teachers to develop their *curriculum script* for a mathematical topic. This 'script' is an event-structured organisation of knowledge, forming a loosely ordered model of goals, resources and actions for teaching the topic, incorporating potential emergent issues and alternative courses of action; it interweaves mathematical ideas to be developed, appropriate topic-related tasks to be undertaken, suitable activity formats to be used, and potential student difficulties to be anticipated, guiding the teacher in formulating a suitable lesson agenda, and in enacting it in a flexible and responsive way. Finally, the introduction of new technologies may influence the *time economy* within which teachers operate, changing the 'rate' at which the physical time available for classroom activity can be converted into a 'didactic time' measured in terms of the advance of knowledge.

The status of this conceptual framework remains tentative. It prioritises and organises previously disparate constructs developed in earlier research, and has

Table 4 Components of the structuring features framework (Ruthven 2009)

Structuring feature	Defining characterisation	Examples of associated craft knowledge related to incorporation of digital technologies
Working environment	Physical surroundings where lessons take place, general technical infrastructure available, layout of facilities, and associated organisation of people, tools and materials	Organising, displaying and annotating materials Capturing or converting student productions into suitable digital form Organising and managing student access to, and use of, equipment and other tools and materials Managing new types of transition between lesson stages (including movement of students)
Resource system	Collection of didactical tools and materials in use, and coordination of use towards subject activity and curricular goals	Establishing appropriate techniques and norms for use of new tools to support subject activity Managing the double instrumentation in which old technologies remain in use alongside new Coordinating the use and interpretation of tools
Activity structure	Templates for classroom action and interaction which frame the contributions of teacher and students to particular types of lesson segment	Employing activity templates organised around predict-test-explain sequences to capitalise on the availability of rapid feedback Establishing new structures of interaction involving students, teacher and machine and the appropriate (re)specifications of role
Curriculum script	Loosely ordered model of goals, resources, actions and expectancies for teaching a curricular topic including likely difficulties and alternative paths	Choosing or devising curricular tasks that exploit new tools, and developing ways of staging such tasks and managing patterns of student response Recognising and responding to ways in which technologies may help/hinder specific processes and objectives involved in learning a topic
Time economy	Frame within which the time available for class activity is managed so as to convert it into "didactic time" measured in terms of the advance of knowledge	Managing modes of use of tools so as to reduce the "time cost" of investment in student learning to use them or to increase the "rate of return" Fine-tuning working environment, resource system, activity structure and curriculum script to optimise the didactic return on time investment

proved a useful tool for analysis of already available case-records. While it has been noted that "the differing provenance of the five central constructs raises some issues of coherence" (Ruthven 2011b, p. 97), such eclecticism is characteristic of the powerful intermediate theory that effective analysis of issues of teaching requires. However, further studies are now required in which data collection (as well as analysis) is guided by the conceptual framework, so that it can be subjected to fuller testing and corresponding elaboration and refinement. To adequately address issues of professional learning, such studies need to be longitudinal as well as cross-sectional, and to focus on teachers' work outside as well as inside the classroom. Likewise, the current reach of this conceptual framework is deliberately modest; it simply seeks to make visible and analysable certain crucial aspects of the incorporation of new technologies into classroom practice which other conceptual frameworks largely overlook.

An Example

A study of teachers' use of graphing software to teach about algebraic forms at lower-secondary level used the Structuring Features of Classroom Practice framework to help identify various types of adaptation of teaching practices and development of craft knowledge associated with use of such technology through lesson observations supplemented by post-lesson interviews with teachers (Ruthven et al. 2009).

In terms of *working environment*, many of the aspects observed were not specific to graphing software. Relocation of lessons from the normal classroom to the computer suite required teachers to modify their managerial routines, notably those concerned with handling the start of lessons, to include getting students seated appropriately, and their computer workstations and resources opened for use. Equally, adaptation was required to routines for securing the attention of students during periods of independent work, so as to make important points to the class as a whole. Teachers also had to develop fallback strategies to cope with any non-functioning of components of the technological infrastructure.

Typically the *resource system* for lessons consisted of graphing software and printed worksheets: the latter set out tasks and often provided a means of recording results by hand. Making students' use of graphing software functional required teachers to develop strategies to familiarise them with (and later to review) core techniques, and to allow students to explore (and then to share their discoveries of) a wider range of technical possibilities. Teachers themselves were developing expertise regarding the forms of technico-mathematical guidance that students might require: such as explaining how to enlarge a point to make it more visible, or how to enter x^2 in the equation editor; helping students to understand why their graph was a horizontal line rather than the expected sloping one (as a result of entering $y=5+4$ rather than $y=5x+4$), or a straight line rather than the expected curve (as a result of entering $y=x+2^2$ rather than $y=(x+2)^2$); prompting students to drag the displayed image to expose more of a particular graph, or to pursue the limiting trend of a graph.

In terms of *activity structures*, a distinctive type of activity format was emerging for individual or paired student work on a new type of 'target practice' task which capitalised on the interactivity of the software to centre investigative activity around a process of trial and improvement of posited solutions. For example, in two investigations of this type, students were tasked with using the software to find equations – of straight lines in the first investigation, quadratic curves in the second – passing through some specified point or pair of points. In a similar way, teachers had adapted a conventional whole-class exposition and questioning activity format to incorporate use of the software to provide immediate feedback on student suggestions, for example through students 'taking the stage' to use the projected computer to test their predictions.

These preceding elements of adaptation had all been interwoven into teachers' *curriculum scripts* for the topic of algebraic forms. At the core of these scripts, teachers had had to find or devise tasks (such as the 'target practice' type already alluded to) which productively employed graphing to investigate the topic of algebraic forms. On the basis of classroom experience of the ways in which these tasks played out in the classroom, teachers were both refining them and developing a repertoire of strategies to support students in tackling them, concerned with prompting strategic action and supporting mathematical interpretation. One example involved prompting students to zoom out on the displayed image of $0.00000009x^2+x+1$ to test whether it was a straight line (as it had appeared to be to students), then introducing the comparison with $0x^2+x+1$. Another example involved supporting a student who had graphed $x=-yx$ and wondered why it looked the same as $y=-1$, by helping him to rearrange and simplify the first equation. Such examples illustrate how a gradual accretion of teachers' expert knowledge, and its organisation within their curriculum script, takes place through their responding to, and reflecting on, classroom incidents. There was also evidence of certain technology-supported lines of questioning becoming invariant elements of teachers' curriculum scripts for the topic. A recurring pattern across one teacher's lessons arose when, after examining graphs of the form $y=x+c$, she consistently posed the question "How would you draw the diagonal line going the other way… from top left to bottom right?" with a view to using the software to test student responses.

While they had to devote time for students to learn to graph both by hand and by machine, teachers reported that use of the software helped to ease and effect the production of graphs and so to accelerate such activity and elevate students' attention to focus on the mathematical relationships involved. In particular, teachers considered that having students make use of graphing software made investigative lessons much more viable. These changes in *time economy* had required corresponding adaptation of curriculum sequences on this topic and recalibration of their timing.

Commonalities, Complementarities and Contrasts Between the Frameworks

The title of this chapter refers to the 'expertise' rather than the 'knowledge' that underpins successful integration of digital technologies into everyday teaching practice. This is a deliberate choice to emphasise that – put another way – much of the knowledge that teachers use is 'tacit' and resides in schemes of perception and action which they are typically unable to articulate, and may even be unaware of. Nevertheless, an important contribution that researchers can make to the enterprise of professional education and development is to identify such expertise and provide means of representing and analysing it. Typically they have done so by refining techniques of observation-based analysis supported by introspective interview that support inferences about such expertise. The resulting findings are particularly valuable when then taken up and used in teacher education for purposes of structuring and scaffolding the reflexive appropriation and development by teachers of the expertise that has been identified.

There are three conceptualisations of the relations between pairs of perspectives that I find particularly illuminating. The first relates to the contrasting models of knowledge or expertise underlying the Technological, Pedagogical and Content Knowledge framework and the Structuring Features of Classroom Practice framework. If we look back at the descriptors used for elements of TPACK (shown in Table 1), while the term 'knowledge' predominates, there is also reference in the entries under TK to 'skills' and 'ability', under TPK to 'ability' and 'understanding', and under TPCK to 'understanding' and 'techniques', indicating that the TPACK model does acknowledge such broader components of expertise. By comparison, the way in which the examples of 'craft knowledge' are formulated in the Structuring Features model (shown in Table 4) frames these as practical competences without seeking to differentiate either between tacit and articulate knowledge or into technology-, pedagogy- and content-based categories. Perhaps, then, the crucial difference between these frameworks is that the organising concept for the TPACK model is one of epistemological demarcation between different classes of knowledge relevant to teaching, whereas the organising concept for the Structuring Features model is one of how material-cultural factors structure the functional organisation of teaching expertise.

The second illuminating comparison is between the Technological, Pedagogical and Content Knowledge framework and the Instrumental Orchestration framework. The forms of teaching expertise implied by the Instrumental Orchestration framework are those related to the management of the collective instrumental genesis of a class of students. Because this construct is used in a manner that emphasises the way in which development by students of mathematical content knowledge is, to some significant degree, intertwined with development of knowledge of the mediating technology, the process of classroom instrumental genesis is taken as having the growth of students' TCK at its core, even if some components of the knowledge to be developed might be classed as simple CK or TK alone. As well as this technological content knowledge linking topic and tool, the teacher must also have the pedagogical knowledge necessary to manage its development by students. This includes knowledge of how to coordinate the introduction and use of particular features of the tool with a task sequence capable of supporting an effective learning trajectory (as shown by the example of Trouche's 'Customised calculator' orchestration – which might be classed as TPCK) – and of how to exploit a range of more generic classroom configurations in enacting the various stages of such a sequence – (as shown by Trouche's other orchestrations which might be classed as TPK).

The third illuminating comparison is of the Instrumental Orchestration framework and the Structuring Features of Classroom Practice framework. The Structuring Features framework provides a more differentiated characterisation of several key aspects of Instrumental Orchestration. First, it highlights the matter of incorporating a new tool into the resource system (e.g. Establishing appropriate techniques and norms for use of new tools to support subject activity). Alongside that, there is the matter of adapting activity structures to better support the development and use of this tool (e.g. Establishing new structures of interaction involving students, teacher and machine and the appropriate (re)specifications of role). Finally, there is the matter of devising task sequences and associated narratives to incorporate use of the tool within the curriculum script for a topic (e.g. Choosing or devising curricular tasks that exploit new tools, and developing ways of staging such tasks and managing patterns of student response). Equally, the different types of instrumental orchestration identified by Drijvers et al. (shown in Table 3) all correspond – in the terms of the Structuring Features framework – to specific activity formats that exploit a particular resource (sub)system. However, Trouche's instrumental orchestration for development of the limit command (shown in Table 2) corresponds – in the terms of the Structuring Features framework – to customisation of a specific part of the resource (sub)system linked to development of an innovative pathway within the curriculum script for the topic. Moreover, the network of teaching possibilities for a topic that makes up the curriculum script – in the Structuring Features framework – underpins both the advance planning of a 'lesson agenda' – linked to 'didactical intention' in the Instrumental Orchestration framework – and its interactive enaction and adaptation by the teacher – linked to 'didactical performance' in Drijver's extension of the Instrumental Orchestration framework.

In their current state, then, each of these three frameworks provides an overarching set of 'top level' constructs that reflects a particular orientation towards the phenomenon of technology integration in subject teaching. By comparing these differing systems of base constructs I have sought to provide a more coordinated overview that shows how their different perspectives on technology integration in subject teaching are inter-related. I have also highlighted how each of these frameworks provides a more tentative listing of elements and examples at the more concrete level necessary to support the operational use of its main constructs as analytic tools. This points to a crucial need for fuller and more systematic investigation of the phenomenon of technology integration into subject teaching at this intermediate level. Indeed, close examination of each of the studies presented here as an example of the application of a particular framework in use has shown that it required supplementation by other ideas in order to generate illuminating findings. More intensive research work at this more concrete level could serve to better operationalise the existing frameworks or to fuel the development of a single more powerful one. My own view is that any more powerful framework is likely to be organised along functional lines closer to those of Instrumental Orchestration and Structuring Features, but in a way capable of incorporating intermediate level elements from all three existing frameworks. A synthesising framework of this type would provide an overarching system of constructs driven by the need to organise systematically a much richer and fuller inventory of the kinds of intermediate level elements that these three frameworks have started to identify.

References

Association of Mathematics Teacher Educators. (AMTE). (2009). *Mathematics TPACK (Technological Pedagogical Content Knowledge) Framework*. Retrieved February 8, 2012 from http://www.amte.net/sites/all/themes/amte/resources/AMTETechnologyPositionStatement.pdf

Artigue, M. (2002). Learning mathematics in a CAS environment: The genesis of a reflection about instrumentation and the dialectics between technical and conceptual work. *International Journal of Computers for Mathematical Learning, 7*(3), 245–274.

Cuban, L. (1989). Neoprogressive visions and organizational realities. *Harvard Educational Review, 59*(2), 217–222.

Drijvers, P., Doorman, M., Boon, P., Reed, H., & Gravemeijer, K. (2010). The teacher and the tool: Instrumental orchestrations in the technology-rich mathematics classroom. *Educational Studies in Mathematics, 75*(2), 213–234.

Groth, R., Spickler, D., Bergner, J., & Bardzell, M. (2009). A qualitative approach to assessing technological pedagogical content knowledge. *Contemporary Issues in Technology and Teacher Education, 9*(4). Retrieved February 8, 2012, from http://www.citejournal.org/vol9/iss4/mathematics/article1.cfm

Guin, D., Ruthven, K., & Trouche, L. (Eds.). (2005). *The didactical challenge of symbolic calculators: Turning a computational device into a mathematical instrument*. New York: Springer.

Guin, D., & Trouche, L. (2002). Mastering by the teacher of the instrumental genesis in CAS environments: Necessity of instrumental orchestrations. *Zentralblatt für Didaktik der Mathematik, 34*(5), 204–211.

Koehler, M. J., & Mishra, P. (2009). What is technological pedagogical content knowledge? *Contemporary Issues in Technology and Teacher Education, 9*(1). Retrieved February 8, 2012 from http://www.citejournal.org/vol9/iss1/general/article1.cfm

Mishra, P., & Koehler, M. J. (2006). Technological pedagogical content knowledge: A framework for integrating technology in teacher knowledge. *Teachers College Record, 108*(6), 1017–1054.

Rabardel, P. (2002). *People and Technology: a cognitive approach to contemporary instruments*. Retrieved February 8, 2012 from http://ergoserv.psy.univ-paris8.fr/

Richardson, S. (2009). Mathematics teachers' development, exploration, and advancement of technological pedagogical content knowledge in the teaching and learning of algebra. *Contemporary Issues in Technology and Teacher Education, 9*(2). Retrieved February 8, 2012 from http://www.citejournal.org/vol9/iss2/mathematics/article1.cfm

Ruthven, K. (2002). Instrumenting mathematical activity: Reflections on key studies of the educational use of computer algebra systems. *International Journal of Computers for Mathematical Learning, 7*(3), 275–291.

Ruthven, K. (2009). Towards a naturalistic conceptualisation of technology integration in classroom practice: The example of school mathematics. *Education & Didactique, 3*(1), 131–149.

Ruthven, K. (2011a). Conceptualising mathematical knowledge in teaching. In T. Rowland & K. Ruthven (Eds.), *Mathematical knowledge in teaching* (pp. 83–96). New York: Springer.

Ruthven, K. (2011b). Constituting digital tools and materials as classroom resources: The example of dynamic geometry. In G. Gueudet, B. Pepin, & L. Trouche (Eds.), *From text to 'lived' resources: Mathematics curriculum materials and teacher development* (pp. 83–103). New York: Springer.

Ruthven, K., Deaney, R., & Hennessy, S. (2009). Using graphing software to teach about algebraic forms: A study of technology-supported practice in secondary-school mathematics. *Educational Studies in Mathematics, 71*(3), 279–297.

Trouche, L. (2004). Managing the complexity of human/machine interactions in computerized learning environments: Guiding students' command process through instrumental orchestrations. *International Journal of Computers for Mathematical Learning, 9*(3), 281–307.

Trouche, L. (2005). Instrumental genesis, individual and social aspects. In D. Guin, K. Ruthven, & L. Trouche (Eds.), *The didactical challenge of symbolic calculators: Turning a computational device into a mathematical instrument* (pp. 197–230). New York: Springer.

Wilson, S., Shulman, L., & Richert, A. (1987). '150 different ways' of knowing: Representations of knowledge in teaching. In J. Calderhead (Ed.), *Exploring teacher thinking* (pp. 104–124). London: Cassell.

Summary and Suggested Uses for the Book

Alison Clark-Wilson, Ornella Robutti, and Nathalie Sinclair

Abstract This chapter provides an overview of the book's content in relation to the 'grain size' of the focus and analysis of the different methodologies contained within the constituent chapters. In addition it offers some classification in terms of static, dynamic and more evolutionary approaches to researching teachers' uses of digital technologies in classrooms, whilst emphasising the importance of the different approaches. The chapter ends by suggesting some possible approaches to the use of the book's content for academic teaching scenarios, particularly those that involve practising mathematics teachers. The examples that are provided give ideas on how to engage teachers in both reflective thought alongside the provision of use of theoretical constructs that may support the ongoing development of their classroom practices with technology.

The current research interest in teacher education (which Sfard has called the *era of the teacher*) is a relatively new domain when compared to other research themes relating to mathematics content, curriculum, students, learning, cognitive processes, policy and equity. This era marks an important milestone in the evolution of mathematics education, toward the teacher having an important role to play within the classroom. No theoretical approach, whether it is constructivist or participationist

---

A. Clark-Wilson (✉)
London Knowledge Lab, Institute of Education, University of London,
23-29 Emerald Street, London WC1N 3QS, UK
e-mail: a.clark-wilson@ioe.ac.uk

O. Robutti
Dipartimento di Matematica, Università di Torino, Via Carlo Alberto 10, Turin 10123, Italy

N. Sinclair
Faculty of Education, Simon Fraser University, 8888 University Drive,
Burnaby V5A 1S6, BC, Canada
e-mail: nathsinc@sfu.ca

and no innovation, whether it concerns new digital technologies or new forms of assessment, can shade the importance of the teacher in current and future research.

Sfard (2005) stresses that the advent of *the era of the teacher* has brought about a re-conceptualisation of the relationship between the teacher and the researcher, arguing that in most of the international research studies, the question is not *what* is taught in classrooms, but *how* it is taught. The chapters in this book bear witness to the role of the teacher, and many of them look at the different activities, both within professional development and in the daily activities in the classroom. In this role, the teacher organises and anticipates the main actors in the classroom, which includes the students and the digital technologies. Without the coordination of the teacher, it is hard to develop students' mathematical understanding as, however sophisticated the digital tools, putting them into the students' hands does not automatically or transparently result in meaning. The teacher's intervention is necessary at different levels: designing activities with instruments; planning teaching practices according to the activities; developing them in the classroom; and making choices about and observing students' learning.

Within the classroom, researchers have choices to make with respect to the 'grain size' of their focus and analysis, which can range from the individual analyses of teachers/lessons involving particular technologies, to studies of the evolution of teaching/learning processes over time.

In this book, readers can find different types of analyses: Bretscher's sociocultural approach distinguishes between the use of software and hardware, contrasting teacher-centred and student-centred uses of technology, offering some reasons that explain these differences. Mason's chapter, by suggesting three different web-based *e-screens*, invites the reader to engage in challenging mathematical experiences. These experiences can be useful ways for pre-service and in-service teachers to reflect on how using digital technology shapes and changes the personal mathematical experience.

Trigueros and colleagues examine different uses of technology that can occur in the classrooms, as *replacement, amplification* and *transformation* activities, and relate them to aspects of the role of the teacher in terms of communication of mathematics, interaction with students, validation of mathematical knowledge, the source of mathematical problems, and the actions and autonomy of students. Drijvers and colleagues use the notion of *instrumental orchestration,* combined with elements of TPACK, to describe seven different orchestrations that teachers may develop when they use digital technologies in their classrooms. The metaphor of the teacher as the conductor of an orchestra, where students are the players of instruments under his or her coordination, is useful to help focus attention on a teacher's responsibility in planning and managing the use of digital technologies in a classroom. Gueudet and colleagues use the same frame of orchestrations at the kindergarten mathematics classroom level and find new orchestrations related to this school context. Their chapter shows the need for theory development that is attentive to the age and grade level of the students. Haspekian, using the instrumental approach within the context of teachers' professional development, introduces *instrumental distance* and *double instrumental genesis* to analyse the use of spreadsheets as a mathematical and pedagogical resource. Also Abboud-Blanchard, adopting a meta-level approach in empirical research concerning the use of a computer algebra system, dynamic geometry program and web-based resources, shows how it is possible to identify the characteristics of ordinary teachers' uses of technology, classifying them in terms of the different components of their practice, with the use of the *double approach*. These approaches give deep insights into the varying ways that teachers integrate digital technologies in the classroom, plan activities with them and use them with students. Thomas and colleague, with the help of PTK, consider the obstacles to secondary teachers' use of digital technologies, and introduce some indicators of teachers' progress in their implementation, namely, proficiency and understanding of the techniques required to build didactical situations incorporating digital technologies.

All the chapters described above take into account the complexity of the teaching/learning situations from the point of view of different modes of teaching, designing activities, and using technologies to improve the learning of mathematics. The next two chapters, in contrast, still focused on the classroom, inquire into how particular events change the course of a lesson. Clark-Wilson and Aldon use the instrumental approach to analyse unexpected situations in the classroom and teachers' responses to them. Their approaches are similar, but give different insights: Clark-Wilson uses the notion of a *hiccup*, as perturbations experienced by the teachers during the lesson that triggered by the use of a digital technology, to highlight discontinuities in teacher knowledge. Aldon uses the notions of *didactic incident* (an event of the didactic system that modifies the dynamics of the teaching situation) and *perturbation*, which is what follows the incident. Hiccups and incidents are theoretical constructs that help to explain the dynamic relationship between teaching and learning, and their evolution and change over time. These constructs help reveal the complexity of teaching/learning processes in terms of choices, changes, and discontinuities.

Other chapters concentrate on the process of the evolution of teachers' professional development, teaching practices, relations with institutions, use of technology and interactions with colleagues and researchers. The chapter by Goos presents the construct of *teachers' pedagogical identities* to direct attention to the process of developing teacher identities when teachers begin to use digital technologies in their classrooms. To explain how these beginners are able to develop practices with technology, the author uses a socio-cultural approach (Valsiner's zone theory) and introduce the notions of a teacher's *zone of free movement* and *zone of free action* as part of the complex system that overlaps with the teacher's *zone of proximal development*.

Bellman and colleagues examine teachers' progression as they use a particular digital technology that makes students' assessment more transparent and their longitudinal observations provide evidence of the increased teacher control of the technology over time.

At university level, it is also possible to analyse the progression in the use of digital technologies, as Buteau and Muller do, in the context of an undergraduate mathematics course. These authors describe the design and implementation of a course that supported their students experience with, and use of, digital technologies.

Arzarello and colleagues describe the complexity of teacher education in relation to institutions, and analyse the evolution over time of the teaching processes, seen from the point of view of teacher education and professional development. The variables involved in this analysis are explained in the context of a model, the *Meta-Didactical Transposition*, through which it is possible to take account of the complexity of teacher education as a process, in which communities of teachers have to relate with institutions on the one hand, and on the other, with the community of researchers as well as that of teacher educators/tutors.

One interesting conclusion that can be drawn from the book in its entirety is that it offers evidence for the emerging cumulative knowledge of teachers' professional development concerning new technologies, and discusses the impact that the different variables might have on the personal development process.

The chapters of the book, summarised here, evidence and try to account for the great complexity of teaching processes, both in terms of teaching practices in the classroom and teacher education, that involve the use of digital technologies. This complexity can be explained using different frameworks, and can be showed in either a static or dynamic way, according to the framework chosen and the research aims. The first set of chapters described in this summary examine the complexity in a more static way, highlighting features, indicators and elements of the context of teaching with digital technologies that are important. The next pair of chapters direct attention to the decisions teachers make when they encounter unexpected tensions with respect to the use of digital technologies. The last group of chapters describe the complexity in terms of evolution in teaching processes and in professional development.

These three main axes of research can be seen in corresponding theoretical frameworks, as exemplified in the last chapter by Ruthven, which demonstrates how varying frameworks function as research tools. The frame of TPACK enlarges the widely-used PCK introducing the technological variable (T) and gives insight on the knowledge that mathematics teachers need to teach. With TPACK, it is possible to consider technology as a component with others, to have a "coarse-grained tool for conceptualising and analysing teacher knowledge" (p. 380), and to describe teaching with digital technology in a static way, depicting its main features. The Instrumental Orchestration approach (which provides a more fine-grained analysis), based on the instrumental approach, can be used (alone or in conjunction with TPACK – see Drijvers) for a classification of teaching practices either in the classroom or in teacher professional development (see Clark-Wilson). Another approach – the Structuring Feature of Classroom practice – has a different purpose, which is to support the identification and analysis of teaching-with-technology expertise, but offers a more differentiated classification of some of the key features of the instrumental orchestration approach.

In terms of future research, we anticipate that more work will be carried out to refine current theories so that they are more useful to particular contexts that have been less researched, such as the teaching of mathematics at the elementary school level using digital technologies. Along these lines, most studies are done in the context of particular kinds of digital technologies and we anticipate the new developments, such as touch-screen devices, will also affect the kinds of practices we see in the classroom, the incidents or hiccups they give rise to and the evolution of their integration over time.

Given the international flavour of this book, it is also important to consider how particular studies conducted in particular countries might inform the work of colleagues in other countries. The ReMath project (http://remath.cti.gr/), which focused on studies of learning mathematics using digital technologies, clearly showed that contextual factors, including culturally-specific interpretations of theoretical constructs and goals of education, made research results difficult to carry over from one context to the other. However, the opportunities that current and future international projects provide for collaborative research and resource development should not be underestimated.

As digital technological tools become more ubiquitous in the classroom, many national educational systems are beginning to grapple with the complexities of scaling students' access through different approaches to wide-scale implementation. Experiences suggest that, although many successful classroom implementations are reported within small scale studies, further research is necessary to build knowledge concerning the best conditions for scalabilty and the retention of epistemological fidelity as practices become more mainstream.

This book has been compiled and edited as an essential text for any teachers and researchers interested in the field of digital technologies within mathematics education, with particular emphasis on the teacher's role. As such, the editors envisage that there are a number of ways in which the reader or groups of readers might engage productively with the text. These are offered as suggestions and they do not constitute an exhaustive list.

Interactions Between Teacher, Student, Software and Mathematics: Getting a Purchase on Learning with Technology by Mason is placed early in the book as an orienting text that will serve to ground the reader, by requiring him or her to engage in mathematical thinking within a digital environment. By considering the personal experience, it is anticipated that some criteria can be established for the worthwhile use of technology within mathematics education. Consequently, whether working individually or as a small group, there is a productive sequence of activities whereby groups come together to: discuss their experiences; debate Mason's structuring ideas and discuss the implications of these on the nature of students' activities with digital technologies within (and outside of) the school mathematics setting.

Ruthven's chapter (Frameworks for Analysing the Expertise That Underpins Successful Integration of Digital Technologies into Everyday Teaching Practice) would be a useful addition to a course reader (or compilation of papers) as it offers a concise summary of the main theoretical constructs that are relevant to mathematics classrooms which incorporate digital tools. This could provide a helpful bridge to the original research papers, whilst also encouraging students and teachers to engage with the constructs in more depth.

Alternatively, selected chapters in the book could be used in a chosen sequence to form part of an academic or professional course in which students or teachers are encouraged to explore their own practice with a view to critiquing and expanding their repertoire of technology use for both teaching and learning mathematics. For example, Fig. 1 illustrates such as pathway.

This pathway might be suitable for a group of practicing teachers who have some experience of using technology for teaching and learning mathematics and who need support to reflect on their existing practices from different perspectives.

Fig. 1 A pathway through the text for practicing teachers

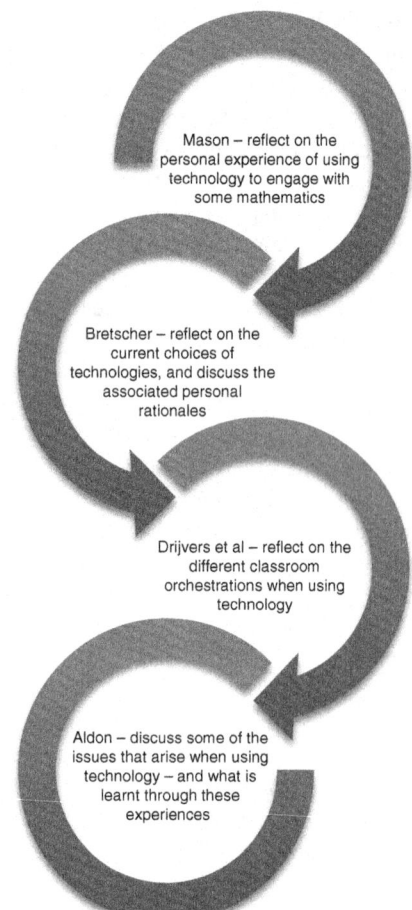

An alternative route is shown in Fig. 2, which might be to focus initially on a particular pedagogic construct, such as Bellman, Foshay and Gremillon's 'adaptive and differentiated instruction' as a means to challenge current practices with technology, whilst offering participants a framework against which they can audit and develop their practices.

In this context the chapter contributions from Aldon and Clark-Wilson both provide an insight into secondary mathematics classrooms in which opportunities for the formative assessment of students' learning have been provided by the technology.

These are just two possible pathways, however the book's index and the accompanying online Glossary (http://extras.springer.com*) offer a supportive mechanism to construct some relevant pathways for different academic and professional purposes.

---

*Log in with ISBN 978-94-007-4638-1

Fig. 2 An alternative pathway through the text for practicing teachers

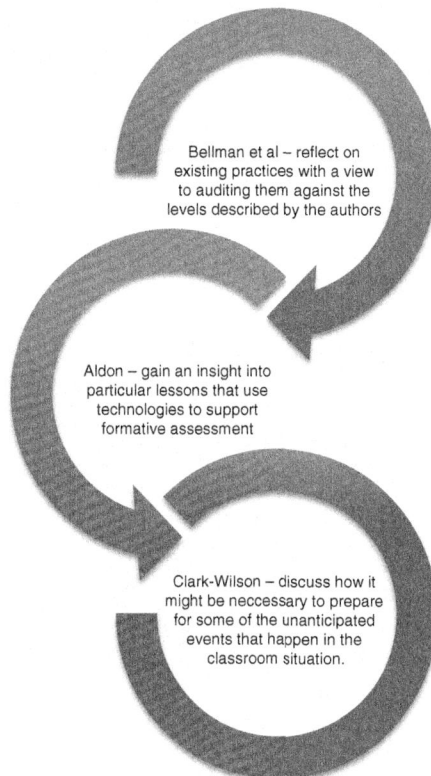

Reference

Sfard, A. (2005). What could be more practical than good research? On mutual relations between research and practice of mathematics education. *Educational Studies in Mathematics, 58*(3), 393–413.

Glossary

Broker Brokers make connections across communities of practice, enable coordination, and open new possibilities for meaning. Brokers facilitate the transition of mathematical concepts from one community to the other.

Community of Inquiry Group of individuals involved in a process of empirical or conceptual inquiry into problematic situations.

Curriculum Script An event-structured organisation of knowledge, forming a loosely ordered model of goals, resources and actions for teaching the topic, incorporating potential emergent issues and alternative courses of action. It interweaves mathematical ideas to be developed, appropriate topic-related tasks to be undertaken, suitable activity formats to be used, and potential student difficulties to be anticipated, guiding the teacher in formulating a suitable lesson agenda, and in enacting it in a flexible and responsive way.

Didactic Incidents An event of the didactic system that modifies the dynamics of the situation.

Didactical Performance A didactical performance involves the ad hoc decisions taken while teaching on how to actually perform in the chosen didactic configuration and exploitation mode: what question to pose now, how to do justice to (or to set aside) any particular student input, how to deal with an unexpected aspect of the mathematical task or the technological tool, or other emerging goals.

Documentational Genesis The transformation of a resource in a document.

Double Approach The didactic and ergonomic approach, which analyses teachers' practices by the mean of five components: cognitive; mediative; institutional; social; and personal.

Double Dialectic The first dialectic between the personal students' meanings of a didactic situation and its scientific meaning and the second dialectic between the teachers' personal interpretation of the first dialectic and the researchers' interpretation.

Double Instrumental Genesis From the same artefact, two instrumental geneses lead to two different instruments, one as a result of teacher's professional genesis, the other as a result of personal genesis.

Hiccup The incidents within lessons where teachers experience perturbations, triggered by the use of the technology, which seemed to illuminate discontinuities in their knowledge.

Instrumental Distance The set of changes (cultural, epistemological or institutional) introduced by the use of a specific tool in mathematics 'praxis'.

Instrumental Genesis The process through which humans transform artefacts into instruments.

Instrumental Orchestration The intentional and systematic organisation and use of the various artefacts available in a learning environment by the teacher in a given mathematical task situation, in order to guide students' instrumental genesis.

Instrumentalisation The process through which humans transform tools into instruments.

Instrumentation The process through which humans learn to use technological tools.

Meta-didactic Transposition The model that describes the dynamic process that occurs, during a teacher education programme, in the dialectical interactions between the community of teachers and that of researchers and their evolution over time, in the context of the institutions involved.

Mutualisation The construction of shared knowledge in the classroom that results from individual or group work and often involves joint action between teacher and students.

Pedagogical Technology Knowledge The construct of pedagogical technology knowledge (PTK) is used as a lens for examing crucial variables related to teachers' use (and non-use) of technology in mathematics. It includes the need to be a proficient user of the technology, but more importantly, to understand the principles and techniques required to build didactical situations incorporating it, and to enable mathematical learning through the technology.

Praxeologies The tasks, techniques, and justifying discourses that develop during the process of teacher education.

PTK See Pedagogical Technology Knowledge.

Structuring Features Framework A framework design to support the identification and analysis of teaching expertise. It contains the following elements: working environment; resource system; activity structure; curriculum script; and time economy.

Teachers' Pedagogical Identities The embodiment of a teacher's knowledge, beliefs, values, life history, and experiences of participating in diverse professional communities that influence how that person has learned 'how to think', 'how to act', and 'how to be' as a teacher.

Technical, Pedagogical and Content Knowledge (TPACK) Originally Technological Pedagogical Content Knowledge [TPCK], now Technology, Pedagogy and Content Knowledge [TPACK] represents an extension of the now classic conceptualisation of teacher knowledge PCK to include Technology.

Theory of Didactic Situations A model that describes the dynamics of the interactions between teacher and students in the classroom as a 'game' where teachers and students win when students learn.

Zone of Free Action The zone of free action (ZFA) refers to the set of activities, objects, or areas in the environment that promote an individual's actions.

Zone of Free Movement The zone of free movement (ZFM) describes the constraints that structure the ways in which an individual accesses and interacts with elements of the environment.

Zone of Proximal Development The zone of proximal development (ZPD) refers to the current extent to which an individual can develop new knowledge, beliefs, goals and practices.

Index

A
Abacus, 234
  children and teacher interactions, 221–222
  Chinese abacus, 217, 235–236
  digital projector, 217, 218
  interactive white board, 217, 218, 220–221
  laptops, 220–221
  lessons, 218–219
  software, 219–220
Abboud-Blanchard, M., 6, 7, 297–315, 397
Abrahamson, D., 168
Actions-Processes-Objects-Schemas (APOS), 168
Adaptive and differentiated instruction, three-level model, 4
  ability grouping/setting, 93
  definitions, 92
  developmental progression, 96
  expert level model
    definition, 97
    PCK, 101
    pedagogy, 100
    TPACK and TI-Nspire Navigator, 104–109
  formative assessment, definition of, 92
  immediate (entry-level) model
    definition, 96–97
    PCK, 99–100
    pedagogy, 99
    TPACK and TI-Nspire Navigator, 105
  master-level model
    definition, 97–98
    PCK, 103
    pedagogy, 101–102
    TPACK and TI-Nspire Navigator, 106–107
  phases, 94–96
  requirements, 92–93
  supplemental educational services, 93
Adler, J., 49
Advanced Mathematical Thinking and the Computer, 168, 181
Ainley, J., 22
Aldon, G., 7, 319–342, 397, 400
Animation
  fractions, 123–124
  metric units of volume, 120–121
  Rolling Polygon
    affordances, 30–31
    aligning student attention, 30
    extensions, 31
    implications for teaching, 32
    mental imagery, 13–14
    phenomenal mathematics, 30
    variation theory, 30
    ways of working, 31
Anthropological Theory of Didactics (ATD). *See* Meta-didactical transposition (MDT) model
Apple Classroom of Tomorrow (ACOT) project, 96
ArAl project, 358, 366–367
Argyris, C., 49
Artigue, M., 263, 311
Arzarello, F., 7, 245, 262, 347–370, 398
Association of Mathematics Teacher Educators (AMTE), 377
Assude, T., 47
Autograph, 54

B
Baker, J., 243
Ball, D.L., 215, 349, 370
Bartolini Bussi, M.G., 351
Bass, H., 349, 350
Becker, H.J., 45, 84
Bellman, A., 4, 91–109
Ben-El-Mechaiekh, H., 170
Bennett, J., 20, 21
Besamusca, A., 5, 189–210
Blondel, F.-M., 244
Board-instruction orchestration, 201–202, 204, 209
Boon, P., 5, 189–197, 210
Borba, M., 140, 149
Bretscher, N., 3, 43–68
Brokering, 7, 356–357
Brousseau, G., 72, 85
Bruillard, E., 244
Bruner, J., 18, 19
Bueno-Ravel, L., 6, 213–245
Burtch, M., 174
Buteau, C., 5, 166, 171, 173, 180

C
Capacity measures' program, 119–120, 135
Capponi, B., 245, 262
Carifio, J., 79
Chae, S., 169
Chevallard, Y., 7, 348, 351, 352
Chinnappan, M., 72
Clark, D., 356, 357, 369
Clark-Wilson, A., 1–9, 51, 108, 215, 221, 258, 277–294, 325, 395–401
Communities of practice, 194–195, 350, 357
Community Documentational Genesis (CDG), 195
Community of inquiry, 86, 356, 367
Computer algebra systems (CAS), 72, 165, 166, 170, 180, 182, 252, 298
Concerns Based Adoption Model (CBAM), 96
Coulange, L., 245
Cuban, L., 68
Curriculum scripts, 48, 215, 314, 386, 387, 389, 391
Cusi, A., 7, 347–370

D
Data projector
 in English mathematics classrooms
  accessibility, 52–53
  frequency and perceived impact score of, 56–59
  individual level factors, 60–61
  work-and-walk-by orchestration, 192
Didactic incidents. *See* EdUmatics project
DidaTab project, 245
Differentiated instruction model. *See* Adaptive and differentiated instruction, three-level model
Digital Mathematics Environment (DME), 196, 206, 383
Digital projector, 217–218, 228
Digital technology
 constraints and obstacles
  calculators, 74
  computer use, 73
 NCTM's position, 190
 pedagogical technology knowledge
  epistemic value, 77
  instrumental genesis, 76, 77
  MKT, 75–76
  PCK, 75
  professional development, role of, 84–87
  PTK and TCK, 75
  ROG, 76–77
  teacher orientations, 76–84
  vs. TPACK, 75–76
 primary school teachers, role of (*see* Role of the teacher)
 in secondary school mathematics (*see* Secondary school mathematics)
 teachers' professional development (*see* Professional development (PD))
Discuss-the-screen orchestration, 192, 201, 202, 214, 220, 230, 237, 384, 385
DME. *See* Digital Mathematics Environment (DME)
Documentational genesis, 7, 324–326, 338
 communities of practice, 195
 at kindergarten level, 215
Doorman, M., 5, 189–210
Double approach, 7, 252, 397
 cognitive component, 302
 institutional component, 303
 mediative component, 302
 personal component, 302–303
 social component, 303
Double dialectics, 357, 366–367
Double instrumental genesis, 3, 6, 247, 254, 256–258, 279, 396–397
Drijvers, P., 5, 6, 189–210, 214, 215, 225, 226, 228–230, 232, 298, 311, 315, 384–386, 391

Index

Dubinsky, E., 168
Dynamic geometry, 54, 61, 306

E

EdUmatics project, 7
  collaborations, 319
  context and methodology, 326–328
  dialogues, teacher and students, 335
  documentational genesis and incidents, 320, 324–326
  French situation, analysis
    ascendant analysis, 329–330
    classroom observation, 330–334
    descendant analysis, 328–329
  milieu
    didactic bifurcation, 324
    didactic situation, 321, 322
    Egyptian fractions, 323
    game, 320–321
    lower didactic levels, 323
    naturalized knowledge, 321
    nil-didactic situations, 323
    noospherian situation, 322
    ordinary classrooms, 321
    structuring of, 322
  syntactic incident, 335–336
  teacher development
    documentational genesis, 338
    researchers and teachers collaboration, 327, 336–338
  TI-Nspire technology, 335
Electronic-Exercise-Bases (EEB), 300
  double approach framework, 301–303
  emerging 3-axis structure (CPT), 303–305
  French project, 300
  paper and pencil environments, 305
  teacher's role, 307–308
  tutoring environment, 300
Electronic screens. *See* E-screens
Electronic whiteboard, 118–119, 121
Elliott, P., 168
Email, 364, 367
  access to, 54
  frequency of use, 59
Emotion, 16–17, 32, 33
Enactivism
  learning, 113
  use of digital technology, teachers' role in (*see* Role of the teacher)
Enciclomedia, 136
  capacity measures' program, 119–120, 135
  evaluation of, 112
  fractions, animation, 123–124

  intention of, 112
  'The Number Line,' 125–127
Erdogan, E.O., 44, 45
Erfjord, I., 233
Ervynck, G., 178, 179
E-screens, 3
  action, 20–21, 25
  activity, structure of
    motivation, 21–22
    resources, 22
  applet, secret places
    affordances, 34
    complexity, 36–37
    extensions, 35–36
    mathematical reasoning, 32
    narrative, 32–34
    teacher-led exploration, 14–15, 32
    ways of working, 34–35
    website, 15
  attention, 24
  human psyche
    awareness, 16
    harnessed emotion, 16
    rational justification, 17–18
    responsibility, 17
    socio-cultural-historical milieu, 18
    trained behaviour, 16–17
  learning
    experience, 23–24
    maturation process, 18
    MGA, 19–20
    SEM, 18–20
  Rolling Polygon animation
    affordances, 30–31
    aligning student attention, 30
    extensions, 31
    implications for teaching, 32
    mental imagery, 13–14
    phenomenal mathematics, 30
    variation theory, 30
    ways of working, 31
  screencasts
    affordances, 26–27
    commentary, 27–28
    example construction, 29
    human powers, mathematical use of, 29
    mathematical themes, recognition of, 29
    questions, 26
    similar problem, 13
    technique-bashing mode, 25–26
    ways of working, 28–29
    website, 13

Eurydice, 242
Explain-the-screen orchestration, 192, 201, 204, 214, 219, 225, 230, 385

F
Forgasz, H., 46, 50, 74
Foshay, W.R., 4, 91–109

G
Gadanidis, G., 140, 149
Gardiner, A., 29
Garuti, R., 7, 347–370
Gattegno, C., 16–18, 30
GeoGebra, 54, 196, 206, 362
The Geometer's Sketchpad, 45, 54
Glover, D., 46
Goos, M., 5, 139–159
Graham, C.R., 210
Graphic calculator (GC)
 in English mathematics classrooms
  accessibility, 52–53
  frequency and perceived impact score of, 56, 57
 PTK, 77–79
 TPACK, 380
Gremillion, D., 4, 91–109
Grounded theory, 4, 94
Gueudet, G., 6, 47, 48, 195, 213–245
Guide-and-explain orchestration, 201, 202, 206, 208

H
Habermas, J., 17
Haspekian, M., 6, 241–272
Heidegger, M., 22
Heid, M.K., 72
Hennessy, S., 45, 48, 63, 66–68, 71, 84, 298
Hiccups, 6, 258, 289–290, 292, 293
Hollingsworth, H., 356, 369
Holton, D., 179
Hoyles, C., 169, 233
Hughes, J., 115
Hyde, R., 46, 49, 51

I
ICMI. *See* International Commission on Mathematical Instruction (ICMI)
Individual Link-screen-book orchestrations, 202, 209
Information and Communication Technologies (ICT), 3
 English mathematics teachers' use of (*see* Teachers' ICT practices)
 spreadsheet (*see* Spreadsheet)
 teacher-guided use of
  action, 20–21
  activity, structure of, 21–22
  applet, secret places, 14–15, 32–37
  attention, 24
  human psyche, 16–18
  learning, 18–20, 23–24
  Rolling Polygon animation, 13–14, 30–32
  screencasts (*see* Screencasts)
Inquiry Based Science Education (IBSE), 364
Instrumental distance, 3, 247–249, 252, 264–266, 396–397
Instrumental genesis, 380–381
 at kindergarten level, 215
 pedagogical technology knowledge (*see* Pedagogical technology knowledge (PTK))
 spreadsheet (*see* Spreadsheet)
Instrumentalisation, 75, 195, 215–216, 227
Instrumental orchestration, 5, 209, 396
 Algebra Arrows tool, 383
 classroom teaching and learning, 381
 didactical configuration, 191, 192, 381, 382, 384
 didactical performance, 191
 discuss-the-screen orchestration, 192
 DME, 383
 explain-the-screen orchestration, 192
 exploitation modes, 191, 381, 382, 384
 instrumental genesis, 380
 at kindergarten level
  choices, 215
  didactical performance, 214
  documentational genesis, 215
  instrumental geneses, 215
  instrumentalisation, 215
  mathematical package project, 216–217
  passenger train software (*see* Passenger train software)
  types, 214
  virtual abacus (*see* Abacus)
 link-screen-board orchestration, 192
 notion of, 190–191
 Sherpa-at-work orchestration, 192, 381
 spot-and-show orchestration, 192
 teaching expertise, 390
 technical-demo orchestration, 191
 technology-mediated teaching and learning, 380

Index

tool-adaptive form, 383
typology of, 384, 385
work-and-walk-by orchestration, 192
Instrument utilisation scheme
perpendicular functions, 283–285
prime factorisation, 281–283
Interactive whiteboards (IWBs), 3–4
in English mathematics classrooms
accessibility, 52–53
frequency and perceived impact score of, 56–60
individual level factors, 60–64
school level factors, 55–56
self-reported pedagogic practices, 64–65
instrumental orchestration, kindergarten level, 217–218, 220–221, 228
International Commission on Mathematical Instruction (ICMI), 164, 169, 348
Internet, 65, 116, 174, 197, 376
Interpretative flexibility, 48, 67
Italian teacher education programmes
ArAl project, 366–367
MMLab-ER project
fundamental mathematical activities, 364
IBSE, 364
praxeology, evolution, 364–365
researcher praxeologies, 365
SCK, 366
semiotic mediation, 365
teacher knowledge, 365
theoretical discourse, 365
M@t.abel project, 368
brokers, 359, 360
cognitive apprenticeship, 359
GeoGebra software, 362
internal/external components, 363
Matematica per il cittadino, 358–359, 368
Ministry of Education, 358
MIUR, 359
platform, 362
shared praxeologies, 362–363
teacher's logbook, 361–362
tutors, 360–361
IWBs. *See* Interactive whiteboards (IWBs)

J
James, W., 17, 23
Jaworski, B., 23, 86, 352, 366, 369
Jorgensen, R., 154, 155
Joubert, M., 232

K
Kendal, M., 297, 305
Keynes, H., 164
King, K., 169
Koehler, M.J., 194, 376

L
Laborde, C., 306
Lagrange, J.-B., 44, 45, 262, 315
Lavicza, Z., 165, 182
Lenfant, A., 303
Leontiev, A.N., 302
Levy, S.T., 232
Link-screen-board orchestration, 192, 204, 209, 230, 384, 385
Logo, 54, 58–60
Lowrie, T., 154, 155
Lozano, M.-D., 4, 111–137

M
Malara, N.A., 7, 347–370
Manipulating-the-familiar, getting-a-sense-of, and articulating (MGA), 19, 20
Marshall, N., 170, 171
Martignone, F., 7, 347–370
Martinovic, D., 173
Marton, F., 23
Mason, J., 2–3, 11–37, 165, 167, 173, 366, 396, 399
Mathematical Association of America (MAA), 164–166, 181
Mathematical Beliefs Questionnaire, 144, 145, 151
Mathematical knowledge for teaching (MKT), 75–77, 349–352, 355, 369–370
Mathematical package project, 216–217
Mathematics Integrated with Computers and Applications (MICA), 180, 181
conjecturing task, 173–174
Exploratory Object, 173
Mandelbrot Set, 171–172
numerical and graphical investigation, 177
Observations and Findings section, 172
Special Thanks and Credit section, 171
final projects phase, 178
lab sessions, 175–177
lectures, 173–174
microworlds, 170
preliminary task analysis, 171
program principles, 170
student development process model, 171

McDonald, M., 168
Meta-didactical transposition (MDT) model, 7–8, 398
　ArAl project, 366–367
　brokering, 356–357
　double dialectic, 357
　institutional aspects, 352–353
　internal and external components, 355–356
　MKT model, 351–352, 355, 369–370
　MMLab-ER project (*see* MMLab-ER project)
　M@t.abel project (*see* M@t.abel project)
　praxeology, 353–355
　social/political change, 347
　teacher education and Italian context, 348–351
MICA. *See* Mathematics Integrated with Computers and Applications (MICA)
Miller, D., 46
Ministry of Education (MIUR), 351, 352, 358, 359
Mioduser, D., 232
Mishra, P., 194, 376
Mitchell, D., 165
MMLab-ER project
　fundamental mathematical activities, 364
　IBSE, 364
　praxeology, evolution, 364–365
　researcher praxeologies, 365
　SCK, 366
　semiotic mediation, 365
　teacher knowledge, 365
　theoretical discourse, 365
Monaghan, J., 298, 310, 315
Morselli, F., 174
Moss, G., 44
Motivation, 21–22
M@t.abel project, 368
　brokers, 359, 360
　cognitive apprenticeship, 359
　GeoGebra software, 362
　internal and external components, 363
　Matematica per il cittadino, 358–359
　Ministry of Education, 358
　MIUR, 359
　platform, 362
　shared praxeologies, 362–363
　teacher's logbook, 361–362
　tutors, 360–361
Muller, E., 5, 167, 170–174
Multi-commented transcripts (MT), 367
Multi-representational technology, 6
　artefact, 278
　concept of variance and invariance, 278
　design and orchestrate classroom activities, 280
　double instrumentation, 279
　hiccups, 289–290, 292, 293
　human-instrument interactions, 278
　lesson bundles, 280–281
　perpendicular functions, 283–285
　prime factorisation, 281–283
　professional development setting, 278
　situated learning, 290–292
　students' handheld screens on public display, 288, 289
　Subject-Instrument-Object triad, 279
　teachers' epistemological development, 278
　teaching and pedagogic knowledge, 280
　transformations of functions, 287, 288
MyMaths, 51, 65
　accessibility, 54
　mean frequency score, 58–60
　preparation time, 61

N
Norton, S., 266
Noss, R., 167, 168, 182

O
O'Brien, T., 32
Olson, A., 164
Omnigraph, 54

P
Palmer, J.M., 4, 71–87
Pampaka, M., 50
Passenger train software, 224
　autonomous-use, 226
　digital projector/IWB, 228
　'discovering' mode, 223
　instrumented teaching techniques, 223
　'learning' mode, 223
　lessons, 224
　score sheet, 223
　software, 224–225
　supported-use, 226–227
PD. *See* Professional development (PD)
Pedagogical content knowledge (PCK), 75, 193, 194, 349, 350
　adaptive and differentiated instruction
　　expert level model, 101
　　immediate (entry-level) model, 99–100
　　master-level model, 103

PTK, 75
TPACK, 374–379
Pedagogical knowledge (PK), 193, 194, 233, 369
Pedagogical technology knowledge (PTK), 4
　epistemic value, 77
　instrumental genesis, 76, 77
　MKT, 75–76
　PCK, 75
　professional development, role of
　　calculators, use of, 84–85
　　community of inquiry, 86
　　informal interaction, 84
　　teachers' mathematical knowledge, 86
　PTK and TCK, 75
　ROG, 76–77
　teacher confidence, 76–84
　vs. TPACK, 75–76
Pédauque, P.T., 338
Perla, R.J., 79
Piaget, J., 17, 22
Poincaré, H., 17
Poisard, C., 6, 213–245
Pólya, G., 29, 33
Popham, W.J., 92
PowerPoint, 54, 58–61, 66, 67, 280
Pratt, D., 22
Professional development (PD)
　data analysis, 200–201
　EdUmatics project (*see* EdUmatics project)
　method, 195
　　classroom intervention design, 196–198
　　DME, 196
　　geometry module, exemplary online task, 197
　　learning management system, 196
　　linear equations module, exemplary online task, 198
　　Moodle, project environment in, 196
　　orchestration chart, 199
　　participants, 197, 198
　　research instruments, 198–199
　pedagogical technology knowledge
　　calculators, use of, 84–85
　　community of inquiry, 86
　　informal interaction, 84
　　teachers' mathematical knowledge, 86
　results
　　community, influence of, 207–208
　　Computer-paper-classroom, 207
　　Degree of difficulty, 207
　　ICT questionnaire, 206
　　orchestration frequencies, 204
　　Planning the module, 207
　　post-project questionnaire, 208
　teachers' orchestrations, 201–203
　TPACK knowledge and skills, 205
　theoretical framework
　　communities of practice, 194–195
　　instrumental orchestration
　　　(*see* Instrumental orchestration)
　　TPACK perspective, 193–194
PTK. *See* Pedagogical technology knowledge (PTK)

R
Rabardel, P., 256, 278, 293
Ralph, B., 167, 170
Rasmussen, C., 357
Ravitz, J.L., 45
Reed, H., 199
ReMath project, 399
Remillard, J.T., 3, 47, 48
Response repertoire, 293, 336, 338
Resources, Orientations, and Goals (ROG), 76–77
Robert, A., 262, 301, 302
Robutti, O., 1–9, 347–370, 395–401
Rogalski, J., 262, 301, 302
Rojano, T., 245, 262
Role of the teacher, 4–5
　EEB, 307–309
　in secondary school mathematics (*see* Secondary school mathematics)
　use of technology in primary school, 122, 128, 133
　　amplifier, 115
　　The Balance' program, 128–132, 135
　　capacity measures' program, 119–120, 135
　　communication of mathematics, 114, 132
　　computer tools, use of, 113
　　electronic whiteboard, 118–119, 121
　　formal training, need of, 136
　　fractions, animation, 123–124
　　interaction with students, 114, 132–133
　　learning, 113
　　limitations, 134–135
　　mathematical knowledge, validation of, 114, 133–134
　　mathematical problems, source of, 114–115
　　metric units of volume, 120–121
　　'The Number Line,' 125–127, 135
　　replacement, 115
　　research questions, 115
　　research tools, 117–118
　　states and schools, 116

Role of the teacher (*cont.*)
  students' actions and autonomy, 115, 134
  teachers, 116–117
  transformation, 115
  worksheet, 124–125
Roschelle, J., 109
Rossi, G., 244
Ruthven, K., 8, 9, 45, 47, 48, 63, 66–68, 71, 84, 98, 108, 210, 215, 298, 314, 315, 373–392, 398

S
Sabena, C., 7, 347–370
Sabra, H., 195, 325
Sahl, K., 140
Saljö, R., 23
Sandoval, I., 4, 111–137
Saxe's cultural model, 298
Schoenfeld, A.H., 76
Schön, D.A., 49, 350
Schurrer, A., 165
Screencasts
  affordances, 26–27
  commentary, 27–28
  example construction, 29
  human powers, mathematical use of, 29
  mathematical themes, recognition of, 29
  questions, 26
  similar problem, 13
  technique-bashing mode, 25–26
  ways of working, 28–29
  website, 13
Secondary school mathematics, 5
  ICT, teachers' use of (*see* Teachers' ICT practices)
  research design and methods
    lesson cycles, 145
    questionnaire, 144
    semi-structured scoping interview, 144
    snowballing methodology, 144–145
    teachers' general pedagogical beliefs, 144
  teacher's role in
    assessment task, 152, 154
    computer software, 156
    external curriculum and assessment requirements, 149
    Flying Gonzo game, 152, 153
    graphical approach, 152
    graph matching activity, 146
    interview and lesson observation data, 154
    Mathematical Beliefs Questionnaire, 145, 151
    mathematical performance approach, 146
    modeling tasks, 146, 148, 149
    motion detector, 146
    Newton's Law of Cooling spreadsheet, 146–149, 159
    pedagogical identities, 149–151, 155–157
    people-environment relationships, 156
    professional development workshop, 152, 154
    quadratic functions, 152
    regression function, 152
    school's organisational culture, 156
    student-centred views, 145
    technology metaphors framework, 146
    TI-Interactive software, 154
    timetabling practices, 150
  technology-enriched mathematics teaching
    sociocultural perspectives, 141, 143
    teacher learning and development, 142–143
    teaching and learning roles, 141–142
    Valsiner's zone theory, 142–144
Second Information Technology in Education Survey (SITES), 46, 67–68
See-experience-master (SEM), 18–20
Semiotic mediation, 365
Sfard, A., 348, 349, 396
Shared praxeology, 354, 355, 361, 367, 368
Sherpa-at-work orchestration, 192, 228, 233, 384
Shulman, L.S., 75, 193, 349, 369
Sinclair, N., 1–9, 395–401
SITES. *See* Second Information Technology in Education Survey (SITES)
SMILE, 54, 58, 59
Social utilisation scheme, 282, 284
Specialised Content Knowledge (SCK), 350, 366
Spiral learning, 18–19
Spot-and-show orchestration, 192
Spreadsheet
  algebra, 244, 245
  benefits of, 244–245
  class level, 260
  constrained, 258–259
  data, 254–255
  didactic and ergonomic approach, 243
  DidaTab project, 245
  domain changing, 260–261
  expert teachers, 243, 265–266

Index 415

instrumental approach, 242
  algebraic culture, 250
  artefact/instrument distinction, 246
  cell variable, 250
  computer transposition, 251
  instrumental distance, 247–249
  psychological and socio-cultural theory, 246
  mathematical and instrumental content, 260
  Newton's Law of Cooling, 146–149
  paper and pencil mathematics, 261
  professional genesis, 255–256
  professional instrumental genesis, 265
  professional websites, 246
  pupils' instrumental geneses, 256–260
  research study
    expert teachers' practices, 261–264
    reducing instrumental distance, 264–265
  school learning, 241
  teachers' double instrumental genesis, 256–258
  teachers' practices
    didactic and ergonomic approach, 252–253
    professional instrumental genesis, 253–254
Stacey, K., 297, 305
Stein, M.K., 48, 60, 67
Structuring features model, 398
  activity structures, 389
  components of, 386, 387
  curriculum scripts, 389
  lower-secondary level, 388
  material-cultural factors structure, 390
  resource system, 386, 388
  teacher craft knowledge and thinking, 386
  time economy, 389
  working environment, 386, 388
Sugden, S., 243
Sullivan, P., 353, 354
Supplemental educational services (SES), 93
Sutherland, R., 245, 262
Swift, J., 17

T
Tabach, M., 210
Tacoma, S., 5, 189–210
Tahta, D., 22
Tall, D., 168, 169, 181
TCK. *See* Technological content knowledge (TCK)

Teachers' ICT practices
  curriculum materials, Remillard's perspective of, 47–48
  hardware use
    accessibility, 52–53
    frequency and perceived impact score of, 56–58
    teachers' choice of, 45–46
  individual level factors, 47, 60–64
  quantitative and qualitative gap in, 44–45, 65–66
  school level factors, 47
    mean school support scores, 55–56
    overall school support scores, 55
  self-reported pedagogic practices
    computer suite, 65
    espoused theory, 49
    IWB, whole-class context, 64–65
    'theory-in-use,' 49
  software use
    accessibility, 53–54
    mean frequency of, 58–60
    teachers' choice of, 45–46
  survey instrument, sample and data analyses, 49–52
Teachers' pedagogical identities, 5, 149–151, 155–157, 397
Technical-demo orchestration, 191, 202, 219, 220, 225
Technological content knowledge (TCK), 194, 390–391
  PTK, 75
  TPACK, 75, 376–379
Technological knowledge (TK), 194, 376–377
Technological, pedagogical and content knowledge (TPACK), 8, 208, 210
  adaptive and differentiated instruction
    expert level model, 106
    immediate (entry-level) model, 105
    master-level model, 106–107
  AMTE, 377
  definition of, 104
  dynamic algebra/geometry software, 377
  ecological adaptation and cognitive economy, 374
  intersections/interactions, knowledge domains, 375
  knowledge and skills, 205
  participating teachers, 378–379
  PCK, 374–376
  professional development programme, 378
  vs. PTK, 75–76
  school-based lesson-study group, 380

Technological, pedagogical and content knowledge (TPACK) (*cont.*)
  teachers' professional development, 193–194
  technological and pedagogical components, 377
  two-and three-way interactions, 375
  Venn-diagram metaphor, 375
Technological pedagogical knowledge (TPK), 75, 194, 376, 378, 379
Technology integration
  beginner teachers' practices
    double approach framework, 301–303
    emerging 3-axis structure (CPT), 303–305
    pre-service mathematics teachers, 300
    professional dissertations, 301
    teacher's role, 308
    trainees, 301
  Computer Algebra Systems, 298
  educational research, 297
  EEB
    double approach framework, 301–303
    emerging 3-axis structure (CPT), 303–305
    French project, 300
    paper and pencil environments, 305
    teacher's role, 307–308
    tutoring environment, 300
  experienced teacher's practice
    double approach framework, 301–303
    dynamic geometry software, 299
    emerging 3-axis structure (CPT), 303–305
    non-technology-based lesson, 299
    paper and pencil environments, 305
    teacher's role, 307, 308
  mini-classes, 310–311
  primary school teachers, role of (*see* Role of the teacher)
  professional constraints, 315
  quasi-disappearance, collective phases, 311
  Saxe's cultural model, 298
  in secondary school mathematics (*see* Secondary school mathematics)
  time of teaching and learning, 311–313
  Valsiner's zone theory, 142–144
Theory of didactic situations (TDS), 320–324, 339
Thomas, M.O.J., 4, 71–87
TIMSS, 44, 46, 349
TI-Nspire, 279, 335
  adaptive and differentiated instruction, 95
  expert level model, 106
  immediate (entry-level) model, 105
  master-level model, 106–107
  transformations of functions, 287, 288
TPACK. *See* Technological, pedagogical and content knowledge (TPACK)
TPK. *See* Technological pedagogical knowledge (TPK)
Trigueros, M., 4, 111–137
Trouche, L., 5, 47, 48, 191, 195, 252, 259, 356, 382, 384, 397

U
University mathematics education, 5
  constructionist approaches, 169
  cooperative work, 169
  creativity, technology and educating students, 178–180
  experimental approaches, 169
  ICMI studies, 164, 169
  mathematics departments, 166–169
  MICA I, roles of tutors in
    conjecturing task, 173–174
    Exploratory Objects, 170–173, 177
    final projects phase, 178
    lab sessions, 174–177
    lectures, 173–174
    microworlds, 170
    preliminary task analysis, 171
    program principles, 170
    student development process model, 171
  motivate explanations, 169
  student thought processes, 169
  technologies in, 165–170

V
Valsiner, J., 24, 142, 149, 150, 157
van den Heuvel, C., 5, 189–210
van der Veer, R., 24
van Hiele-Geldof, D., 24
van Hiele, P., 24
Verillon, P., 278, 293
Virtual abacus. *See* Abacus
Voogt, J., 210
Vygotsky, L.S., 6, 25, 302

W
Wenger, E., 143, 194
Windschitl, M., 140
Wong, Y.T., 45
Work-and-walk-by orchestration, 192, 202, 206, 208, 220, 226, 311

Z

Zbiek, R.M., 47
Zone of free movement (ZFM), 149, 151, 155, 156, 158, 397
   individual accesses and interacts, 142
   organisational culture, 156
   teaching actions, 143
   ZFM/ZPA complex, 142, 150
Zone of promoted action (ZPA), 5, 149, 155, 157, 158, 397
   formal professional development, 150, 156
   participation, 150–151
   people-environment relationships, 156
   teaching approaches, 143
   university research projects, 150
   ZFM/ZPA complex, 142, 150
Zone of proximal development (ZPD), 5, 143, 149, 150, 155, 156, 158, 397
   goals and actions of participants, 142
   social setting, 142
   students' actions, 25

CPSIA information can be obtained
at www.ICGtesting.com
Printed in the USA
LVOW13*0916010718
582385LV00007B/62/P

9 789400 746374